# Green Phoenix

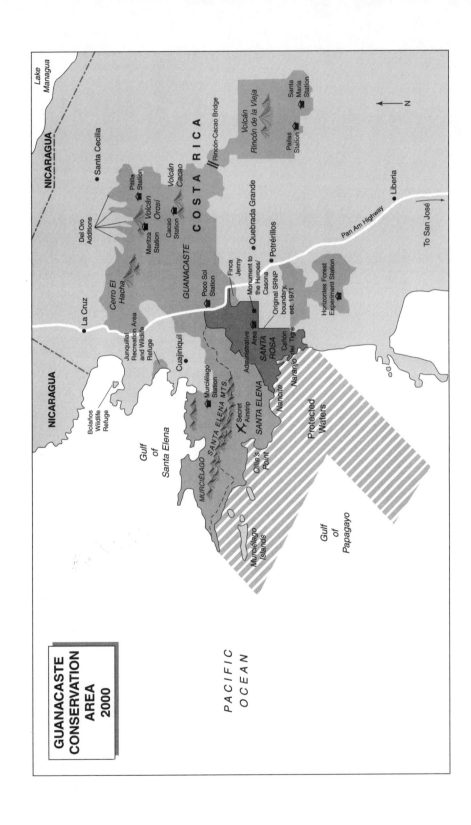

GUANACASTE
CONSERVATION
AREA
2000

Lake Managua

NICARAGUA

COSTA RICA

Santa Cecilia

Del Oro Additions

Pitilla Station

Volcán Orosí

Volcán Cacao

Maritza Station

Cacao Station

GUANACASTE

Cerro El Hacha

La Cruz

Junquillal Recreation Area and Wildlife Refuge

Cuajiniquil

Poco Sol Station

Finca Jenny

Monument to the Heroes/ Casona

Original SRNP boundary, est. 1971

Administrative Area

SANTA ROSA

Cañon del Tigre

Nancite

Naranjo

Rincón-Cacao Bridge

Volcán Rincón de la Vieja

Santa María Station

Pailas Station

Quebrada Grande

Potrerillos

Pan Am Highway

Liberia

To San José

Horizontes Forest Experiment Station

N

NICARAGUA

Bolaños Wildlife Refuge

Gulf of Santa Elena

Murciélago Station

SANTA ELENA MTS.

MURCIÉLAGO

Secret Airstrip

SANTA ELENA

Ollie's Point

Protected Waters

Gulf of Papagayo

Murciélago Islands

PACIFIC OCEAN

# Green Phoenix

Restoring the Tropical Forests of Guanacaste, Costa Rica

**William Allen**

OXFORD
UNIVERSITY PRESS
2001

# OXFORD
UNIVERSITY PRESS

Oxford New York

Athens Auckland Bangkok Bogotá Buenos Aires Calcutta Cape Town
Chennai Dar es Salaam Delhi Florence Hong Kong Istanbul
Karachi Kuala Lumpur Madrid Melbourne Mexico City Mumbai
Nairobi Paris São Paulo Shanghai Singapore Taipei Tokyo Toronto Warsaw

and associated companies in
Berlin Ibadan

Published by Oxford University Press, Inc.
198 Madison Avenue, New York, New York, 10016

Oxford is a registered trademark of Oxford University Press

Library of Congress Cataloging-in-Publication Data
Allen, William, 1952–
Green Phoenix : restoring the tropical forests of Guanacaste, Costa Rica /
by William Allen.
p. cm.
Includes bibliographical references (p. ).
ISBN 0–19–510893–0
1. Guanacaste National Park (Costa Rica)
2. Restoration ecology—Costa Rica—Guanacaste National Park.
3. Rain forest ecology—Costa Rica—Guanacaste National Park.
I. Title
SB484.C8 A44 2000
333.75'153'0972866—dc21

                                        00–036729


For permission to reprint parts of previously published articles written
by the author, grateful acknowledgment is made to:
The American Institute of Biological Sciences, for "Biocultural restoration of a tropical forest," *BioScience* 38, no. 3 (March 1988), and "The
varied bats of Barro Colorado Island," *BioScience* 46, no. 9 (October
1996).
The New York Academy of Sciences, for "Last Stand in Guanacaste,"
*The Sciences* 29, no. 4 (July-August 1989).


9 8 7 6 5 4 3 2 1

Printed in the United States of America
on acid-free paper

*To my wife, Debra Ewanic Allen*

*and*

*To my mother and father, John and Mary Allen*

# Contents

# Acknowledgments

I deeply appreciate the assistance of the following people, who went beyond the call of duty during the decade-plus of interspersed work on this book. Some have moved on to other places: Kirk Jensen, my very skillful and understanding editor at Oxford University Press, and Jeanne Hanson, my always chipper literary agent; at the *St. Louis Post-Dispatch* and Pulitzer Inc.: Virginia Baldwin Gilbert, John Carlton, Ron Cobb, Patricia Corrigan, Laszlo Domjan, Robert Duffy, Linda Eardley, Peter Hernon, George Landau, James Maloney, Nancy Miller, Jim Mosley, Jerry Naunheim, Jr., Teak Phillips, Richard Weil, Richard Weiss, and, most of all, William Woo, now of Stanford University; at the Massachusetts Institute of Technology: Paula Anzer, Martha Henry, Victor McElheny, Peggy McNally, and especially Barbara Goldoftas and my fellow 1996–97 Knight Science Journalism Fellows Mark Grossi and Madeleine Drexler; the students in the 1996–97 Tropical Ecology course at Harvard University, and teaching fellow Christopher Dick and lecturer Mark Leighton; at the University of Florida: Stephen Mulkey and Kaoru Kitajima; at the International Center for Tropical Ecology, University of Missouri-St. Louis: Bette Loiselle, Greg Basco, Bob Marquis, and Eduardo Silva; at the Missouri Botanical Garden: Peter Raven, Barry Hammel, Michael Grayum, A. J. Dickerson, Sr., James Trager, Jim Miller, Janine Adams, and Connie Wolf; at the World Resources Institute: Mary Houser, Kenton Miller, Walter Reid, and Janet Abramovitz; at the University of

Illinois: Carol Augspurger, Jeff McMahon, Robert Reid, Sandra Brown, May Berenbaum, Judy Rowan, Fred Mohn, Larry Smarr, and Jeremiah Sullivan; John Bouseman of the Illinois Natural History Survey and Stanley Changnon of the Illinois State Water Survey; at the University of Missouri-Columbia: James Carrel, Richard Loeppkey, and Robert Logan; at the Institutes for Journalism and Natural Resources, Missoula, Montana: Frank and Maggie Allen; at the University of Pennsylvania: Daniel Janzen, Winnie Hallwachs, and Pauline Moslemi; many at the Organization for Tropical Studies, including Donald Stone, Jonathan Giles, and Barbara Lewis; Thomas Eisner of Cornell University; George Godfrey of Haskell Indian Nations University; Elisabeth Kalko of the University of Tübingen; Meredith Lane of the Academy of Natural Sciences, Philadelphia; Sally Pobojewski of the University of Michigan; Martin Quigley of Ohio State University; John Sterling of *Genetic Engineering News*; and several staff members at Conservation International, the Nature Conservancy, World Wildlife Fund, and Sierra Club.

In Costa Rica, I especially want to thank Don and Goldie Marrs, who shared their home with me on several occasions and helped me begin to find my way. Fellow journalists Diane Jukofsky and Chris Wille of the Rainforest Alliance's Tropical Conservation Newsbureau assisted in many ways, as did Rod Hughes, Brian Harris, and other staffers at the *Tico Times*. Also, sincere thanks to Oscar Arias, Rodrigo Gámez, Pedro León, and Alvaro Ugalde for unusually kind help.

I deeply appreciate the efforts on my behalf of many scientists and conservationists I met in Guanacaste Province, including Mariamalia Araya, Mónica Araya, Róger Blanco, Camilo Camargo, Colin Chapman, Lauren Chapman, María Marta Chavarría, Roberto Espinoza, Randall García, Ian Gauld, Guillermo Jiménez, Alan Journet, Frank Joyce, Patricia McDaniel, Sigifredo Marín, Alejandro Masis, Johnny Rosales, Guiselle Méndez Vega, Brad Zlotnick, and several park guards.

Many thanks to the librarians who assisted me at Harvard, Illinois, MIT, Michigan, the Mercantile Library of St. Louis, Missouri Botanical Garden, and the public libraries of St. Louis, St. Louis County, and Urbana, Illinois. Thanks also to Alice Serrano, an instructor at the Meramec campus of St. Louis Community College, who lent me many items from her collection of tropical forest books and articles.

I am indebted to the many people who read and offered comments on this book or parts of it, including Debra Allen, John Allen, J. T. Allen, Mary Allen, Greg Basco, Kim Bell, Doug Boucher, Randy Curtis, Tom Eisner, Rodrigo Gámez, Ian Gauld, Anna Gillis, George Godfrey, Barbara Goldoftas, Mark Grossi, Winnie Hallwachs, Dan Janzen, Frank Joyce, Diane Jukofsky, Ramona Kurek, Keith Leber, Bette Loiselle, John McPhaul, Steve Mulkey, Lyle Prescott, Alice Serrano, Alvaro Ugalde, and Chris Wille. Errors that may still exist are solely my responsibility.

Many thanks to Samantha Burton for her chapter opener drawings, to Dave Gray and Caroline Huth of xplane.com for the map, and to Elisabeth Kalko for her bat photo.

I deeply appreciate the help of other friends and family members who provided information, logistical support, and encouragement on the long road to completing this book: Barbara Allen, Chuck Allen, J. T. Allen, Kathryn Clark Allen, Larry Askew, Wendy Bachhuber, John Burness, Joe Bynum, Mary Creswell, Keay Davidson, Catherine Ewanic, Peter Ewanic, Larry Hoffman, Phyllis Hoffman, Staci Kramer, Dale Jones, Bob Lobonc, Kathy Lobonc, Rebecca O'Connor, Gail Paul, Letie Rangel, Rick Rangel, Robert Rebholz, John Wotal, and the people of the late, great Hotel Galilea in San José.

A special thanks to my parents, John and Mary Allen, and their "Allen Family Foundation" for providing room, board, and a great place to write at several critical times; to my children, Katie, John, and Robyn Allen, for their lifetimes of patience with a distracted father; and above all to the one person whose dedication and countless sacrifices most made this journey possible: Debra Ewanic Allen, my wife.

William Allen

*St. Louis, Missouri*
*May 2000*

# Prologue: The *Contrafuego*

Julio Díaz, tall and thin, darted along the edge of a stand of jaragua grass, dropping little balls of flame from a wand. The balls burst alive and crackled through the grass, six feet tall and dry as tinder. Fingers of fire reached forward fast, then widened, joining in a mad dash to the center of the field. A sudden gust of wind whipped the blaze into a thirty-foot-high frenzy. With this explosion of intense heat, Díaz and the men near him turned their heads away. Some of them smiled and shouted, as if over the deafening rush of a great waterfall.

The wave of fire raced through the jaragua stand, then pushed into the adjacent tropical forest, a mix of green-leafed, brown-leafed, and bare trees. This parched mosaic is typical of the forest in northwestern Costa Rica during the dry season. In this patch of forest, without the jaragua grass as fuel, the fire subsided to a slow, knee-high march through fallen leaves, in contrast to the powerful assault through the jaragua. It moved back and forth along an uneven front, overrunning twigs and stumps, devouring saplings, and charring only the bottoms of the larger trees. It seared bromeliad plants, tearing them as if with bayonets. Clear juices leapt out of steaming cracks in the succulent fronds and trickled down stems to the ground, making dark stains in the soil.

Often, encouraged by a fresh wave of wind and a pile of leaf litter at the base of a trunk, flames climbed a dead tree, and branches crashed in surrender. The burning produced a medley of pops, bangs, peeps, crackles, hisses, and high-pitched whines. It was an eerie juxtaposition: this zany noise, this insidious destruction. The fire left behind a carpet of ashes and glowing embers, and high above the canopy, dozens of birds flew erratically around the rising column of ash and white smoke, seeking insects flushed by the fire.

Díaz moved like a deer as he set these fires. An affable man in his mid-twenties, he wore a white yachtsman's hat and an olive green fatigue shirt hung over his blue jeans. He jumped and paused, jumped and paused on the path, delivering the balls of flame every few feet from the tip of the wand, which was connected to a small red canister of diesel fuel. The other men teased him about his stern expression, a reflection of his determination to perform with skill and care.

Díaz was no ordinary arsonist. He was, in the most literal sense, fighting fire with fire. A guard with the Costa Rican National Park Service, he was setting a *contrafuego*—a backfire—meant to burn its way toward a larger fire, one approaching dangerously close to Santa Rosa National Park, in Guanacaste Province (pronounced gwahn-ah-KAHS-tay).

Whipped along by high winds, the main fire that day, February 24, 1988, was burning on Finca Rosa María toward Santa Rosa. This ranch, and several like it, generated many devastating fires, usually when cattle ranchers annually burned their pastures to clear unwanted growth. These conflagrations swept through pastures and into the trees, ignoring property lines. They regularly nipped and cut at the national park, as the Rosa María fire threatened to do.

On its front lines on the ranch the day before, the guards had fought the fire hand to hand, trying to beat it out with straw brooms dampened in buckets of water. But they couldn't gain control. The *contrafuego*, set just outside the Santa Rosa boundary, was the next logical countermeasure. They hoped it would burn through a forested section of private property that stood between the park and the main fire. By consuming fuel there, the larger fire might be kept from the forests of Santa Rosa.

While tracking the main fire on the neighboring ranch, the fire fighters also had to monitor the *contrafuego*, to make sure it didn't spread beyond the intended path. They had to stay alert to wind shifts that could cause it to circle back and outflank them. They raced through the woods, panting and sweating, wielding brooms, machetes, chainsaws, and water packs in an attempt to mold the destructive force they had unleashed. Theirs was a strenuous, if not dangerous, game, and one with high stakes.

The Rosa María fire that day threatened Santa Rosa's special treasure: one of Latin America's few significant remnants of tropical dry forest. This remnant—actually an archipelago of forest islands surrounded by a sea of jaragua grass—was under constant attack by fires. Like a wave washing a little higher on a beach, each new blaze cleared the way for more of the aggressive, exotic grass. Meanwhile, thousands of species of animals and plants lingered in the shrinking forest islands near the brink of extinction. The Rosa María fire, potentially, was yet another assault on this biological richness, another inward push of the biologically impoverished, but powerful, sea of African grass.

The fire also threatened a bold, new crusade to restore the fallen forest. Known then as the Guanacaste National Park Project, this crusade sought to raise the forest, like a modern-day phoenix, out of the charred and tattered aftermath of centuries of ecological destruction by ranching, farming, burning, logging, and hunting in northwestern Costa Rica. One of the project's goals was to expand the forest islands embedded within Santa Rosa's roughly 42 square miles into a vast canopy covering seven times that much, and complete with healthy populations of all the species that once thrived there. The main method to accomplish this was more active prevention and control of fires set by people, which would allow the forest to grow back on its own, to reweave its intricate ecological quilt. Another goal was to infuse the value of the forest into the culture of Guanacaste Province so that future generations would want to conserve it.

The Guanacaste project was the brainchild of Daniel Janzen, a University of Pennsylvania professor long based at Santa Rosa, and Winnie Hallwachs, his biologist wife. In the 1970s and 1980s, Janzen and Hallwachs had witnessed the accelerated demise of the region's forests. In the mid–1980s, by what many saw as the sheer force of will, they launched this counterattack. The project had its roots in the establishment of Santa Rosa National Park by biologists and conservationists such as Kenton Miller and Alvaro Ugalde. Following in their footsteps, Janzen and Hallwachs assembled a set of ecological, political, cultural, and financial insights that, used together, constituted a revolution in tropical forest conservation.

Although the size of the Guanacaste project was impressive, its concept of restoration was what made it a momentous event in the history of science and the environment. To understand the event, it's important to understand that deforestation of the tropics became a mainstream scientific concern only in the late 1970s. It took on a new sense of urgency in the mid–1980s, largely because its rate had accelerated. Fueled by scientific reports, conservationists and journalists helped deforestation achieve popular attention as an environmental issue. Tropical deforestation was a dramatic phenomenon, defined by a complex set of causes and consequences described elsewhere by several authors. For the purposes of this book, suffice it to say that by the mid–'80s, when I first visited Santa Rosa, the most widely quoted global rate at which tropical forests were disappearing was about 50 acres a minute, or 27 million acres a year, an area slightly larger than Ohio. In some parts of the world, such as Southeast Asia, Central America, West Africa, and coastal Brazil, Colombia, and Peru, surges of population and economic development had exerted especially great deforestation pressures. There was a deep-seated feeling that once gone, the forests were gone forever.

In parallel with their clarion call about deforestation, biologists cautioned that the ensuing mass extinction of plants and animals would cause irretrievable loss of unexplored natural resources, and the loss of the economic, scientific, and aesthetic benefits they could bring. If millions of species were wiped out, they, too, would be gone forever, drastically reducing the potential for new pharmaceuticals, crops, scientific knowledge, and countless other goods and services. As for aesthetics, researchers spoke of the tiniest insect as a marvel that made the finest human works of art look pathetic by comparison, "once you get to know it," as Stanford biologist Paul Ehrlich put it. Deforestation, Ehrlich said, was like "flame-throwing several hundred thousand 'Mona Lisas.'"

"Once you get to know it"—that was a key phrase. I gradually realized that a knowledge of organisms and the way they interacted with each other and the environment gave biologists a special appreciation of what tropical forests contained. This appreciation led to a deep sense of frustration about what was being lost, and it dawned on me that it was the same frustration articulated so beautifully by the late, great U.S. conservationist and wildlife biologist Aldo Leopold, who wrote that "one of the penalties of an ecological education is that one lives alone in a world of wounds."

Those with an "ecological education" were often field biologists whose training and experience gave them an ecological perspective, a perspective that revealed the elegant and fragile processes governing life on Earth. The wounds in the "world of wounds" were, in one sense, the acts of destruction that humans inflicted on nature. In another sense, the wounds were the frustrations felt by those with an ecological education when they realized that their fellow citizens did not understand how subtle disruptions in a complex ecosystem can have profound and lasting negative effects. To the ecologically educated, the wounds were compounded by public nonchalance, or even disbelief.

As I talked with tropical biologists in the 1980s and '90s, it was clear that the world of wounds weighed heavily on their minds and souls. Leopold wrote, "An ecologist must either harden his shell . . . or he must be the doctor who sees the marks of death in a community that believes itself well and does not want to be told otherwise." Janzen, Hallwachs, and their colleagues had seen the marks of death and hardened their shells. But Janzen and Hallwachs, like a few other biologists of their time, scrambled to take action to stem the destruction of life around them. What distinguished them was their audacity.

Until that time, the deforestation crisis had gone through two phases: one, awareness of the problem, and two, action to protect threatened areas before they were damaged or destroyed. The leaders of the Guanacaste project went a step beyond. They sought not just to protect the forest from invasion, a defensive tactic, but to actually take back the forest, an aggressive, offensive maneuver. Restoring a tropical forest on such a scale, and with such scope, would be unprecedented. The attempt alone was a defiant act against great odds.

If this conceptual leap proved feasible, the success might ignite new hope among those who sought to save forests elsewhere in the world. Likewise, if the worldwide assault on tropical forests was ever halted or reversed, historians might well look back on the Guanacaste project as a watershed.

Despite its promise, this nascent effort was endangered. The surviving islands of forest, like shrinking reserves elsewhere in the tropics, were under constant attack. The charge by the runaway fire in Rosa María was just one such assault.

In North America, popular awareness about the fate of tropical forests centered on the Amazon rain forests of South America. As I would soon learn, the forests of Central America were every bit as interesting as those of the Amazon, and even more threatened.

In June 1986, I joined George Godfrey, a mentor and friend, at Santa Rosa. Godfrey was an entomologist with the Illinois Natural History Survey, in Champaign, where he had directed me in field and lab work when I attended the University of Illinois at Urbana-Champaign in the early 1970s. A veteran of North American field biology, he was considering expanding his research beyond the United States. I wanted to find science stories to write. Formerly a reporter with the City News Bureau and United Press International in Chicago, and soon to be with the *St. Louis Post-Dispatch*, in 1986 I covered science for the public information office at the University of Illinois and on my own time was trying to build a career as a freelance science journalist.

I went in search of a unique, important story in the conservation wars, and I found more than I was looking for in the Guanacaste project. During that first trip and six others over the next decade, I followed the people, politics, and biology of the project. I got to know Dan Janzen and Winnie Hallwachs and tracked the project from afar. I talked with Rodrigo Gámez, Alvaro Umaña, and other key Costa Rican proponents of the project, as well as many of its international corps of scientists and volunteers. I also talked with some of its enemies.

I watched the project grow from idea to reality as its leaders walked a cultural and political tightrope, gained international media recognition, negotiated for land, and raised millions of dollars in donations and grants. I came to know and, frankly, admire the Costa Rican people, their society, and their history.

I wandered the region where the Guanacaste project unfolded, from the cloud-fogged summits of volcanoes to crescent Pacific beaches. Along the way, I encountered the life of the tropical forest—plants, animals, and their ecological interactions over vast and varied landscapes. "These are living creatures, not just museum specimens," Ian Gauld, a biologist with the Natural History Museum in London, reminded me one night as we collected insects in a misty forest on the side of a volcano. "If you look at their biology, and know about their life history, you can see the most incredible relationships. Each one is a story."

Thanks to Gauld, Godfrey, Hallwachs, Janzen, and other scientist-naturalists, I learned some of these stories. The larger lesson these men and women taught me was that a forest constitutes a living entity—a complex combination of seemingly infinite processes, from individual details of a creature's life to broad connections across mountains, lowlands, and bodies of water. The forest is a dynamic system, constantly self-renewing even while intact. It is beautiful. It is fascinating. It is filled with intrinsic value.

Discovering and appreciating these stories provided a potent intellectual experience. It also helped me understand many of the biological, political, and cultural themes of the Guanacaste project—itself a kind of complex ecological interaction.

The account in this book unfolds in three parts. In Part I, we trek the Guanacaste project landscape while encountering some of the inner workings of the tropical dry forest, from the burst of leaves and caterpillars with the onset of the rainy season to the ebb and flow of animal migration to the interplay of predators and prey. Along the way, we meet snakes, angry wasps, and a jaguar, and I begin to deal with my fear of the forest, a fear fueled by books and movies that unrealistically portray its risks. We place the project in its historical setting, from the conquistadors to William Walker to the Nicaraguan contras to the cattle boom and subsequent bust that made it possible for the conservationists to acquire large tracts of land. We learn about the nearby operations at Oliver North's secret airstrip built to aid the contra rebels in Nicaragua, and we explore the dimensions of the prime threat to the region's forests—fire—and some of the fire-fighting lore at Santa Rosa. We introduce key players, including Janzen, whose shift from hard-core research scientist to Messianic conservationist set the stage for the project. It was one of the many "peculiar accidents" that made this historic ecological restoration possible. We cover details of the restoration plan and the first steps by Janzen and Hallwachs in pursuing it.

Part II presents an on-the-ground view of how the restoration effort unfolded during its early years. Much of the action centers around the visit to Santa Rosa by President Oscar Arias, a major milestone for the Guanacaste project politically, ecologically, and financially. Arias signed several decrees that aided the project, including one that continued the decade-old expropriation case of Santa Elena, the ranch with the secret airstrip. The message sent by the

Arias visit helped move the project forward in its efforts with local landowners and international donors. It also tipped the balance in the struggle with some in the Costa Rican National Park Service who viewed the project as a threat. We report the attempt to kick Janzen out of Costa Rica and the $24.5 million fund-raising bonanza assisted by traders at Salomon Brothers in New York. We describe the mechanics of restoration, both ecological and cultural, and how Janzen and other leaders reaped success in their dealings with the likes of Jimmy Swaggart, Prince Philip, and the British Broadcasting Corporation. We reveal the largely hidden, yet powerful, role played by Hallwachs in the Guanacaste project. Meanwhile, there's more about the natural history of Guanacaste—its birds, sea turtles, and insects—and the ecological matrices that bind them together.

In Part III, the story jumps a decade ahead, with a journey through what the Guanacaste project has become. Evidence of the project's success includes the stunning surge of new forest and the huge expansion of protected land and water, then known as the Guanacaste Conservation Area. We report events since the project's early years, among them the victory in wresting more local control for conservation areas and the development of a professional fire-fighting unit. We describe parataxonomists, the National Biodiversity Institute, the All Taxa Biodiversity Inventory, and other related developments in Costa Rican conservation science. In the late 1990s, several problems remained; for example, the Santa Elena acquisition was still mired in an ugly expropriation dispute, and the conservationists were having trouble restoring rain forest on the wetter volcano slopes. The Epilogue assesses lessons learned in the Guanacaste project and prospects for using its approach elsewhere in the world.

As time progressed from the mid–1980s through the end of the century, the Guanacaste National Park Project led by Janzen, Hallwachs, and others went by a few different names and acronyms. To avoid cluttering the book, I frequently use the terms "Guanacaste project" or "the project" instead. Although other conservation projects existed in Guanacaste Province, none is mentioned in this story without its specific name.

Much of the information for this book came from documents. Much came from personal observation, meaning the story was gathered and written with some of the standard biases of a North American, a "gringo," to use the vernacular of Latin America. And much came from interviews, which always suffer the problem of the interviewee's bias and selective memory. This may have been all the more problematic in Costa Rica, where telling "little stories"—spinning fabrications in one's self-interest—is said by many to be common practice, although I have my doubts that this is a trait unique to Costa Ricans. Few of the people involved in the Guanacaste project, regardless of nationality and even though they are honorable men and women, were above stretching the truth, or even avoiding it from time to time. That left me with the challenge of sorting through the "little stories"—and my gringo biases—to get as close to reality as possible. I offer this truth: I tried my best.

The reader should know that for years I have been deeply influenced by the writings of Aldo Leopold, especially his essays in *A Sand County Almanac* and journal entries compiled in *Round River*. Leopold's perspectives have touched generations of professional ecologists and wildlife biologists as well. Many of these scientists especially called my attention to a theme Leopold termed the "land ethic." It is a complex idea involving ecology, evolution, and the role of humans in the landscape, but it boils down to these words of his: "A thing is right when it tends to preserve the integrity, stability, and beauty of the biotic community. It is wrong when it tends otherwise." I was struck by how often this theme emerged during my journeys through Guanacaste Province, as did many other themes Leopold espoused. Although the Guanacaste project did not arise directly out of Leopold's literature, this genre of thinking, by him and others before and since, has done much to shape conservation in the United States and Costa Rica, including the project.

There are many noble efforts in the conservation wars, and this is the story of just one. It is neither a complete record of events, debates, and decisions that comprised the Guanacaste project so far nor is it an official history. Rather, it is a journalistic story, told by an independent outsider. Unbounded though he may be, this outsider has been impressed by the spirit of the people who set this *contrafuego*, who tried to bring the forest to life again, to get the phoenix to fly.

# Green Phoenix

# PART I

# In the Place of the Tree with Ears

# The Nucleus

Behind the Casona, a large house that served long ago as the residence and base of operations for Hacienda Santa Rosa, in north-western Costa Rica, a rough stone stairway is dug into the side of a hill. The stairs are sturdy, but rain has washed holes in the mortar, and muddy puddles often gather in the low spots. It's easy to get lost in admiring the chaotic patterns in the way the stones fit together, and that's just what I was doing when a foot-long snake, black with red lengthwise stripes, shot off the step in front of me and slipped into a crack in the stairway.

It was the summer of 1986, and I had just arrived for my first visit to Santa Rosa National Park. No sooner had I dropped my bags at the crumbly old cabin where I would stay than my friend, biologist George Godfrey, urged me to walk with him to the Casona, and to a hilltop monument behind it which offered a great view of the park and the landscape for miles around.

Now, carefully detouring around the step where the snake had disappeared, I finished trudging up the hill as a breeze rustled the leaves of the surrounding trees. We came out of the trees at the top, and there, on a small, grassy clearing, was the Monument to the Heroes of 1856–1955. The monument, a huge brick and concrete "H" structure, was built to honor several generations of Costa Ricans who had fought aggressors from the north.

Builders of the monument had worked thick reinforcing iron into one of the brick walls to form a simple ladder up the 25 feet or so to the horizontal concrete platform. I tamed my fear of heights, and, scaling the wall, I imagined I was climbing the spikes of a telephone pole without a safety strap. As we swung our legs onto the platform, a howler monkey erupted with guttural barks in the forest below. With no railing to grab for security, I steadied myself against one of the vertical walls and took in the view.

To the east, well outside Santa Rosa's boundary, rose a giant wave of land formed largely by two volcanoes, named Volcán Orosí and Volcán Cacao, believed to have been active as recently as 3000 years ago. Along the top of the wave stretched a long, heavy cap of gray clouds that looked like the foam of a 40-mile-wide tsunami about to break on top of us. Below the foam, the wave itself was the darkest possible shade of green. These were the evergreen cloud forests that adorn the volcanoes.

The cloud forests looked dreamlike, distant, lush, and powerful. The view seemed like an oil-on-canvas by some landscape impressionist, except for the insidious little fingers of light green pasture that reached up the slopes where the forest had been cleared.

A huge strip of pasture covered the plains below the volcanoes. Somewhere between this and the peaks was the Pan-American Highway, which formed much of the eastern border of Santa Rosa. From that border to our feet stretched a thin blanket of forest tattered by patches of sand-brown or green jaragua and small scrub and seedlings. To the south, west, and north this green-dominated scene shifted to a moonscape nightmare, as the jaragua had claimed larger sections of the plateau from the forest. The largely light-brown mosaic had a forbidding quality, with tall, jagged islands of green surrounded by a desert of jaragua. The scene came to an end about 12 miles to the southwest in the deep blue of the Pacific Ocean and in the north and northwest with the towering Santa Elena mountains. Beyond those barren brown peaks, only a couple dozen miles farther north, lay La Frontera—the Nicaraguan border.

Santa Rosa's boundaries at that time protected about 42 square miles of land and marine waters about 12 miles out to the national limit. Within this area were some of the last significant remnants of dry forest on the Pacific side of Latin

America, a fragile mangrove swamp ecosystem, seasonal rivers, and a beautiful and broad set of beaches used as nesting sites by endangered olive ridley, leatherback, hawksbill, and Pacific green sea turtles. Santa Rosa was a mosaic of deciduous forest, evergreen forest, semi-evergreen forest, and old pasture in the process of regenerating into forests. (Evergreen trees in the dry forest are large, broad-leafed species that maintain their leaves all but about two weeks of the year, in contrast with deciduous trees, which drop them during all or most of the dry season.) The park stretched from the Pacific to the highway, rising from sea level to about 1150 feet. In all, the park was home to thousands of insect species and hundreds of species of plants, mammals, birds, reptiles, and other animals. Santa Rosa was one of twenty-nine units in Costa Rica's national park system at that time.

The storm clouds flowing over the volcanoes finally reached us. When thunder rumbled above, we scurried off the platform, like sailors clearing the conning tower of a diving submarine. As we descended the steps to the Casona, a strange noise approached from behind, heading straight for us. At first it sounded like the whining tires of a Mack truck speeding down an interstate highway. We froze. The whine built into a high-pitch buzz. Suddenly, a large swarm of bees passed just above the tree canopy. The dark swarm sped on and the whine faded, leaving behind only the swish of leaves in the wind and a murmur from the clouds.

"Are they Africanized bees?" I asked, a little worried. These are aggressive honeybees known for deadly mass attacks on people who venture near their nests. Only a few weeks hence, a swarm would sting to death a graduate student from the University of Miami at a wildlife refuge a few miles south of here.

"I couldn't tell," Godfrey answered, calmly, thoughtfully. "They could be. Or they could be a different kind of bee or wasp."

After a few moments, Godfrey resumed the descent. I followed, a wave of goosebumps running up the back of my neck. My thoughts turned to the apparent risks that tropical biologists take when they go out into the field every day to do their research, whether at Santa Rosa or the more famous tropical forest research centers of La Selva, on the Atlantic side of Costa Rica, or Barro Colorado Island, in Panama. I'd read about Africanized honeybees; snakes like bushmasters and fer-de-lances; malaria; tree falls; scorpions; jaguars; caimans, a kind of small crocodilian that lives in the rivers near Santa Rosa; and crocodiles, which live in the estuaries, close to the ocean. Some of these were legitimate threats, some not—for a wide range of reasons.

As Godfrey and I walked the rest of the way down the slope, no snakes charged, no bees swarmed from behind a tree. Both are exaggerated, if not silly notions, I later learned. But such are the images of the tropical forest often "revealed" in popular books, magazine articles, and Hollywood movies. I gradually realized that it's best to leave such baggage at the forest door. Give due caution, but don't let the image of a man-eating forest block the path to understanding the true workings of tropical nature.

During the rest of the hike, the only surprise to greet us was a flock of several dozen parakeets whose green feathers shined so brightly against the dark gray clouds that they seemed fluorescent. They criss-crossed rowdily in the air in front of us, chattering like a concert hall full of kids screeching on kazoos. The rain began to fall, and the birds perched in a large tree, continuing their jabber as we strode back to the park headquarters and the shelter of our cabin.

A few hours earlier, Godfrey had picked me up in Liberia, the capital of Guanacaste Province. I had ridden a bus for about five hours from San José, Costa Rica's capital city, just over 100 miles to the southeast. As we drove out of Liberia, we passed the provincial jail, around which several troops from the Civil Guard loitered in olive green uniforms, with black leather holsters glistening in the sun. Godfrey had heard that two American and two French mercenaries were among those held in the jail awaiting trial. They had been captured during a Civil Guard patrol near the Nicaraguan border, and the word was that they had tried to establish a base camp to aid the contras, the U.S.-backed guerrillas fighting to overthrow the Sandinista government in Managua.

We drove north on the Pan-Am Highway, and twenty minutes later, only a few miles south of Nicaragua, pulled into the entrance to Santa Rosa. An enormous field of jaragua, like a straw welcome mat, stretched for several hundred yards on either side of the asphalt entrance road, choking out all but a few spindly trees. It was shocking to see this biological desert at the front door to one of Costa Rica's most celebrated natural areas. The jaragua extended well into the park—more than a hundred yards down the road to where the gatehouse stood.

After the park guard on duty waved us in, we cruised down the curvy road, passing, intermittently, vast fields of jaragua and patches of tall trees that loomed along the roadside. Some 3 miles in, we stopped and hiked off the road to the Mirador Santa Elena, a promontory facing north and west. There, across a great green valley and beyond the park border, were the Santa Elena mountains.

As we walked back to the car through a jaragua field, we saw a faded blue Toyota Land Cruiser coming down the park road. The vehicle stopped, and the driver shouted something through the rolled-down passenger window. He had an angular face set off by dark eyebrows, and gray shocks of hair curled out from his full salt-and-pepper beard and his balding head. He was clad in a dark green cotton field shirt, and around his neck hung a thick string attached to a pen stuck in his shirt pocket. Godfrey introduced me to Daniel Janzen.

At that time I only knew a few things about Janzen: He was an eminent tropical ecologist based at Penn and Santa Rosa; he had taken much of the cash from the $100,000 Crafoord Prize, the so-called "Nobel Prize of Ecology," won in 1984, and used it to bring telephone and electric service into the Santa Rosa administration area; and he was pushing a plan, known then as the Guanacaste National Park Project, to restore the forests of Santa Rosa and expand the park into a much larger natural area.

Janzen, then forty-seven, told us that he was returning from negotiations to buy some nearby pasture land. He seemed to be barely able to conceal a restlessness—it was perhaps even impatience—beneath a thin veneer of politeness. As I discovered later, he was the quintessential man with a mission, pursuing it with energy and single-mindedness that I have seldom encountered in more than twenty-five years of observing scientists.

Off Dan Janzen sped.

Godfrey continued showing me around Santa Rosa. Like Janzen, he had a full salt-and-pepper beard, was of average-to-thin build and stood a bit under six feet. But unlike Janzen, he carried an air of patience and friendliness. He enjoyed teaching what he knew, and at the most unexpected moments he would reach for a joke, even if only the simplest pun. Godfrey, then forty-two, had done all of his scientific work in North America, and at that time worked for the Illinois Natural History Survey. This was his first time in the tropics, too.

We continued driving the roughly 6–mile road between the entrance and our destination: the administration and living area of the park. We stopped and got out of the car at a place called Bosque Húmedo, a patch of old-growth forest whose moistness stands in marked contrast to Santa Rosa's predominant and newer dry forest. Although it was early afternoon on a sunny day, this part of Bosque Húmedo was darkly shaded and wet. Godfrey had seen monkeys here only a few days before. Sure enough, a group of capuchins moved high in the trees across the road right above us. These fine-boned, black primates with white shoulders and faces picked their way from branch to branch in a steady wave. A small monkey clung to its mother's back. Others leapt in graceful arcs across gaps in the forest canopy. The troop was silent, save for the swish of a branch now and then as it flexed under the weight of a landing monkey.

After watching for a few minutes, what seemed like a loud burp burst from a treetop down the road. We strode toward it and were greeted by a chorus from a small band of howler monkeys. The closer we came, the louder the chorus, until it sounded like a pack of large, wild dogs barking at us. From about 40 yards away, I could see they were larger, darker, and chunkier than their white-faced cousins, and they sat in the branches as if at anchor. Something in the power of their howls pulled us to a stop well before we could get under them and take a good look. Just as well, we later learned. Howlers have a tendency to urinate on humans.

At the administration area, the hard surface gave way to a part blacktop, part rock-and-mud road that looped around a wild clump of trees and shrubs. On the outside of the loop stood a string of simple, small houses and cabins made of cinder blocks and corrugated fiberglass or metal roofs. These housed the park administrator, park guards, and scientists. Other nearby structures included the *comedor*, or dining hall. A few tents were pitched among the houses. It had rained constantly for two weeks, and the area looked as if the floodwaters of a river had just receded, leaving pockets of water, mud, and debris everywhere. Lizards of all sizes lurked in the ditches.

A few minutes after arriving, Godfrey and I saw what was later identified as a coral snake slithering along one of the tents. About a foot long, it had yellow and black bands, with at least one iridescent red band near its head. I remember thinking how odd it was that such a small, beautiful creature could be so deadly. As I learned later, though, it was not odd at all. The coral snake's distinctive color pattern is meant to warn away potential predators, as if to say, "Make no mistake—I am lethal."

Word spread about the sighting, and several people came by to ask questions about it, for the snake was by now gone. They seemed genuinely excited, as if we had witnessed something special, like a shooting star or a president. The only people whose enthusiasm was tempered with a shade of concern were Frank and Katy Joyce, two Cornell graduate students under whose tent the radiant celebrity had slithered.

Godfrey and I shared a room in a cabin that housed two other researchers, a volunteer, and a park guard. The walls were covered with maps and posters, including one calling for action against U.S. military intervention in Central America. Eighteen brown papier-mâché wasp nests swayed head high in the middle of the room, hanging from thick string tied to a beam. Frank Joyce had made these as part of his study of birds that build nests near wasp nests. The bathroom had a dark, moldy shower stall with cold running water, a sink, and a toilet that drained along a pipe to a septic tank in the woods nearby. A cardboard box next to the stool was meant to receive soiled toilet paper, since to place it in the toilet would clog the drainpipe. Placing paper in a container next to the toilet is common practice in many Latin American countries.

The house was, shall we say, "well ventilated," with openings of all kinds, including tattered screens that ran between the top of the walls and the roof. Snakes and insects regularly visited, and in the middle of the night rats scurried up and down the walls. John Thompson, a Canadian biology student who slept in the room next to ours, learned you can't just plop into bed at the end of a long, hard day. One night he was about lie down when, just in time, he noticed a scorpion on his pillow.

As we returned from supper that first night at Santa Rosa, Godfrey found a so-called kissing bug on the outside wall of the house near the door. Called the chinche chupasangre bug by Costa Ricans, Godfrey knew it as *Triatoma*. It had a dark, oval-shaped body about an inch long and a narrow snout that seemed to extend down like another leg. Godfrey explained that this blood-sucking insect, named for its reputation for biting on or near the lips, sometimes transmits a parasite that causes Chagas' disease, a chronic wasting malady found in Central and South America. Some victims have described the bite as feeling like a red-hot needle stuck in the skin, though others say it is painless. The bug transmits disease by brushing parasite-laden feces over the wound as it walks by following its blood meal. Only a few cases of Chagas' disease are reported each year in Costa Rica, but experts estimate it has infected some 20 million people in Latin America, as well as 50,000 people in the United States. They estimate that thousands of people in Latin America die each year from the disease, often from problems caused when the parasite lodges in heart muscle. A smooch from a kissing bug in South America reportedly plagued Charles Darwin after his voyage on the *Beagle*, although some medical historians debate this claim.

Ever the curious naturalist, Godfrey picked up the bug from behind, pinched its wings to its body, turned it over, and took a close look at its underside as its legs rapidly treaded air. After a few moments of quiet study, he tossed it into the quickly darkening dusk, and it disappeared. He must have caught my worried expression.

"Pucker up," he said, smiling.

After dark that first night, Godfrey suggested I join him in a sortie into the forest. As part of his research, he had collected dozens of live caterpillars at Santa Rosa and was raising them to determine their species. The guides that biologists use to determine the species of an animal often focus on the size, shape, color patterns, and other physical characteristics of the creature in its adult stage. So Godfrey needed to rear the caterpillars, known as larvae, into adult moths to identify them. To rear them, he needed to supply them regularly with fresh food, which meant gathering sprigs of the same kind of plant on which they had been captured. Known in biological parlance as the host plant, it was the particular plant species that the caterpillar liked to eat and on which its mother laid its egg. Each species of moth caterpillar has one or more favorite host plant. Now was the first chance Godfrey had had all day to gather some for his captives.

He stuffed plastic bags into a canvas rucksack, and we drove down the blacktop road a couple of miles out of the administration area. He stopped the car near a trailhead, and we walked into the forest, each of us scanning the trail for snakes with the beam of a battery-powered headlamp. I worried about a report of a recent jaguar sighting near the houses, but kept the concern to myself.

It had rained heavily after our walk to the Casona, and another shower had washed through just after supper. The air was warm and wet, and my shirt turned dark from sweat and from the huge drops of water that fell from the forest canopy. The drops smacked the ground all around us, too—so many of them that it created a kind of hissing noise beneath the steady chirp of crickets and frogs and the close-range ring of mosquitoes. Godfrey stopped when he found the host plants he needed.

Godfrey was a lepidopterist, meaning he studied the group of insects that consists of butterflies, moths, and their caterpillars. Within this group he concentrated on a family of moths called the Noctuidae. As the name implies, the adult noctuids are active at night, as are most other moths. During this time, they mate, lay eggs, locate moisture and nectar sources, and avoid predators.

Godfrey had spent countless nights and early mornings in North America collecting moths in an effort to document the diversity of their forms and life cycles. Thus, by definition, he also was a taxonomist, one of the scientific great-grandchildren of Carolus Linnaeus, the eighteenth-century Swedish botanist who started the effort to inventory all the items of nature using a binomial system. In this system, each organism known to science receives a two-part Latin name, one for its genus and one for its species.

The effort to inventory nature in this way is called taxonomy. Taxonomists believe that to understand how organisms relate to one another and their environment, one must first know what they are—must describe and classify them. Classifying them reflects similarities and differences in the physical structures of animals and plants, and, more important, it is an exercise necessary to sorting out their evolutionary relationships. Taxonomy has been described variously as the backbone, nucleus, road map, or central nervous system of the biological sciences, since to find, describe, and name a species is the first step toward understanding it. Taxonomy also is a window through which to observe and think about the diversity of the natural world—not just the diversity of species, but also the diversity of habitats, ecosystems, and landscapes in which they live.

In the muggy darkness, amid the continuous racket of droplets and critters, Godfrey patiently pulled off small branches and leaves of the host plants and stuffed them in plastic bags. The dreamlike scene was interrupted by a flashlight shining through the underbrush at us, then voices. Swift human shadows came down the path.

They were park guards, half a dozen men in jungle fatigues, one of whom carried a carbine. An hour earlier, just after nightfall, they had spotted a light in the forest and, assuming it was a poacher, perhaps in search of the jaguar, formed a search patrol. By the time they got there, the poacher was gone, but when they saw our lights, they came toward us, believing we might be the culprit. The guards quickly recognized us, and, like quiet, powerful antelope, disappeared into the dark forest as swiftly as they had come.

The next morning, Frank Joyce, the Cornell doctoral student, offered to show me his research project. Joyce was tall and lanky, with a disorderly crop of black hair and a wisp of a black goatee. Sweat darkened his olive fatigues. Joyce had first come to Santa Rosa in 1980 as a student in Janzen's three-week tropical biology course at Penn. As part of his Ph.D., he was studying the nesting behavior of the banded wren, rufous-naped wren, and yellow-olive flycatcher. He hypothesized (and later demonstrated) that these birds often build their nests near wasp nests by choice, and that this gives their fledglings a benefit by adding protection from monkeys and other predators, which avoid wasp nests and the possibility of inadvertently disturbing the wasps and unleashing their ire. The birds nested near both real and artificial nests. When I asked for more details, he offered to demonstrate.

Things proceeded innocently enough, at first. Joyce took me to a tall tree not far from the house. He pointed to a wasp nest about 15 feet above. It looked like a large, gray pasta shell glued to the trunk, and it blended in well. As he described the nest, he stared at it alertly, as if he had just snuck up behind a sleeping guard dog. I couldn't reconcile his subtle smile with the look of caution in his eyes.

"O.K. Here we go," he said. "Get ready to run."

Huh? I looked at him in puzzlement and was about to ask him to explain. But suddenly he slapped the tree trunk hard and fast three times, then sprinted away, his boots pounding on the forest floor.

"Go! Go! Go!" he shouted.

I started running after him and within a few steps realized why he had issued the order. A huge swarm of angry wasps flushed out of a hole at the top of the nest with a loud whoooosssssshhhhhhhHHHHHH. It was somewhere around the time when the small "h's" became large "H's" that I passed him.

One of the wasps caught up with Joyce and landed on his back. It stung his hand fiercely as he tried to brush it off, but the only sound he made was a barely distinguishable hiss, as if he had just slapped on some cologne after a particularly rough shave. In half a minute, we had outrun the wasps, and Joyce stopped to examine the quickly rising, painful welt on his hand. That, he said, is how tropical wasps react to creatures that threaten their nests.

My pleasant smile, eager eyes, and breathless "Uh huh" covered a deep inner suspicion: Was this some kind of hazing ritual for journalists? But Joyce had none of that in his eyes, nor in his attitude. It was clear that he had intended no harsh initiation. Just the purest form of learning by doing. I followed him for the rest of the afternoon as he wandered through the forest with an aluminum step ladder, removing the young from bird nests at one of his study sites and charting their growth by weighing and measuring them.

At Santa Rosa, Dan Janzen lived in a house about 50 yards from ours, fronted by a large guacimo tree. On the ground near the base of the tree lay piles of large seed pods and the weathered skull and pelvic bone of a horse. I often saw him leaning over two students seated at a weathered picnic table outside the front door, directing them in the finer points of processing insects. These usually had been collected somewhere in the park and piled in a box on the picnic table. Janzen himself would dart into and out of the house attending to various tasks, or sit at the table gently but efficiently picking up moths from the mass of insects and carefully placing them on a board to spread their wings.

After dinner on the day of Joyce's wasp demonstration, I stopped by Janzen's house to ask him for an interview about the Guanacaste National Park Project. He continued his busy movements about the house, saying he was swamped right then but might have an hour sometime in the next few days. He seemed distracted and gruff. It caught me by surprise because he had seemed friendly, even outgoing, when we had first met near the Mirador Santa Elena the day before.

I returned to my cot and pulled out a package Janzen had sent in February. Under the 60–watt glow of a bare light bulb, and while Godfrey tended his caterpillars, I looked over the seventy pages of documents inside. Night fell outside the house, and a symphony orchestra of frogs and toads tuned up and began to play. The orchestra featured a range of instrumentalists, from the most profound basses, which issued thunderous booms, to piccolos, which cried out with screeches of the highest possible octave. Their patterns of play ranged from simple strokes on the E-string, to tones that sounded like the pull of a hand saw or the space-laser "boing-boing-boing" of a video arcade game.

Three thoughts struck me in particular about what Janzen had sent. For one, in the cover letter, he had written, "George Godfrey tells me that you are likely to be coming to visit Santa Rosa this spring. If so, there is a thing happening that you should know about. It is the other half of what field biologists do, or had better be doing if they want to have anything to do biology in in the future." What an uncharacteristically drastic, even threatening statement from a scientist.

The second striking thought was contained in the beginning of a manuscript marked "rough draft": "There was once as much dry tropical forest in Central America and tropical Mexico [referred to herein as Mesoamerica] as there was rainforest. Today there is still enough Mesoamerican rainforest that we can argue over what to save and what to cut. However, of the original 550,000 square kilometers [about 212,000 square miles] of dry forest on the Pacific side of Mesoamerica, equal in area to about five Guatemalas, today only 0.09 percent [181 square miles] has conservation status and at least 98 percent of the remainder has been cleared or so severely altered that it and its species have only relictual status." One-fourth of that 181 square miles was within Santa Rosa alone.

By conservation status, Janzen meant the land was part of a national park, forest reserve, or some other designation that protected it, at least on paper, from use by loggers, poachers, and others. By relictual status, he meant that the forest, and its species, were restricted to such small, isolated pockets that they could be considered mere remnants of animal and plant populations, on the precipice of local extinction, like broken pieces of pottery from some ancient culture, crumbling away in the wind.

This dire situation hadn't developed overnight. In fact, waves of destruction had eroded the shores of the dry tropical forest for centuries. "The conservation battle over large blocks of Mesoamerican dry forest would have had to have been fought in the 1800s, if not earlier," he wrote. "The first Spanish haciendas in Mesoamerica were established in the 1500s on dry forest land already subject to indigenous agriculture. Today, many of Mesoamerica's breadbaskets, cotton fields, and cattle pastures occupy what was once the dry forest habitat."

The third striking thought in the material was Janzen's proposal to use Santa Rosa as a nucleus around which to create the first large tropical forest restored by man with secure numbers of all the species native to the area at the time of the Spanish arrival. The new preserve would be five times larger than Santa Rosa and contain far more habitat types than it already had. It would be the first national park in the New World tropics not only based on restoring forest, but designed from the outset to serve multiple uses as a research station, tourist site, and education center, with its managers living in the park. It would be called Guanacaste National Park, after the host province and Costa Rica's national tree, the guanacaste.

This was a revolutionary idea. Although elsewhere in the world there had been efforts to restore individual species of animals and plants to forests and other habitats before, like rearing an endangered bird or mammal in captivity and returning it to the wild, no one had yet attempted to restore a whole forest, a whole region of habitats, of interconnected ecosystems. In that sense, the Guanacaste project was unique. In the global war against deforestation, it represented a brand new and uncharacteristically offensive thrust, in the military sense of the word.

It was time for sleep, and Godfrey and I were now treated to a biodiversity performance that we came to appreciate every night. It was set up by the fact that all evening, insects in the forest around us had been answering their behavioral imperatives, enticed by our light bulb to join us in our tiny room. They snuck in through cracks around the opaque window above Godfrey's cot, through tears in the screen near the eaves, and through countless other rifts, most of them too small to notice. There were moths, beetles, flies, wasps, flying ants, crawling ants, fireflies, and spindly creatures of all kinds. Once inside, they spun about the light bulb, climbed all over the walls, floor, and furniture, or came to rest on bright surfaces. They constantly rained on our backs and heads. It was a harmless rain, and we accepted it as part of the territory, like sweat, mud, and the shrill symphony of frogs.

When Godfrey switched off the light, the real fun began. Put simply, the insects inside the room went crazy. Deprived of light from the bulb, their incandescent homing device, they scrambled for the new light source. That, of course, was now the relatively faint moonlight outside, which they perceived through the window. To them, the weak light was no subtle halo, no barely perceptible gray glimmer. It was an aurora borealis of the bug world, and it must have punched on the impulse to flee. The reaction of the insect mob was as if one of them had suddenly shouted, "Fire!" and all had panicked and instinctively fled.

Alas, the insects weren't quite as good at getting out as they had been at getting in. So for more than an hour, the room was a madhouse of buzzing, circling, crashing insects. They smashed into screens, windows, walls, and clothes that hung on lines stretched above our cots. A typical flight pattern was traced in the dark by a firefly with a bright turquoise light that flashed every second or so. Repeatedly, it circled the room twice, banged into the window, bounced away, recovered, circled the room twice, and banged into the window again.

The beetles buzzed loudest and hit the windows with the most percussive cracks. A particularly noisy, low-pitched beetle circled with the greatest arc short of hitting the walls. We called it "the B–17." Moths flew, too, although almost silently. The only clue to their activity was the soft flutter of delicate wings as their flight paths passed my ear.

As we lay there in our undershorts in the hot, humid air, brushing insects off our bodies, I remembered Frank Joyce's words that afternoon while he watched his hand balloon to twice its normal size. "I hate to sound masochistic, but I really *did* want to know what the sting of this particular wasp felt like," he said, and he pulled a small bottle of pain-killer out of his pants pocket and sprayed it on the hand. What he meant was that by knowing the sting, he was coming to know the forest better. And what I realized, laying there that night in the house, was that Joyce chose to know the forest better by becoming part it. Since the wasp was part of the forest, and the sting was part of the wasp, and the swelling was part of the sting, then Joyce had become part of the forest. Godfrey and I, caught up in the whirl of insects, had become part of the forest, too. We accepted the frenzy, and laughed.

Eventually the creatures made their escapes or otherwise calmed down, with a few exceptions. The firefly stopped circling and buzzed at the window in tandem with an ichneumon wasp, a kind of small, slender wasp whose name is derived from the Greek word for tracker. Through all the insect commotion that night, I felt nary a bite, but sometime in the early morning, after the room had cooled and I had slipped under a sheet and fallen asleep, I was awakened by a sharp nip on my right buttock. Thoughts of a misguided kissing bug rushed into my head, and I quickly reached down and crushed the pumpkin-seed-size critter between two fingernails. I picked up the carcass and put it on the floor beneath my cot, thinking I could identify it in the morning. I rubbed the smarting skin and dozed off.

By dawn the mystery bug had disappeared from the floor, its body carried off, perhaps, by the rats, or by ants. The firefly, too, had departed. But the ichneumon wasp, now excited by the sun's powerful rays, still buzzed at the window.

■ *2* ■

# *Cañón del Tigre*

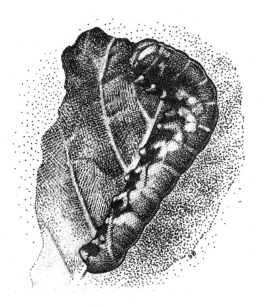

The caterpillar swung back and forth in front of my eyes as I strode through a patch of forest in Santa Rosa National Park. I had been concentrating on the path to avoid some rocks, and the caterpillar seemed to suddenly dart into my face. About the size of a matchstick, brilliant green with black stripes running the length of its body, it had been hanging from a high branch by a thin silk, which became attached to the brim of my hat when I walked by. It startled me at first, but after a moment, I laughed at the idea of a caterpillar hanging by a four-inch string in front of my face. Like some obsessive hypnotist reincarnated as an inchworm, it swayed to and fro with the rhythm of my gait. We walked along together like that for several minutes, until the silk snapped and the caterpillar fell into the greenery along the path.

George Godfrey and I were hiking from the Santa Rosa administration area to Playa Naranjo, the Pacific beach that formed much of the western boundary of Santa Rosa. It was a 15-mile round trip, and it took us about eight and a half hours. The trek, on June 10, 1986, provided more insight into the makeup of Santa Rosa, its creatures, and their interactions, and into how a remote natural area can become a stage in an international conflict.

Carrying packs and cameras, we had departed at 7:30 a.m. The air at that hour was already hot and humid, and the sun's rays seemed to pierce my shirt. We walked slowly but deliberately down a muddy, two-track road through the old pastures, flanked on either side by shoulder-high jaragua grass, brown and dry even in the midst of the rainy season. We sweated profusely, and I found myself taking great gulps from my water bottle.

Dozens of turtles wallowed in mud puddles, and a few horses grazed on the pastures. We came into and out of forest patches, walking over rolling hills and across shallow streamlets. We saw and heard deer crashing through the woods, and lizards, frogs, butterflies, and birds darted about everywhere. I was "on alert" for snakes and peccaries, the Central American version of a warthog. The hyped anecdotes about dangerous animals in a few tropical backpacker adventure books I had read before going to Costa Rica were, unfortunately, still on my mind. I was a fairly experienced North Woods hiker, but this was the *jungle*, man, according to these writers. Nature very red in tooth and claw. So against the background of our gently thudding footsteps on the wet ground, any sudden noise came as a shock. When I heard the grunt of a peccary around a bend in the path and saw a few furry hulks in the jaragua ahead, I froze in fear and looked for the nearest tree to climb. Godfrey pulled up silently, too, but he seemed to enjoy the event, and he watched the peccaries studiously. After a few moments, they moved off the path, and he calmly stepped forward.

Godfrey pointed out the fresh prints of a cat's paw among the horse tracks in a muddy section of the road. A rather large cat, actually. We stopped to examine them, discussing where the cat might have gone after it passed this spot, and how long ago. I looked around at the wall of jaragua, wondering where the big cat was. I didn't know at the time that jaguars are rarely seen in Guanacaste.

The forest calmed my concern, a fresh, rich scent wafting out of the woods. It came, I think, from the previous night's heavy rain, which soaked into the web of life as if into the roots of a rose, releasing whole ranges of musky bouquets that blended to become one earthen aroma. There was humor in the forest, too, if you chose to look at it that way. The caterpillar swinging from its silk was a case in point. Each time we entered large clumps of forest, it became necessary to dodge a vast obstacle course of caterpillars hanging on translucent strings. And if you stopped for a moment to listen, you could hear, above the periodic calls of the birds, above the ring of the mosquitoes, a subtle but steady rain of frass— caterpillar droppings.

Frass rain was just another sign that the dry tropical forest of Santa Rosa had recently reawakened. Triggered by the onset of the rainy season in May, plants had burst to life, and vegetation flourished. That, in turn, had triggered a riot of animal life. Godfrey and I had walked right into the middle of a dramatic display of the tropical dry forest's seasonality.

Most tropical forests—even rain forests—have *some* seasonality. In northwestern Costa Rica, the dry season lasts about six months. Seasonality of rainfall was "a dominant ecological force," as biologists Peter Murphy of Michigan State University and Ariel Lugo of the U.S. Department of Agriculture put it in a 1986 paper about tropical dry forests around the world. Most organisms of the forest synchronize their reproduction and growth to the availability of moisture.

The dry forests in and around Santa Rosa contain some evergreen species. The crown of old-growth dry forest can be between 20 and 50 percent evergreen, but one of the prevalent characteristics of younger dry forests, one of which we were walking through, is that between 60 and 75 percent of all the tree species are deciduous, putting out a new set of leaves around the time the rains come in May. Many species wait until the rain falls to produce flowers and leaves, but for many others, flowering and leaf flush occur weeks or months before the rain begins.

Although the forest bursts with greenery several times during the year, the biggest burst centers around the annual coming of the rains. It is a time when the forests reach out with their greatest mass of greenery to capture energy from the sun, through photosynthesis. Quiet yet omnipotent, obvious yet so often overlooked and underappreciated, this silent biochemical industry of leaves is really the basis for all life on land, including human life. Without photosynthesis, and without the complicated chain of ecological interactions it supports, we simply would not exist. Photosynthesis lifts life off the landscape, giving plants the energy to grow.

Vegetation-eaters, or herbivores, are the next link up the chain of life. They become extremely active and abundant with the flush of new leaves, timing their new generations to take advantage of the bounty provided by plants.

The explosion of growth had filled the air with little green worms, and it seemed that at least half the shrubs and trees along the road had pieces chomped out of their leaves. More often than not, when we stopped for a minute to look, we could see at least one caterpillar, and usually more, clasping a leaf and chewing away like a hungry child on buttered sweet corn.

In scientific parlance, when this eating binge is severe it is called a "defoliation event." Santa Rosa experiences many defoliation events, and a long cast of characters takes part in these dramas: hundreds of plant species, more than 3000 species of caterpillars that munch away on leaves from the forest floor to the crowns of the tallest trees, and hundreds more kinds of animals that eat parts of living plants—or prey on caterpillars.

A typical example of a defoliation event was one studied by Janzen during the 1983 rainy season, when a moth known as *Aellopos titan* assaulted trees of the genus *Randia* in a patch of Santa Rosa's forest known as Bosque San Emilio. He described it in a 1985 paper, "A Host Plant Is More Than Its Chemistry."

From May 28 to June 4, thousands of adult moths flew through the forest, darting among the small trees and shrubs and landing on branches and buds. If a moth found a *Randia* tree, it sometimes laid a small, pale green egg on one of the plant's thorns, twigs, or newly expanding leaf buds. It then darted to another part of the same plant, or to a neighboring plant, and did the same.

The result was that most *Randia* plants in San Emilio carried dozens or thousands of *A. titan* eggs. Meanwhile, the enemies of the moth began an assault. Various kinds of ants carried off the eggs. Birds and pirate bugs fed on them as well. In five or so days when the remaining eggs hatched, more predators attacked the emerging caterpillars: ants, birds, wasps, reduviid bugs, spiders, and beetles. But the moths won the battle with numbers. They kept laying eggs, more eggs kept hatching, and by the end of the first week of June small caterpillars, their pale green color blending well with the new leaves on the plants, wandered and gorged on their hosts, sometimes only one or two per plant, but sometimes by the hundreds.

By June 18, the surviving caterpillars reached the largest stages of larval growth, known as the penultimate and ultimate instars. Some crawled or dropped off the plants and onto the forest floor, where they began the next stage in their life cycle by forming pupal cases in the litter. Other species, such as the ones that rappel down silks, feast in other parts of the forest and use different methods to descend and begin the pupal stage. Or, they may not descend at all, preferring instead to pupate somewhere up in the plant.

The *Randia* trees had been defoliated to various degrees, including many that were totally stripped of their leaves. Thousands of the larger caterpillars had fallen victim to still more birds and predaceous insects, as well as a new level of predator that included scorpions and unidentified mammals, possibly a gray fox or small cat. (One scat specimen likely to have come from such an animal contained the head capsules of twenty-six *A. titan* caterpillars.) Disease killed more. Despite the slaughter, many moths hatched and flew out of the area, presumably to mate and lay eggs for the next generation in another location, where predators may not have been so primed. By late August, however, the *Randia* trees were producing a new crop of leaves.

The story of *A. titan* and *Randia* is replayed constantly by thousands of other species of animals and plants in this forest. And although to the naked eye of a passer-by each story may look like the simple tale of an animal eating a plant, it is never just so. Hidden behind each is an intricate interplay of phenomena like host-plant defoliation, moth population cycles, and predator-prey balance. These stories are all just a part of the stunningly complex drama that is the tropical dry forest.

When they hear about a forest in the tropics, most people think of "rain forest," as in "save the rain forest." This is at least in part due to the success of conservation groups getting their message out in the 1970s and '80s about the crisis of tropical deforestation. It's really quite remarkable how the idea of the rain forest has changed in recent decades. Before then, it was called "the jungle," and its popular conception was the dangerous, disease-ridden, mysterious hell on Earth described by authors like Joseph Conrad in *Heart of Darkness*, or the leech-and-mosquito-infested sweatbox that assaulted Humphrey Bogart and Katherine Hepburn in *The African Queen*. Recently, a more benign view of "the rain forest" has emerged, a perspective that modern writers fill with vivid colors, rare orchids, exotic animals, and warnings of impending loss. Adoration and understanding have replaced fear.

Long before all this, the tropical dry forests dropped out of view, removed for fields and pastures. Dry forest may well be the most invaded and least studied of the widespread tropical habitats. Compared with rain forest, the tropical dry forest, often called tropical deciduous forest, had until the mid–1980s been somewhat neglected by scientists and environmentalists. Perhaps the rain forest's reputation for richer species diversity has given it an advantage. Perhaps rain forests seem more remote, pristine, or mystifying. Or perhaps the general public doesn't really distinguish at all between dry forests and the three basic categories of rain forest—rain forest, wet forest, and cloud forest. In other words, the term "rain forest" is used for all four types in popular culture. The nonscientist, in most cases, would be hard-pressed to see a difference when walking through these forest types during the rainy season.

What is a tropical dry forest? Technically it is any forest in the tropics that annually averages rainfall of between 10 and 80 inches and has four to eight months of dry weather. Santa Rosa averages about 60 inches. By comparison, a tropical rain forest averages between 80 and 160 inches.

Dry forests are generally shorter and less complex than rain forests in their three-dimensional structure, that is to say, the structure of the trees and other vegetation. Their leaf area is typically less, too. Yet dry forests are richer in seed and animal biomass than rain forests and usually have a comparable number of vertebrate species. Despite running second in diversity, they are still incredibly diverse. An acre of dry forest can be home to 90 species of trees and thousands of species of animals and other plants. Santa Rosa's 42 square miles contain an estimated 700 species of plants, 115 species of mammals, 170 species of resident birds, and well over 30,000 species of insects.

Perhaps the most crucial distinction of tropical dry forests is their threatened status. Humans have long had more of an affinity for tropical dry forest than rain forest, or at least for the land on which these kinds of forests grow. In their 1986 paper, biologists Murphy and Lugo attributed this affinity to historical patterns of settlement, richer soils, the relative ease with which dry forest can be cleared, and the fact that cattle and many crops favor the drier climate. Malaria and other human diseases may pose less of a threat in a dry environment, and the dryness makes these forests more easily removed with fire.

We descended from Santa Rosa's sun-drenched upland mosaic and into the dry forest of *Cañón del Tigre* (Canyon of the Jaguar). Tall trees and brush flanked the rocky road, forming a deep curtain of greenery that you could stare into for a few dozen feet before the view disappeared in dark shadows. Heavy rain had washed out at least one of the road's tracks along extended stretches, gouging ruts that sometimes dove five or six feet into the ground. I watched these dark subterranean trenches with fear of snakes. They never produced any, but the woods did. We frequently heard things crashing through them. Once, a snake about the thickness of a baseball bat and 10 to 12 feet long slithered away through a tangle of brush to our right. It stuck its head up behind a fallen log about 15 feet away, watching us a bit brazenly, I thought.

We paused every so often to talk about caterpillars on the shrubs and saplings along the trail. Mosquitoes the color of shiny blue steel swarmed us; these were the "blue devils," so named for their propensity to vex anyone standing still while trying to work in the forest. They can spread yellow fever, although this did not occur in Santa Rosa at that time. As biologist C. L. Hogue put it in *Costa Rican Natural History,* "They are strongly attracted to man and indomitable in their attacks, during which they emit an irritating high-pitched whine."

I certainly found them irritating, but Godfrey seemed unperturbed, even serene. He focused on his interest: How caterpillars deal with their leafy environment. In North America, he knew, the jaws of some caterpillars change as they break out of their eggs and go through a series of growth stages. He wanted to learn how the jaws of caterpillars he found in Costa Rica compared with those of related ones in the temperate zone. The information, he believed, could shed light on feeding behavior and evolutionary relationships.

At one of our stops, Godfrey squatted like a baseball catcher taking photos of a bush infested with caterpillars, and I stood 12 feet up the path, looking out into the forest. All was quiet. I switched my focus to a cluster of saplings at my feet. One leaf had a chunk of material chewed out of one edge near the tip. Others had all sorts of holes, small and large, heart-shaped, oval, or in completely random patterns. Some leaves were eaten from within, some along the edges. Some had shriveled, some appeared shredded, the torn parts brown and curled. Some had been "mined," a process in which the insect gets inside and eats the tissue between the top and bottom layers of the leaf, leaving clear, cellophane-like patches.

Several other insects toiled on or around the leaves: flies of many sizes, small black wasps, spiders that had stretched webs among the branches, a small yellow beetle. Every so often a large wasp, yellow with black bands around its abdomen, landed on a leaf, walked around, and took off.

For a minute or so I watched a tiny caterpillar gouge one of the leaves, its black head swinging back and forth like a gold miner's pick striking the mother lode. It was clearing a wide swath on its way toward the leaf tip; other miners, now gone, had cut different routes in nearby leaves: straight lines, S-curves, and chaotic paths. I lifted the leaf to watch from below. A row of white eggs as perfect as a string of pearls stretched along the ridge of the central vein.

A dark shadow crossed into my field of view, interrupting. I looked up from the saplings and watched as the forest produced a large cat, trotting nonchalantly through the brush parallel to the road about 10 feet below me. As it happened, I was straddling the intersection of an animal trail and the road, and the cat, somewhat mechanically following the trail, turned toward me, looked up, and froze.

"O.K., George, here we go," I said to Godfrey, in a loud but still conversational tone. Only a few feet away, the cat stared with deep, black eyes. It seemed about the size of a large timber wolf. It looked somewhat gray in the shade, although there was a tinge of yellow on its flanks, which were covered by a beautiful pattern of open, butterfly-shaped black rings with a dot in the middle. The tail stuck lazily out behind. The paws were hugely out of proportion to the body, and the ears were rounded and erect, the mouth tightly closed.

Those eyes. Those deep, black eyes. There was something strange and wild in them, something that said, in a rather regal way, "This—is—my—forest." There was something shrewd in those eyes, too, and I knew that behind them a brain wired with an ancient and very efficient intelligence was calculating the answer to a basic question: "Kill or run?"

Later, back at the Santa Rosa administration area, Godfrey and I talked with other scientists and determined that the cat was a jaguar, known as *el tigre* in Spanish. It must have been a female or a young adult male, since it didn't seem nearly as heavy as the 220 pounds some males are said to weigh. They pointed out that we must have been downwind of the cat, or it would have smelled us and never approached. The jaguar is the largest and most powerful carnivore in Central America and stands at the top of the food chain, eating carrion, crippled animals, garbage, fruit, and now and then some healthy prey. It also is known to kill cows, pigs, goats, and dogs. Its jaws have such strength that it can crush small crocodiles and splinter turtle shells. As biologist Adrian Forsyth once wrote, "a jaguar would find the human skull about as challenging as a ripe cantaloupe." The fearlessness of this predator earned it a prominent place as an icon in indigenous cultures throughout tropical America, although some scientists today believe this tradition probably passed to the jaguar after larger prehistoric cats that truly earned that distinction disappeared.

We perused C. B. Koford's entry on jaguars in *Costa Rican Natural History* and found that they had once been common in the region but were now an endangered species throughout the New World tropics, or neotropics. "Because of its conspicuous tracks, the high value of its pelt, its reputation as a stock killer, and its vulnerability to hound pursuit and still hunting, this cat is now rare except in parts of large unhunted reserves," Koford wrote. Hunting was outlawed when Santa Rosa became a national park, but before then it had taken a heavy toll on jaguars. A Nicaraguan hunting in the bottomland area just below the *Cañón del Tigre* in the mid-1960s, for example, reportedly shot sixteen jaguars and mountain lions in one year. Despite export bans, the skins of jaguars shot in Costa Rica were still smuggled into Nicaragua as late as the early 1970s. The skins fetched a poacher $200 apiece. Habitat loss due to deforestation, however, was the main reason for the jaguar's decline.

Thomas Belt, in his 1874 memoir *The Naturalist in Nicaragua*, described how while hunting bird specimens in the forest one day (this would have been about 100 miles northeast of Santa Rosa) a jaguar emerged from some bushes. It was a large cat, and Belt's first impression was that in its jaws he would have been "nearly as helpless as a mouse in those of a cat. He was lashing his tail, at every roar showing his great teeth, and was evidently in a bad humour." With the jaguar about sixty feet away, Belt's thoughts turned to self-preservation. He recalled, "I had not even a knife with me to show fight with if he attacked me, and my small charge of shot would not have penetrated beyond his skin, unless I managed to hit him when he was very near to me." Nonetheless, Belt dropped to a knee and took aim. With that movement, the jaguar crouched as if about to spring, but then turned and disappeared into the bushes. "I half regretted I had not fired and taken my chance," wrote Belt, who followed the cat for a few yards, "greatly chagrined that in the only chance I had ever had of bagging a jaguar, I was not prepared for the encounter, and had to let 'I dare not' wait upon 'I would.'"

In what struck me as an interesting bit of phrasing, Koford wrote that jaguars appear "not to avoid the scent of man." They may follow a human, but unprovoked attacks are rare, if they actually occur at all. In one recent attack in Panama, a jaguar was said to have charged a man who toted a bag of captured birds. The question is, was it after the man or the birds? Some scientists flatly deny there's any record of a jaguar attack on a human. Such attacks—or at least tales about them—apparently were more numerous in centuries past. A Jesuit missionary in South America in the eighteenth century wrote a detailed account of how a jaguar kills a human, including licking the blood gushing from the severed throat. More modern biologists also have been captivated by jaguar mythology. Aldo Leopold wrote in *A Sand County Almanac* of his visit to the tropics, "We saw neither hide nor hair of him, but his personality pervaded the wilderness; no living beast forgot his potential presence . . . those massive paws could fell an ox, those jaws shear off bones like a guillotine." As Koford put it, a jaguar usually dispatches an animal "with a bite at the nape, the canines breaking the prey's neck or penetrating its skull." Although he was not portraying a human kill, I pondered this description long and hard later that night.

Such ponderings were nowhere near my state of mind that afternoon in the *Cañón del Tigre*. The jaguar's brain may well have been close to making a decision, but the circuits in my brain, perhaps too long disconnected from the forests where my ancestors once lived, were calculating nothing. I was neither about to kill, nor to run. Beyond the rather lame if not useless reflexive act of drawing my hunting knife, I merely stood there, slightly crouched but fully stupefied.

Then, as abruptly as the jaguar had rambled into my life, it turned, jumped, and ran away, paws pounding heavily on the forest floor. Godfrey, thinking that I had seen another snake, had walked toward me, and apparently this had been enough for the cat to choose to run. Only then did I realize that I was holding the knife, gripping it tightly in my right hand, waist-high. I had a difficult time loosening my grip and returning it to the sheath. Godfrey had seen the cat, too, and as we continued our hike, we discussed the details of the still-fresh image. The significance of the encounter sunk in, and my knees began to tremble. A cold, wrenching shiver twisted up through my stomach. It produced no nausea, and it lasted only a few seconds before it soared out of my body and into the warm forest around me. In the end I was left with the glow of an emotion that I can only describe as respect.

My knees had stopped shaking by the time we came down out of the hillside forest of the *Cañón del Tigre*, forded the Río Poza Salada, and crossed an expanse of bottomland forest. We had reached an area called Naranjo. Squatters and other previous inhabitants here had cleared trees for farms and cattle pasture; some had used the wood to fuel fires in extracting salt. We moved through a mangrove swamp, surrounded by thousands of crabs with bright red legs, yellow claws, and metallic blue shells as far as the eye could see. They crouched cautiously near small holes, their bodies flashing like firecrackers against the background of dark brown mud. Soon we splashed through shallow, marshy pools, me watching warily for crocodiles. We pushed through sand and reached the edge of the forest near the beach, where a huge ctenosaur, a lizard similar to an iguana, eyed us warily.

This was Playa Naranjo, a spectacular beach several miles long with large cliffs at either end and completely deserted. Just offshore, jutting out of the lovely blue Pacific, was a huge, rough-hewn, sedimentary rock dated at 60 million years old. This was Peña Bruja, also known as Roca Bruja, or Witch's Rock. Peña Bruja is a favorite of surfers, most of them from North America. They say the waves that move past it have one of the best breaks in Costa Rica, which has become something of a surfer's paradise.

We sat on the beach gazing at the wild waves. No surfers were visible this day, nor could we see something even more interesting going on at that moment just about 6 miles to the northwest, in the shadow of the Santa Elena mountains: Employees of a secret, illegal U.S. effort to resupply the contra guerrillas were trying to free a C-123K Provider cargo plane from a muddy, clandestine airstrip.

The airstrip, on the Santa Elena peninsula, had been renovated and expanded from a much shorter runway built in the 1940s but which had ceased operation in the mid–1970s. Shortly after he took office in May 1986, President Oscar Arias Sánchez ordered the airstrip closed down and asked the U.S. government to stop using it in its war against the Sandinista government. However, by June, unbeknownst to Arias, the airstrip was back in use again. That fact would not become public until the Costa Rican and international press broke the story, bringing Santa Elena to center stage in a bitter dispute between the Arias and Reagan administrations, a dispute that, by many accounts, both aided and hampered the Guanacaste project.

To understand this dispute and how it affected the project, it's important to understand some of the basics of the Iran-contra affair, and why Santa Elena became part of the U.S. government's secret effort to open a southern front in the contra war. The airstrip, only about 12 miles from the Nicaraguan border, was a key component in that effort. It was part of an operation run by the National Security Council staff to give extensive military support to the contras, against the wishes of the U.S. Congress. The resupply operation was coordinated by Lieutenant Colonel Oliver North with the support of the Central Intelligence Agency and State Department. The operation was known as "Project Democracy" by North and as "the Enterprise" by arms dealers who worked with him, including retired U.S. Air Force General Richard Secord. By the summer of 1985, the Enterprise was using Ilopango Air Force Base in El Salvador to fly supplies to contra troops in northern sections of Nicaragua. North and his partners strove to get supplies dropped by air to another group of contras in southern Nicaragua.

According to the 1987 *Report of the Congressional Committees Investigating the Iran-Contra Affair*, North asked for and obtained the help of Lewis Tambs, the newly appointed U.S. ambassador to Costa Rica, and Joe Fernández, also known as Tomás Castillo, the CIA chief in Costa Rica. On August 10, 1985, North flew to Costa Rica to discuss with Tambs and Castillo the details of setting up an airstrip. The next week, North sent back his courier, Robert Owen, to scout the Santa Elena site. Owen and Castillo surveyed Santa Elena, and North concluded that it was ideal.

In December, Admiral John Poindexter, President Ronald Reagan's newly appointed national security adviser, traveled to Central America with North to discuss the Santa Elena airstrip and other issues, including how to get more cooperation from Costa Rica in supporting the contras. Poindexter later briefed Reagan about Santa Elena.

Luis Alberto Monge, the president of Costa Rica at that time, later said that he had been pressured by North and company to permit the airstrip or face losing the millions of dollars a year in aid Washington was giving Costa Rica. The North Americans argued that in the event of a U.S. invasion of Nicaragua, the Sandinista government would attempt to internationalize the conflict by invading Costa Rica and destroying the airport at Liberia, thus creating a need for an additional airstrip to support Costa Rican defenses. Monge allowed the Santa Elena project.

The operatives leased the land from Joseph Hamilton, a wealthy textile manufacturer from North Carolina who ran clothing assembly factories in Costa Rica and who headed the group that owned Santa Elena. Local contractors constructed the facility. North pored over the details of the project, including on-site storage of fuel, construction of guard quarters, and even instructions on using the bulldozers. His men arranged for the airstrip to be protected by guards working for a Costa Rican Civil Guard colonel. To keep inquiring citizens away, a cover story was created that the area was being used for mortar and rifle practice.

The idea of the airstrip was to allow cargo planes, which had a limited range, a place to land and refuel after flying down from Ilopango and dropping supplies to the contras in southern Nicaragua. The airstrip was dubbed "Point West," a play on words suggestive of the U.S. Military Academy at West Point, New York, but also perhaps referring to the fact that the Santa Elena Peninsula is the westernmost point of mainland Costa Rica.

Flight records show that two of the Enterprise's own twin-engine cargo planes—a C–7 Caribou on May 21 and a C–123K on June 6—flew from Ilopango to Santa Elena and returned. On June 9, a C–123K with 10,000 pounds of munitions, uniforms, and medical supplies took off from the El Salvador airbase and tried to make a drop, but couldn't find the contras. To make matters worse, the plane got stuck in the mud when it landed at Santa Elena. This was the rainy season, after all.

It was still stuck in the mud the next afternoon, June 10, when 6 miles to the southeast Godfrey and I ate lunch at a picnic table near the beach under the watchful eye of the ctenosaur. Because our water supply was so low, we decided not to linger. As we headed out of the beach area, three coatimundis—raccoon-like animals—jumped down out of a tree near us and crashed off into the brush. Our walk back up to camp, however, was most memorable for extreme heat, humidity, fatigue, and thirst. Even so, Godfrey paused several times to observe and collect caterpillars.

I got dehydrated and a bit lightheaded and grouchy. Sensing this, and to take my mind off my discomfort, Godfrey told a long joke about two turtles who had argued in a tavern but settled it and decided to drink sodas as a sign of friendship. The sodas each cost a nickel, so one turtle started home to get his money. The second one had a nickel, so he ordered a glass and waited. Days later, after valiantly fighting temptation but with all but a few drops of his soda evaporated, the second turtle decided to take a sip. As he leaned over to drink, the tavern door swung open, and the first turtle stuck his head around and said, "All right, if you're not going to wait, then I'm not going home for my nickel." I laughed deliriously.

After we returned to our cabin and guzzled enough water, we told the scientists and volunteers about the jaguar sighting. There were two reactions. There was a pure, shared joy that someone—anyone—had experienced one of the great encounters of the forest. The people who expressed this talked and quizzed us almost as if they were friends offering congratulations on the birth of a first child. The other reaction was excitement tinged with jealousy. Those who expressed this—and there were but two or three—made an effort to sound happy, but their eyes revealed their true thoughts: They had spent years in the forest without seeing *el tigre*, likely would spend many more years without seeing one, and some tenderfoot journalist waltzes in and gets lucky on his second day in country. It was a fundamental urge among neotropical naturalists to see one. Thomas Belt himself had gone back the next day to the spot where he'd seen his jaguar, only this time he took a rifle with heavier shot. The cat eluded him. "Although afterwards I was always prepared," he wrote with regret, "I never met with another."

The only exception to these reactions was that of Dan Janzen. I don't remember his exact words, but I do remember their nature. He expressed disgust, almost anger, that I had reached for the knife instead of my camera. I *could* have gotten a photo of the jaguar, and *that* would have been useful. I was stunned. I was embarrassed. His words and their tone made me feel stupid, and for a moment, the Irish in me flared, and I felt a burst of indignation rise into ire. I said nothing, respecting my place as a newcomer, but I had learned that Janzen was quite capable of demeaning someone. Even so, as Godfrey and I walked away from the front of Janzen's house, I remember thinking, "You know, he's absolutely right."

# · 3 ·

# The Ant and the Acacia

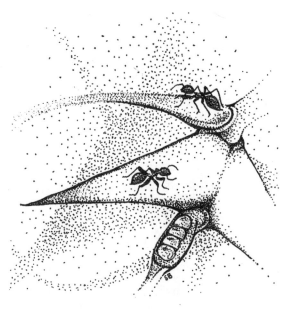

"Now, here you can see a worker stinging the back of my hand in the usual manner."

The man speaking on the movie screen looks a bit disheveled, dressed in an unbuttoned red plaid wool shirt pulled over another shirt and hanging over the belt of his brown pants. Black hair, receding from his forehead, curls back around his ears before falling low on the nape of his neck. He is clean-shaven, yet appears to have a five o'clock shadow. His eyes seem tired, his mouth relaxed.

The camera cuts in close on his hand, where an ant bites him with its jaws and swings its abdomen down to zap him with its stinger. He doesn't flinch, and, rather blandly, considering the circumstances, he continues his monologue.

He is a young Dan Janzen, and he stands in the dim light behind a potted acacia tree in the studio. This is *Ants and Acacias: An Ecological Study*, a twenty-four-minute, 16–millimeter instructional film produced in 1973 by the Open University and BBC-TV. The format of the film is somewhat drab, primarily what is called these days the "talking head." Still, the story of the closely knit lives of the ant and the acacia is fascinating, and the one-man narrative is broken up with close-ups of thorns on the tree, ants, and their enemies. In what appears to be a single take, with no script and no editing, Janzen reveals one marvelous detail after another.

At the time, Janzen was on the faculty at the University of Michigan, where he had already gained repute among students for his unusual field courses and his lectures combining color slides with tales of ecology and natural history. Janzen had greater aspirations, though, and the BBC film would not be the last time he sought to reach a broader audience with an ecological message.

"Back when I was a beginning graduate student, back in 1962, I was down in southern Mexico, in the state of Veracruz, which is right here, collecting insects for the California Insect Survey," Janzen says, pointing to a map of Mesoamerica. "And I was there because I wanted to look for tropical ecological interactions to do my thesis research on. And one day I was walking down a road, and a small bush more or less like this one, but perhaps a little taller, was standing there, and a beetle flew by and landed on one of the leaves. The instant that the beetle landed, an ant ran up and grabbed the beetle. The beetle jumped off the leaf and flew off. And I saw all this happen, but it really didn't sink in, and I walked on down the road, and the little wheels began to turn and after two or three hours it suddenly hit me that there must be some reason why the ant was there. So I turned and walked back to the bush, and there were ants walking all over the surface of the plant. So I picked up a few insects and threw them on the plant, and the ants attacked them. They jumped off, and then the immediate question came to mind as to why the ants were there and what they were doing there."

Thus began Janzen's study of acacia ants and ant-acacia trees, a study that was to influence modern field biology in several significant ways.

The tree in the film was a small version of the acacia trees common in Santa Rosa. Acacias are rather spindly, almost awkward-looking in the way their branches jut out from the trunk in different lengths and directions. All along the branches sprout scores of thorns. Some of the thorns come to a sharp point an inch or so from their base, which is wide and swollen where it connects to the branch. Some of them join in pairs, curving slightly upward into two sharp points. Thus follow the common names of the tree: the swollen-thorn acacia or bull's-horn acacia.

As fragile and flimsy as it looks, the acacia has powerful defenses besides its thorns, thanks to small, rust-colored ants garrisoned there. The ants are constantly on patrol, darting up and down branches and attacking anything that touches the tree. To grab a bull's-horn acacia is to invite a squadron of ants onto your hand within seconds, followed by quick, potent stings. I once carelessly brushed against an acacia while walking in Santa Rosa and was shocked by a fierce burning sensation on my arm. Suspecting I'd just been zapped by a wasp or a bee, I looked down to see an otherwise diminutive ant. Entomologist Justin O. Schmidt once described the acacia-ant sting as "a rare, piercing, elevated sort of pain," like "someone has fired a staple into your cheek." The sting produced a red hole surrounded by a small, white welt, whose irritation persisted for half an hour or so.

Such stings are potent indicators of two scientific concepts key to understanding tropical forests: mutualism and coevolution. Mutualism means cooperation among species, and coevolution is the process by which two organisms evolve together in a gradual, interactive way—not along single, independent lines. In other words, natural selection has favored the advantageous changes that they developed together.

The sting of acacia ants, in this case belonging to the genus *Pseudomyrmex*, is but one of the cooperative services ants provide to trees of the genus *Acacia* in Central America. The sting helps the ants keep the tree free of most insects that would defoliate it. The ants also bite away foliage that touches the tree or grows beneath it. This keeps other plants from competing for light and additional resources.

In return, the tree's swollen thorns house the ants. Large nectar-producing glands, called foliar nectaries, protrude from the leaf stalks, providing a flow of liquid sugar that is "about the consistency of Karo syrup," as Janzen once described it. Another food source comes from modified leaflet tips known as Beltian bodies, which are named for the nineteenth-century English naturalist Thomas Belt, who described them as an appendage that "looks like a golden pear." These contain protein, fat, and vitamins. Belt reported that the ants "are continually employed going from one to another, examining them. When an ant finds one sufficiently advanced, it bites the small point of attachment; then, bending down the fruit-like body, it breaks it off and bears it away in triumph to the nest."

Belt, in his book, *The Naturalist in Nicaragua*, was the first of many scientists to describe parts of the ant-acacia system, contributing the observation that "these ants form a most efficient standing army for the plant." Over the next century, several more scientists described other dimensions, and in a 1966 paper in the journal *Evolution*, Janzen brought together these and other details and proposed that the process by which this mutualistic relationship had come about was coevolution. Only two years before, in a paper in the same journal, the word coevolution had been coined by two young biologists at Stanford University: Paul Ehrlich, future author of the best-seller *The Population Bomb*, and Peter Raven, who now leads the Missouri Botanical Garden. Ehrlich, a butterfly specialist, and Raven, a botanist, suggested that coevolution was a major force in creating the overwhelming diversity of tropical plants and insects.

As part of Janzen's doctoral work in biology at the University of California at Berkeley in the 1960s, he conducted an intensive study of ants and acacias in Central America, primarily at a site in eastern Oaxaca, in southeastern Mexico. The field experiments involved observing an impressive 5000 acacia trees, from many of which Janzen removed the ants by spraying insecticide or clipping thorns.

In the *Evolution* paper that resulted from this unusually exhaustive work, he concluded that certain acacia ants and certain ant-acacias depended on each other not just in a fortuitous way, but in a complex system essential to the survival of each. He called this a "fully developed interdependency," and he outlined a scenario by which this relationship evolved. The characteristics developed by the ants and the acacias had brought them down an evolutionary road together to the point where one could not survive without the other. Furthermore, they had evolved several characteristics that set them apart from similar species. For the acacia ants such characteristics included aggressive behavior, greater speed and agility, twenty-four-hour vigilance on the tree, more ants patrolling the tree, large colonies, and a propensity to maul any vegetation that touched the tree. For the ant-acacia trees they included the nutritious Beltian bodies, year-round leaf production, and large foliar nectaries that produce a heavier nectar flow.

The basics of this coevolutionary process, Janzen argued, applied to other mutualistic relationships between animals and plants as well. Another famous example is the complicated story of the figs and fig wasps, which he elaborated on more than a decade later. This story is set inside the fig "fruit," known as the syconium. It is often mistaken for a fruit but actually is a cluster of tiny male and female flowers enclosed in an inside-out stem—as if the inside of a flowered stem had been pulled out and over like a sock. Some of the many amazing, intricately timed aspects of this work of nature are as follows: The male wasp impregnates the female before the female hatches out of her pupal chamber; the brothers cut a hole in the wall of the syconium, out of which the pregnant female crawls; within a short time the female locates another fig tree in flower, crawls into a syconium through a natural hole at the end, finds flowers, pollinates them, lays eggs in them, and dies. The pollen the female brings is from the male flowers of her birth syconium; it's carried in special pouches to the female flowers in the next syconium.

The ant-acacia paper became the first, best field-demonstrated example of coevolution, and the ant-acacia relationship quickly became the poster child for the process. Other researchers suddenly saw examples of coevolution all around them, and not just in the tropics. The paper also settled a long-standing controversy over whether the ants were merely exploiting the acacias or, in fact, giving protection in return.

A third impact of Janzen's *Evolution* paper was to open a new approach to field biology. In those days, the concept of doing experimental biology "in the field" was quite novel. Except for agricultural scientists, field biologists generally didn't do experiments. They were more like naturalists, wandering, collecting, writing down their observations, painting a picture of nature piece by piece. Experimental biology usually meant indoor laboratory work—petri dishes, test tubes, lab benches. With the ant-acacia paper, experimental biology took on a new tone, and researchers began to think about setting up experiments with a wide range of organisms in their habitats. This kind of field biology grew rapidly, and its practitioners have made great strides in understanding natural systems, even as they disappear.

Since the ant-acacia paper, Dan Janzen has become a larger-than-life figure in science and conservation, one of the world's most aggressive and successful conservation biologists. Within him are blended many qualities rare among scientists. Most important to this story are his abilities to grasp pieces of natural history and biology, assimilate them into a larger picture, develop truly audacious conservation ideas, and then make them happen.

Although others played key roles in conceiving and executing the Guanacaste project, Janzen was by far its main character. The project simply would not have commenced without his ingenious and energetic articulation of it, without his daring push. Nor, once it was launched, would it have traveled the same rapid path through the political, economic, and social milieus of Costa Rica and internationally in the final decade and a half of the twentieth century.

Janzen was born in Milwaukee, Wisconsin, on January 18, 1939. His father, also named Daniel, worked for the U.S. Fish and Wildlife Service, eventually rising to become its director, and his mother was a homemaker, artist, thinker, and, as Janzen once put it, "very strong supporter of those near her." Janzen was educated in the Minneapolis Public Schools and spent much of his childhood tromping through the Minnesota woods, passing hours hunting, fishing, and trapping. He started collecting moths at age nine and during his high school years avidly collected butterflies and other insects on long trips with his parents through Mexico. He served in the U.S. Army Military Police in the late 1950s, with basic training at Fort Leonard Wood, Missouri, and assignments to stateside posts. As an undergraduate at the University of Minnesota, Janzen supported his wife, Karen, and their first child, with scholarships, hunting, and trapping. "I caught all the meat we ate in the last two years," he told a writer for *Smithsonian* magazine. (He and Karen, who had two children, later divorced; he and his second wife, Caroline, also divorced.)

Janzen started by majoring in engineering at Minnesota, but when he accidentally wandered into a zoology building exhibit of mounted birds, he realized he might make a living pursuing his passion of collecting insects. He earned a bachelor of science degree in 1961, then packed his collection into a trailer and drove to Berkeley, where he finished a doctorate in 1965. Edward O. Wilson of

Harvard University remembered buying a "hungry-looking" Janzen dinner during a visit to Berkeley in the '60s. Wilson encouraged Janzen on his ant-acacia project. "He was then as he is today: a wild man, highly entrepreneurial and original," Wilson recalled in an interview. "He charted his own course."

Janzen held faculty positions at the University of Kansas (1965–68), University of Chicago (1969–72), and Michigan (1972–76) before settling in at Penn, where he still teaches a double load one semester a year. He spends the rest of the time in Costa Rica. He won several teaching and research awards, served on the editorial committees of half a dozen scientific journals, and became a member of nearly a dozen scientific societies. He taught field courses in Venezuela, Mexico, Puerto Rico, and Costa Rica, where he helped formulate the educational strategy for the Organization for Tropical Studies (OTS) in 1965. OTS is a consortium of U.S. universities formed in the early 1960s to deepen the exposure of North American scientists to tropical biology.

In 1963, Janzen was among the first graduate students to take a course in Costa Rica taught by OTS's forerunner, Jay Savage's "Advanced Science Seminar in Tropical Biology," funded by the National Science Foundation (NSF). When it began in 1961, only college and university faculty took the course. By 1965 Janzen reorganized it into the "Fundamentals of Tropical Biology" course for OTS, for graduate students, putting "his own stamp of genius" on this and other field courses, as Donald Stone of Duke University and former executive director of OTS put it. In the "competent and energetic hands" of Janzen, the Fundamentals course, with help from others such as Savage, José Sarukhan, John Vandermeer, Rob Colwell, Chip Taylor, Steve Hubbell, C. D. Michener, and Norm Scott, soon earned a reputation as the boot camp of tropical biology. In the course, which is still offered today, participants spend eight intellectually intense and physically exhausting weeks immersed in Costa Rica's main ecosystems, working from dawn to dusk, hearing lectures and conducting research projects in the field, then analyzing, writing, presenting, and discussing the results. This and other courses became OTS's crown jewel and helped it become perhaps the most influential institution introducing U.S. scientists—and now Latin American scientists, too—to tropical biology. However, Janzen's central role faded in the mid–1970s as disagreements mounted between him and the organization's administrative policies.

Janzen eventually studied forests in Africa, India, Southeast Asia, and Australia, besides all of Central America. In addition to his general level of leadership in tropical ecology engendered by his research and publications, he reviewed papers submitted to *Evolution, Ecology, Biotropica, Science,* the *Journal of Tropical Ecology,* and other journals. He gained a reputation as an influential, if not ruthless, reviewer of grant proposals to the NSF and conservation organizations.

After completing his doctorate, Janzen moved all around Costa Rica doing field work. He passed through Santa Rosa many times in the mid–1960s while co-teaching one of the OTS courses and doing research, although at that time it was just another piece of forest to him. Botanist Doug Boucher, then a Peace Corps volunteer at Santa Rosa, recalled how, as the liaison for foreign scientists who wanted to work in the park, he met Janzen in late 1971 at a workshop in Turrialba. "He had done his thesis work in Mexico and Costa Rica, was doing the stuff on seed predators, getting the *Hymenaea* project started, and was looking for a large area in Guanacaste to work. La Pacifica [a dry forest in south-eastern Guanacaste] was small and chopped up. The COMELCO ranch [an acronym for Compania Ganadera El Cortes, in central Guanacaste Province] was big but mostly cattle pasture, and it had an uncertain future. We got to talking, and I talked up Santa Rosa as a place to work."

Janzen went up to Santa Rosa during the dry season of 1971–72, and Boucher showed him around. He told Boucher that he wanted to shift his dry-forest work there. Boucher, who later went to graduate school at Michigan as a student of Janzen's, was struck by the bushy black beard and mustache the biologist sported. "He really looked like a Biblical prophet," Boucher recalled. At a seminar Janzen gave at the University of Costa Rica about that time, the man who introduced Janzen, after listing his background, degrees, and publications, quipped, "So I give you a man of whom it can truly be said, everything he eats turns to hair."

The *Hymenaea* project was Janzen's effort to thoroughly examine the natural history of *Hymenaea courbaril*, or guapinol, a tree famous for producing a resin that is the source of the commercial amber found in the neotropics. Janzen was keeping track of guapinol fruiting and seeding cycles and seed predation, a study he later expanded at Santa Rosa to include other species. He measured growth rates and survivorship of the trees and counted how many fruits and seeds each produced. Each February and March, he had been counting the fruits on nearly 500 of these trees along roadsides from Alajuela, near San José, to Santa Rosa, a distance of a hundred miles or so.

Then disaster struck. Costa Rica's government electric utility constructed a powerline along the Pan-Am Highway, cutting down about half of Janzen's trees in the process. Simultaneously, development caught up with a large forest where he and others conducted research on the COMELCO ranch. The owner had assured them that the forest would be left untouched, forever. "Well, one year we all came back and it had been turned into a big rice field," said Janzen, who lost a huge patch of *Hymenaea* trees there. "I was so flattened by these two events, I said to myself, 'Enough of this research on private land. I'm going to go to a national park.'"

In 1972, he camped beneath the partial roof of a half-completed building at Santa Rosa's administration area. The park, officially dedicated only the year before, was like a huge pasture covered with more than 2000 head of cattle, muddy roads, and small, circular patches of forest. The Rural Guard removed the cattle a few years later, but the jaragua grass pastures remained. "It was like golf course after golf course after golf course, with patches of forest," he recalled. Even so, it seemed a safe place to conduct research, and Janzen launched into it with a single-minded passion.

As he established himself at Santa Rosa, he recalled having two main thoughts: His research would be secure there, and it would help the new park survive. In his younger days, observing the work of his father, he had seen this mutualistic process succeed in U.S. national parks and wildlife refuges. Although Janzen still conducted research in other parts of Costa Rica, by the late 1970s it was done almost exclusively at Santa Rosa.

In my periodic travels to universities, field stations, and research institutes in the United States and Latin America over more than a decade, I witnessed a curious phenomenon whenever tropical biologists discovered I was writing a book on the Guanacaste project. Without my even asking about him, and usually before even talking about the project, they volunteered perspectives on Janzen.

Frequently the perspectives came in the form of a favorite story. Almost everyone who knows Janzen has one, and as they told it, they spoke as if recounting a piece of a legend. The stories were often about something they had seen him do, or heard that he had done. Sometimes they were swashbuckling stories, sometimes oddities, sometimes stories about his darker side. The net effect, if the observer wasn't careful, was that a picture of Janzen as a kind of mythological figure emerged, someone admired and feared at the same time, someone respected and hated, someone with nearly supernatural powers yet pathetically human, even petty. It was one of the most interesting phenomena of the Guanacaste restoration, and it illustrated a key aspect of the project: that Dan Janzen was a man of extremes.

Many of the stories had to do with Janzen's stamina, fearlessness, and determination in the field. Others recalled seemingly superhuman feats. "When I first met him, I remember thinking, here was a guy born too late," said Boucher. "He should have been a mountain man in the 1800s. He has the appearance, the approach to the wild, the spirit of a guy who would go off to the mountains and months later come back with marvelous tales of what he's discovered. To a certain extent that's what he actually is doing, although with more of a scientific perspective."

Another biologist told of a time Janzen became ill in the field with hepatitis but pushed on to complete a research project: "His discomfort and pain threshold is about four times more than others. It's a bit of machismo with Dan."

A former Penn student remembers how, during a dry-season field ecology course in the early 1980s, Janzen was making a point about a flowering tree. He spotted a flower in the tree, scampered up the trunk, and crawled out on the limb. As he reached for the flower, the branch gave way, and he crashed onto a rocky creekbed. "The last thing he did on the way down was to grab the flower he wanted to show us," the student said. "It must have hurt like the devil but he didn't miss a beat. He didn't say 'ouch' or anything. He came up holding the flower in his hand and continued the lecture with tears in his eyes. I watched it and just said, 'My God.' I realized it was just an inconvenience that he hadn't planned on, but which was simply ignored."

Janzen has Ménière's disease, a syndrome in which the inner ear fails to function normally, leading to sudden episodes of dizziness, nausea, and vomiting. The disease has left him totally deaf in his right ear. Despite taking medicine to counteract the vertigo, it struck from time to time. Janzen recalled one particularly violent event. He was giving a seminar before 400 people at the University of Pittsburgh, showing slides and talking about the Guanacaste project. As he looked up at the screen at one of the images, he became intensely sick. He ran for the trash basket and vomited into it. He wiped his chin and, although weakened by the bout, sat on a table and finished the talk.

Having been a man of the deep woods so long, Janzen did things in stride that to outsiders might seem quite strange. While working out his ideas on seed dispersal in the forest, he conducted experiments on whether seeds could grow after passing through an animal's digestive tract. Specifically, he experimented with horses and cows, but supplemented his data by eating seeds himself. He was known to pop caterpillars into his mouth (and swallow) to impress students. "It's just a good way of demonstrating that it isn't harmful," he once told me. "The tissues in the mouth are extremely sensitive to spines."

The bite of a human warble fly sends most people to a doctor for treatment, but not Janzen. This fly, *Dermatobia hominis*, is also commonly (and incorrectly, Janzen notes) called the botfly by biologists. Medical treatment consists of removing the fly larva from beneath the skin. It gets there after a female warble fly catches a mosquito, glues eggs to the mosquito's belly, and lets the mosquito go, after which the eggs hatch and the tiny fly larva burrow into the skin as the mosquito feeds on a mammal. If not removed, a larva grows, twists, and finally emerges from its plasma-slickened burrow—an experience thought to be painful, both physically and psychologically. Once, noticing a fly larva in his lower leg, Janzen decided out of curiosity to leave it there for six weeks until it emerged. The experience was neither painful nor infectious, leading to thoughts that the insect might produce novel antibiotics and painkillers.

He also learned first-hand the answer to a longtime mystery about why the male paca, a kind of large rodent, rears up on its hind legs when courting. He and Winnie Hallwachs, Janzen's common-law wife since 1979, had a paca for a house pet at Santa Rosa, and one day it stood and sprayed urine in his face, marking him as its mate and hoping in vain to cause Janzen to ovulate.

Janzen's North Woods background was reflected in his early, sometimes hungry days in Costa Rica. He once shot a peccary to feed a foodless OTS course, and another time, after observing and photographing a boa constrictor kill a coatimundi, he ate the coati. "Trap-lining," now a common term in tropical ecology, was coined by Janzen to describe how bees and other animals move from flower to flower taking nectar on a daily routine. It referred to the circle of traps set by a trapper, an image from his youth.

Even back in "civilization," Janzen had a reputation for living in his field clothes, often a dark green work shirt and pants, with unlaced hiking boots. A former Penn student told of the first day of school one semester, sitting in a lecture hall waiting for a biology class to begin, when he saw a disheveled man shuffle in, his hair in disarray, his clothes wrinkled. "He looks like one of the vent people," another student whispered, referring to those who spend cold nights near hot-air vents on the Philadelphia streets. It was Janzen. He walked up to the front, turned on the slides, and immediately began to lecture.

"Dan did not provide a lot of background on the first day of class," one student recalled. "Either you kept up, or you were left in his dust. Undergraduates either loved or hated the course, but the vast majority were inspired."

Stone, the former OTS director, remembered a fund-raising talk Janzen gave early in the Guanacaste project to a few dozen high-rollers in the library of the California Institute of Technology Alumni Building, in Pasadena. Stone and Janzen had been out in the field most of the day, as part of a scientific review of the University of California natural reserve system. They arrived at the library in their field clothes. In the men's room, Janzen pulled his dress clothes out of a paper sack. The sport coat had been twisted into the bag, so the sleeves came out coiled, and Janzen spent the night with the cuffs of his coat riding high on his forearms.

Such stories reflect the pleasantly eccentric side of a man more focused on doing science than on the niceties of North American society. Yet Janzen's personality had a distasteful side, too, and those who knew him seemed compelled to talk about it. As they did so, I could see them wrestling with some inner conflict, some love-hate relationship. For many of them it was clearly painful, and I gradually realized that they were struggling with the essential dualism of Dan Janzen, a man who was at once an admirable, crusading genius and an abrasive bully.

Everyone agreed that Janzen was one of those people who never rests, intellectually or physically. "He has more energy than ten locomotives," said a friend. They agreed he was an excellent, productive scientist and skilled public speaker, with great contacts. Some viewed him as a great leader, "like an orchestra conductor who gets the best out of the individual talent around him," as one of them put it.

They praised his knowledge and insight. "He knows more natural history facts than any ten other people," said one biologist. "He is one of the five most influential people in tropical ecology." Another biologist told me that "people enjoy teasing him about his errors because there are so few of them." Another said, "He's done things I would never dream of trying. He's put his money where his mouth is, and he's delivered." Janzen is notorious for shunning such praise. "Dan does not bathe in compliments," an acquaintance of his told me. "I've introduced him a couple of times for talks and it's real tough to get him to shut up long enough for me to say what I want to say in front of a group. He'll try to sort of knock me off stage."

Despite the commendation, Janzen also was viewed as a hurtful egomaniac. "I have a tremendous amount of respect for Dan," said one biologist. "You could tie one hundred of us together, and our contribution to tropical biology wouldn't come close to what he's done. But he can be a complete jerk." Others talked in the same breath of Janzen doing magnificent things but stabbing people in the back, of his intellectual power and his own unwavering belief in it.

Such combinations of traits, of course, are not uncommon among great biologists, nor great people of any field. And such criticisms are not uncommonly offered by students or colleagues who have felt the burn of what they believed to be harsh or cruel criticism, even if it might be the truth.

His strong personality rubbed many the wrong way. "You don't talk to Dan, you listen," observed one researcher, while another told me, "With Janzen, it's the alpha-male. Bottom line: the belly-up thing." A writer for *Rolling Stone* magazine once cracked that Janzen "probably picked up his social graces when he was an MP in the army." My own experience has been that Janzen often—though not always—was so obsessed with a task or series of tasks as to be unapproachable. It took me a long time to realize that this was not personal, but rather a matter of his dedication to his work. "You know the most important thing I learned from Janzen?" Alvaro Umaña, the Costa Rican Minister of Natural Resources in the late 1980s, once asked me. "One night we were at my home having dinner, just the two of us. He said to me, 'Alvaro, I think I have thirty years of productive life left. And I have already decided how I'm going to spend them.'"

Many saw Janzen as overbearing and annoying when it came to discussing details of how nature worked. "Everybody sees something different in a forest," one eminent researcher explained. "One of Dan's problems is, he's not tolerant of the way others see the forest, although he's mellowed with age and is more accepting now." Another scientist, acknowledging that Janzen was "great and brilliant," lamented that he had become "too dictatorial." That was partly because he had had carte blanche from the NSF since getting his Ph.D., the scientist said.

Janzen made plenty of enemies, even among those who respected his ideas. "Everyone has tangled with Dan," one biologist told me, while another said, "If Dan fell, there would be plenty of people who wouldn't mind. But I would never bet against him."

For many in the current generation of tropical biologists, he was, along with scientists like Stone and Savage, "sort of like our Dad," said one researcher. Early in his career, Janzen was considered by some to be a tyrannical father, so hard was he on his doctoral students. "Dan thought everyone should know as much as he did, and of course, when you didn't, he couldn't stand it," one member of the younger generation told me.

On the other hand, I've spoken with several tropical biologists who remember writing to Janzen as high schoolers or undergraduates, asking advice on tropical studies, and receiving a personal, single-spaced typed letter of five or more pages detailing the state of the field, options for study, and ideas for work. They remember the letters as entirely encouraging and supportive.

In February of 1988 I watched Janzen lead a group of students from Penn and Uppsala University, in Sweden, through various habitats in Santa Rosa. At one stop, while the group watched an army ant column advance through the forest, he picked up a twig and unconsciously manipulated it as he talked about the natural history of these foraging ants. When he finished, he absentmindedly tossed the twig away, but when he saw that it had caused a wasp to take flight—a wasp that one of the students was about to photograph—he apologized profusely. It was clear that he felt his act irresponsible and inconsiderate, because it had interfered with the student's interest in the forest.

After his ant-acacia work, Janzen began looking more widely at coevolutionary relationships and literally dozens of other aspects of tropical ecology. He seemed undaunted by the challenge of understanding an entire tropical ecosystem, a challenge that loomed much too large for most natural scientists. Biologists might spend a few weeks or perhaps a few months in the field during summer and winter breaks, collecting or otherwise studying a few plants or animals of choice. Janzen seemed to embrace a total, rough immersion in the forest, and, unlike most scientists, he refused to specialize. Instead, he investigated a wide array of organisms in various habitats, collecting and cataloging them and elucidating the complexities of their interactions. He wrote about his subjects with extreme precision and intensity, yet with a writing style that was more approachable than most standard scientific prose.

The way he wrote, and the way he sometimes built hypotheses on his observations, gained him something of a reputation as a Victorian naturalist, telling, as one tropical ecologist once phrased it, "Just-So" stories. The phrase, borrowed from Rudyard Kipling, has been used before by biologists in other fields to attack what they felt were conclusions made by colleagues without sufficient data.

Janzen published papers at a rapid rate. Their scope was impressive, the topics including the timing of sexual reproduction of trees during the dry season, insect diversity and behavior, cross-pollination of plants by insects, chemical defenses of plants, bird behavior, snakes, rodents, wasps, climate differences

among forest types, rates of forest regeneration, and various aspects of seed predation and dispersal by animals. In essence, he learned the foundations of tropical dry forest function, how the organisms of the forest were inextricably linked
in countless ways, and even how the dry forest itself was inextricably linked in
still more countless ways to other ecosystems across the landscape.

All this brought him to a profound realization that to preserve an ecosystem,
one must preserve not just the *species*, but also the *interactions* among the species.
This perception is now popular among biologists concerned that the U.S.
Endangered Species Act is too narrowly focused on saving individual species. But
Janzen was among the earliest to see it. "If you alter a species' interactions, you
have altered it as much as if you changed its color, diet or teeth," he wrote in
1983, noting that "parks function to conserve interactions, and the conservation
of organisms is a happy by-product." Furthermore, because these interactions
transcended ecosystems—since ecosystems were not isolated worlds unto themselves—to preserve an ecosystem like a dry forest meant one must preserve other
ecosystems across the landscape, too.

By 1985, Janzen's reputation as one of the world's leading tropical ecology
researchers was well established. He had published nearly 250 scientific papers,
an average of more than one a month, considered quite prolific in the critical and
sometimes slow milieu of peer-reviewed scientific publication. The list, which
included articles on ecology, evolution, systematics, and tropical biology, was as
impressive for its impact as for its number. Janzen's gift for writing—not just
research and field observation—became renowned. For many scientists, any
genius displayed in the lab or field is overwhelmed by a kind of intellectual
arthritis when they sit down at a keyboard, and the mental pain of writing slows
them immensely. When Janzen sat down to write, words flowed. Among his
publications with the largest impact was *Costa Rican Natural History*, which he
edited. First published in 1983, the 816-page collection of essays and short
accounts of plant and animal species was written by him and more than 150 other
researchers. The book illustrated Janzen the prolific photographer, too. He
always carried his Pentax and Nikon in the field, and the book's photos were
largely his.

Even those who thought Janzen went too far theoretically credited him for
being interesting and provocative, often propelling the science forward into
uncharted waters. He almost routinely expressed his creative streak. Or was it
iconoclasm? Many papers had anything but the turgid, methods-results-discus-
sion format of traditional scientific publications. The titles alone provide a clue.
Most were of a standard type, such as "Weather-related color polymorphism of
*Rothschildia lebeau*" and "Ecological distribution of chlorophyllous developing
embryos among perennial plants in a tropical deciduous forest." However, some
less-reverent (if not elegant) ones appeared, including "How To Be a Fig," "Mice,
big mammals, and seeds: It matters who defecates what where," and "Why Don't

Ants Visit Flowers?" (There was even "Tool-using by the African grey parrot (*Psittacus erithacus*)," whose senior author was M. J. Janzen—Janzen's parrot, Madge.) It was evidence of a scientist unafraid to challenge dogma, more interested in effective communication of ideas than formality. It also was evidence of a scientist with enough power, contacts, and standing to succeed despite his unorthodoxy.

Janzen, the hard-core researcher, was not blind to environmental destruction in the "laboratories" where he and other tropical biologists did their field work. In the 1970s, he had begun to write of "the uncertain future of the tropics" and "the deflowering of Central America," as the titles of two papers indicate. This reflected, mainly, a broad level of interest. He expressed concern about the impact on Santa Rosa in a 1983 paper in the journal *Oikos*, "No park is an island: Increase in interference from outside as park size decreases," in which he looked at the effects of human development on parks from a researcher's view. He pointed out that an "ocean" of ecologically degraded land around a forest island in Santa Rosa probably upset the natural process of recruitment at a typical tree fall in that patch of forest.

Dispersal, recruitment, and the related phenomena of gaps, seed rain, seed bank, and succession are key to the life of a forest. Dispersal is the word biologists use to describe the process by which seeds come into a disturbed, open area and grow—and are therefore recruited into the forest. This happens when a forest tries to grow back on degraded land, for example, land that was deforested or burned. It also happens as part of natural forest dynamics, such as when wind blows down an old tree, creating a gap, or opening in the canopy. At this point, seedlings from many species of plants begin to compete for light and other nurturing aspects of the gap. The seeds have been deposited by a "seed rain," a constant flow of seeds throughout the forest from a variety of plants near and far from the gap. Depending on their design, seeds are carried by the wind or by animals, including birds, bats, and other mammals that eat the seeds and deposit them randomly when they defecate. A few tropical seeds last for years in a seed bank in the soil. Most happen past just as a gap opens. Typically, fast-growing plant species, called "pioneers," will first colonize this new sunlight-strewn patch of forest floor. Eventually, however, slower-growing species win the competition and take over the gap, returning the forest patch to its original condition—at least resembling the original structure prior to the tree fall, although not necessarily with the same plant species. This process is called "secondary succession," whether it happens naturally in the forest or unnaturally, as when a forest is cut for its timber and left alone. The outcome of the process is determined by which plants arrive at the site right after the tree falls, which arrive later, and how they interact. Owing to centuries of human use, Santa Rosa was predominantly secondary successional dry forest.

In his *Oikos* paper, Janzen outlined his concern that in a small park, the successional process may be unnaturally influenced by large surrounding areas of human-disturbed forest. The problem, put simply, is that the comparatively more intense and unnatural seed rain from the human-disturbed forest may easily swamp out the seed rain of the relatively tiny natural forest. Based on observations of a tree-fall gap created in December 1978 by a blast of dry-season wind in Bosque Húmedo, he concluded that the gap was being colonized in an unnatural way. The interactions of the colonizing species had been altered because of heavy seed rain from a large area of nearby secondary successional forest, an effect that was as destructive to the forest as killing all the large carnivores. "What we always seem to bemoan is the extinction of species from natural areas, when we should be just as worried about the addition of species," he noted.

During the early 1980s, Janzen also began to agitate for a large-scale taxonomic assault on tropical biodiversity. He outlined his idea in an October 1984 letter to Robert M. Adams, then Secretary of the Smithsonian Institution. The letter, a response from Adams, and a further response from Janzen were published in the spring 1985 issue of the *Bulletin of the Entomological Society of America*. The message that came through in this "dialogue," as it was called by the *Bulletin* editors, is an important one, for it helps explain the origins of some of the Guanacaste project's most interesting spinoffs and parallel developments, including the parataxonomists, the National Biodiversity Institute, and the All Taxa Biodiversity Inventory. In the letter, Janzen pleaded for the U.S. Department of Agriculture and the Smithsonian's U.S. National Museum to combine their insect taxonomy resources and launch a "megalomanic push" to collect, identify, and store the entire insect fauna of the New World tropics. The tone of the letter was provocative. Janzen described almost tripping over a taxonomist working on the museum floor because of lack of space. Virtually scoffing at the museum's plans to expand its insect collection, he noted that the effort he envisioned would require one hundred times the space. "That you may not view this as hyperbole, let me just mention that your taxonomists are currently rejoicing in the news that 215,000 insect drawers are currently ordered and on the way. My 'Moths of Costa Rica' project, funded by NSF and fueled by rapidly growing local support, will fill that many drawers alone in 4 years." Adams responded with concern tempered by bureaucratic reality. "Shifts in priorities are inevitably slow," he wrote.

The letter had no effect, Janzen later told me, except to generate a flurry of correspondence to him from individual taxonomists agreeing with his call. At the time, he was "just a professor, in my naive professorial way, seeing that this ought to be happening." The motivation behind it was largely the same one that had driven scientists since Linnaeus: Finding and naming organisms were the first steps toward knowing them.

The dialogue was important because it revealed that Janzen had become increasingly aware of two major trends. In one of his letters to Adams, he wrote: "First even the most avid conservationists do not see how quickly the tropics is sliding down the craw of human consumption; to see it clearly you have to both live in it and be an ecologist excruciatingly finely tuned to how little perturbation it takes to send an ecological interaction into delirious orbit. Second, I see a groundswell of home-grown interest in doing what needs to be done to really understand the tropics, even as it disappears." In other words, he saw the forests suddenly and quickly disappearing, and he realized that residents of neotropical countries, or at least some Costa Ricans, wanted to save and study them.

Even so, by the mid–1980s Janzen's priority was still research, and there was little sign that he would alter course. He told several acquaintances about the stack of papers he intended to publish on tropical ecology before the close of his career. When he traveled to Sweden to accept the Crafoord Prize in 1984, the first of what Janzen called "peculiar accidents" occurred that ultimately led him to change priorities. He encountered a Swedish journalist who pressed him on what he was doing to save tropical forests. Janzen bluntly answered that his job was to study forests, not save them—the latter was the conservationist's task. However, he told her that if she raised any money for conservation, he would channel it to a group that knew how to use it, such as the Nature Conservancy. "My mindset at that time was, 'They're doing their job and I'm doing my job,'" he told me. "We came back over here [to Santa Rosa] and just went on doing what we were doing."

Then came Corcovado. It was the second, and perhaps most important, peculiar accident.

The Costa Rican Park Service, already short-staffed and underfunded, had been hurt even more by the economic crisis of the early 1980s. Some experts speculated that the government might eventually open up the parks to development. When several banana plantations closed, many of the unemployed workers headed for Corcovado National Park on the Osa Peninsula along the southern Pacific Coast to be gold miners. Joining them were unemployed bus drivers, factory workers, farmers, and other sorts of fortune-seekers.

There, many hundreds of them engaged in gold mining. Park officials had tolerated the few local traditional miners who had long worked the park's streams for gold. However, the new wave of destruction alarmed the Park Service and the newly formed National Parks Foundation, established by Alvaro Ugalde and Mario Boza in 1979 as a nonprofit, nongovernmental organization. The main role of the foundation, led by several Costa Rican biologists and conservationists, was to raise money from international donors, purchase land, and donate it to the Park Service. As such, the foundation served as a unified fund-raising entity for Costa Rican conservation and as a check on the possibility that international donations to the national parks might be siphoned off into other units within the Ministry of Agriculture, where the young Park Service was housed at the time.

These Costa Ricans knew that the threat to Corcovado—and how they responded—would test their credibility among conservation organizations and other potential supporters in the developed world. Pedro León, a University of Costa Rica biologist and foundation board member at the time, explained it to me this way: "If Corcovado went, how were we going to explain to our donors that we were believable?" With the backing of the foundation and the World Wildlife Fund, the Park Service commissioned a study of the problem. To lead this study, Alvaro Ugalde, the Park Service director and a foundation board member, chose Janzen, who was appreciated and respected by the Costa Ricans for his scientific work. Janzen had conducted much of his early research in Corcovado and, while working there once in 1980, helped the Park Service by weighing the first gold confiscated in the park on the scale he used for moths. "The reason we used Janzen in 1985 is exactly the same as the one for which I would use him today: he's an excellent person," Ugalde told me. "I admire him, and I trust him."

On short notice, Janzen put a team of researchers together that included Rodolfo Dirzo, a Mexican biologist with the Universidad Autónoma de México; Gina Green, a U.S. graduate student at Oxford University; Juan Carlos Romero, the Costa Rican director of Corcovado National Park; Gary Stiles, a U.S. ornithologist with the University of Costa Rica; Gerardo Vega, a Costa Rican who was a former gold miner and research assistant to Janzen in Santa Rosa and Corcovado; and Don Wilson, a U.S. Fish and Wildlife Service mammalogist.

The team assessed the problem during a June 1985 trip to Corcovado. They hiked through the park, talked with miners, made a low-altitude inspection flight, and held a public meeting one night with miners and residents who lived near the park. In all, they examined eleven different river systems in Corcovado on foot and six more from the air. They also assessed several reports already written by Park Service personnel.

Janzen returned to Santa Rosa and wrote the team's 78–page report to the World Wildlife Fund, "Corcovado National Park: a perturbed rainforest ecosystem." It outlined the devastation to Corcovado caused by the activities of the gold miners, especially the destruction of aquatic ecosystems and virtual elimination of mammal populations in the southern third of the park. The rivers and creeks there had been converted from rich, complex habitats into "liquid deserts." An estimated 800 to 2200 miners and their dependents lived in the park, "by and large hard-working, thoughtful, and respectful people."

The hunting out of mammals, a process begun long before the miner invasion, would alter the forest structure, the report warned. It would lessen the species richness of plants in the Corcovado forest, since animals were a major force in dispersing seeds. This effect would linger long after hunters were removed. "In short, guns and hunting dogs drastically alter rainforest without [anyone] ever touching a chainsaw. And once the mammal populations have been severely reduced, the lightest hunting pressure and human disturbance will keep them at zero even if there is a nearby source area for reinvasion."

In one passage, Janzen described walking up to the same point on the Río Rincón where he had stood twenty years earlier. Before, he remembered, the site was "a crystal clear shade-dappled rainforest river rich in large fish, crabs and crayfish, aquatic mammals, aquatic birds, and insects; that day, it was also decorated with a 3.80 m long [about 11 feet] crocodile. Today it is a sun-baked sterile muddy ditch through a 10 km long [about 6 miles] gravel pit through large scale and productive rice and corn fields."

The Corcovado crisis was only one example of a general challenge to the whole park system at that time, in which people and farms were rapidly marching up to the borders of all the Costa Rican national parks. Often, rural populations entered the parks in search of harvestable resources, such as minerals, trees, and animals. One day this people pressure could overwhelm conservationist arguments about the value of a park's intact ecosystems. They were, the report said, "almost undefeatable contenders" for the resources within the parks, partly because of economic conditions but also because of the need for land on which to live. From the miners' perspective, "the species 'National Park' fits within the genus 'unused State-owned land,' legitimately colonized if there is sufficient 'need.'"

The report cited a body of "concerned and involved personnel" in the Park Service, which it credited with "a level of efficiency, honesty and purpose that is truly outstanding for a neotropical government agency." The Park Service's biggest weakness was a lack of resources in the face of major conservation problems, a weakness exacerbated by the fact that international donors were less attracted to supporting maintenance than to helping buy land. Despite acknowledging these hurdles, the report criticized the agency for "inactivity" and for failing to develop "the kind of hard-nosed managerial drive, budget, and policy to preserve those areas in the face of rising social pressures."

Several steps were needed to reduce the threat to both Corcovado and other parks that, inevitably, would soon face a similar crisis. The report recommended a permanent halt to all mining in Corcovado within a year. It listed several suggestions for evicting the miners, ensuring that they didn't return, and educating the communities around the park about why an intact park was important to them and the nation. Exactly when and how to evict the miners was left up to the Costa Rican government, although in an interesting comment that shed light on political tension in Guanacaste Province at the time, as contra rebels clashed nearby in Nicaragua with Sandinista troops, the panel noted that careful planning was needed if the Costa Rican Rural Guard was to be used in Corcovado, since "given the problems on Costa Rica's northern border, a national emergency could drop on the eviction at any moment." However the eviction was handled, one point was clear to the team: "Corcovado National Park, the Costa Rican National Park Service, and conservation biology in Costa Rica hang in the balance."

A broader policy change was needed as well. The report recommended that the Park Service take steps to deal with growing pressure for timber, land, and other resources in the parks. For one, parks must be viewed not as fortresses but rather as a community resource. That could be accomplished in many ways, including placing explanatory signs on nature trails, training park guides to give lectures in schools and other public venues, and encouraging guards to relocate their families from distant towns to link the guards more closely to the community. This was part of what the authors called "biocultural restoration," defined as "imbedding the parks as firmly in the hearts and minds of Costa Ricans as are its schools and churches." Residents around Corcovado, for example, were "highly rational" and capable of learning about the impact they had on the park and the park's value to them and the country. "There are two ways to save a tapir," the report noted. "Assign it a 24-hour Park Guard with a submachine gun, or expend the same amount of energy rendering it immoral to serve tapir meat to your family and neighbors."

The Park Service would need to increase the amount of money it spent on park maintenance. This kind of funding was "the hardest to obtain, yet absolutely the most critical," the report said. Also, the agency would need to decentralize, since Park Service officials in the central San José office had become too far removed from problems in the individual parks. The authors recommended transferring staffers from the central office into the field, appointing park directors for longer terms, and other structural changes. They also noted a strong, debilitating tendency "for park personnel to deflect decisions upward, to use their superiors as a shield (and therefore to not act when that shield is not in place)."

In mid-1985, Ugalde declared a state of emergency, transferred many Park Service guards to Corcovado, and opened an office in its closest town, Puerto Jimenez. The Park Service spent nine months talking with the miners, basically telling them that they were parked on a yellow curb and would get a ticket if they did not move. When it came time for the well-advertised "big arrest," in March 1986, only 298 of the original 1500 or so remained, and they were only symbolically arrested. After two more years of unrest outside Corcovado, including a hunger strike in San José by seventy miners and picketing at the Legislative Assembly, the crisis ended with government payments to those evicted. Even so, mining continued to be a problem in the park.

The Corcovado affair was important to the Guanacaste project partly because of the impact it had on Costa Rican conservationists. They realized they would have to take much more seriously what was happening outside the borders of parks as well as inside. They created a sister organization to the National

Parks Foundation—the Fundación Neotrópica, or Neotrópica Foundation—whose role was to improve the relationship between parks and residents around them. The Neotrópica Foundation, under the presidency of University of Costa Rica biologist Rodrigo Gámez, also served as a kind of second financial arm of the National Parks Foundation, and both foundations became the chief handlers of a wide range of financial transactions involving the Guanacaste project.

But Corcovado was at least as important for its profound impact on Janzen. Because of its dramatic example of ecological catastrophe, he realized that his study area, Santa Rosa, was just as vulnerable to destructive invasion, unless someone figured out how to have it be "occupied," that is, to have an owner. "He went into Corcovado and he became like a born-again conservationist," León told me. "I think he always was a conservationist, but he'd never been an activist. He'd always taken the easy attitude of many researchers, which is to complain and say you're upset about it, but you're not willing to take any action, because you're too busy with your research."

Said Ugalde: "Of course, it gave Dan a scare. He literally understood that the same thing could happen to Santa Rosa, his baby. He went back to Santa Rosa and he began to dream his project and to figure out how to get Santa Rosa away from that danger."

Even before Corcovado, Janzen and Hallwachs had been frustrated by the lack of management at Santa Rosa, especially as fires were allowed to continue cutting away at the forest islands. Two months after Corcovado, they flew to Australia at that government's request to examine the remnants of the tropical forests that once blanketed the northern third of the country and comment on what could be done with them. This turned out to be the third "peculiar accident," one that sealed Janzen's transformation from pure scientist to avid conservationist.

They found the forests that once covered the region to have been virtually wiped out by fire, along with a vast array of large animals. Janzen called it "the Great Australian Barbeque," and he drew parallels between Guanacaste Province and dry tropical Australia. The Australian grasslands, which stretched for hundreds of miles with only a few or no trees, foreshadowed the Santa Rosa area of the future if the current pattern of fires continued in northwestern Costa Rica. Soon after the visit, Janzen wrote: "Australia is a marvelous example of how what you see in the tropics, species-rich as it may be, can still be but a tattered ecological remnant rather than an evolutionarily fine-tuned ecosystem millions of years old. Yes, there are hundreds of species of plants in dry tropical Australia. But this list is undoubtedly much shorter than it was 30,000 years ago, and it is going to get abruptly shorter as the European-style fire regime eliminates the last relict dry forest pockets."

The chance to view the ultimate consequences of dry tropical forest fires a hemisphere away reinforced the restoration notions then shaping in Janzen's mind. "It was dumb luck," he told me. "Now that the cows were gone from Santa Rosa and the grass was tall, the fires were eating the remaining pieces of these forests, and what we saw in Australia was the outcome of letting it happen for a hundred years. When we got back here [to Santa Rosa], we realized that we had never asked the question of Guanacaste that we had been asked to ask of northern Australia: what do you do with the tropical dry forest?"

Back at Penn for the fall 1985 semester, Janzen and Hallwachs put together the distressing new insights from Corcovado and Australia with notions about forest succession that had arisen from their field research. The product was the startling idea that the forest could be restored on a grand scale. Within a week, they wrote the plan that became the thin, salmon-colored book, *Guanacaste National Park: Tropical Ecological and Cultural Restoration.* It became a classic of environmental restoration, part Bible, part cookbook, part *Common Sense.*

Like an ant harvesting Beltian bodies and suddenly feeling the acacia vibrate, Janzen had decided to turn and defend the plant. In a foreword to *The Naturalist in Nicaragua,* written at about that time, Janzen noted, "Just as Belt exposed some of the outlines of tropical biology, today we have by comparison an army of workers observing and recording the natural history of the tropics. But Belt was working amidst an ocean of nature, while all that remains to us are small and rapidly shrinking ponds. . . . If we do not act now, the remaining small bits of tropical forest that Belt describes and celebrates will be lost to us and to the future, and *The Naturalist in Nicaragua* will cease to be an introduction to the splendors of the tropics and will become an obituary for them instead."

# · *4* ·

# The Tree with Ears

The enormous tree rose up over the landscape, spreading its majestic canopy like a giant umbrella. Resting in the cool disk of shade it cast on the grass underneath, I looked up into this symbol of patriotism and courage—a guanacaste tree.

The tree stood like a sentry in front of the Casona at Santa Rosa. Except for a time when they became extinct locally, guanacaste trees like this one have watched the story of this region unfold, as prehistoric animals roamed the land, as humans settled during successive waves, and as the present battle over the tropical forests engaged. To understand the Guanacaste restoration, one must understand the meaning of the guanacaste tree, and why Costa Ricans feel so strongly about this region.

*48* ∎

On the ground beneath the tree, an odd-looking fruit has fallen. It is lobed and curved round on itself, a deep brown crescent with little bumps where seeds are wrapped in the case. The whole thing looks like, well, an ear. A human ear. In fact, the name guanacaste comes from the word *cuauhnacaztli,* used by the Nahuatl Indians of Mexico to identify the ear-shaped fruit. The guanacaste tree is also known as the ear fruit tree, and a few Costa Ricans told me it was sometimes described as "the tree with ears."

The tree, known taxonomically as *Enterolobium cyclocarpum,* is native to Mesoamerica and South America as far south as Brazil. It is often seen in Guanacaste Province as one of the solitary giants left standing in otherwise barren pastures.

Guanacaste seeds won't germinate unless the seed coat is roughed up by months in the soil. A small rodent, *Liomys salvini,* one of the spiny pocket mice, eats the seeds, or stores them for future meals, both finding the seeds in the fruits below the tree and harvesting them from horse dung. Peccaries and tapirs also eat the seeds, but peccaries usually destroy them and tapirs defecate in the forest where the tree cannot survive, unless washed into an open flood plain far below.

The guanacaste is the national tree of Costa Rica. It's also the namesake tree of the province, and some say it's a symbol of the character of the people who live there. The province covers just over 3800 square miles, about half the size of New Jersey, from the Pacific Ocean eastward to a volcanic cordillera that runs down Costa Rica's backbone, and from the Nicaraguan border southward roughly a third of the length of the country. The volcanic backbone separates Guanacaste from the eastern side of the country, which is constantly soaked by moisture-laden tradewinds. Guanacaste Province lies in the "rain shadow" of the cordillera and for months is quite dry, but when the sun returns from its position south of the equator and shines overhead, the hot air rising off the land pulls in moisture from the Pacific, which itself rises, condenses, and stokes the rainy season.

The people of Guanacaste trace their roots to the *sabaneros*—which means "savanna men," or cowboy ranch hands—whose discipline, competitiveness, and machismo have defined a culture of toughness from peasant to landlord. "The people of Guanacaste are confident," a former *sabanero,* Santos Sihézar Cid, then in his sixties, once told me. "They like challenges, and they're not afraid of anything." Nearly a century ago, lawyer and landowner Victor Guardia Quirós said, "I'll hit whoever says I'm not a Guanacastecan." Such statements reflect a strong regional pride that has long been associated with Guanacaste residents. In a 1988 interview with Marc Edelman,* an authority on the social and

* The definitive study of the province is Edelman's book, *The Logic of the Latifundio: The Large Estates of Northwestern Costa Rica Since the Late Nineteenth Century* (Stanford University Press, 1992). This chapter owes a special debt to the book.

economic history of the province, one resident boasted, "First, I'm Guanacastecan, then I'm Guanacastecan, and last I'm Guanacastecan. [I'm a] Costa
Rican by circumstance, nothing more."

The history of Guanacaste is full of such bluster. It is a history of frontier settlement, of struggles between rich and poor, and of the politics of agriculture,
including cattle ranching. Only relatively recently has the politics of conservation been added to the mix.

Guanacaste trees were among the many kinds of trees that towered over the lowlands of western Mesoamerica when humans first wandered into the region from
North America sometime around 25,000 B.C., scientists estimated recently.
These nomadic people lived among giant prehistoric mammals, known as the
Pleistocene megafauna. Named partly for the geologic epoch in which they
thrived, roughly from 600,000 to 12,000 years ago, the animals included such
wide-ranging beasts as megatheria (giant ground sloths), toxodons (akin to hip-
popotamuses), glyptodonts (imagine two-ton armadillos) and gomphotheres
(mastodon-like animals).

The Pleistocene megafauna were extinguished by a wave of specialized
hunters about 9000 years ago. The disappearance of the megafauna had a profound impact on the guanacaste and dozens of other trees that had evolved to
depend on the giant mammals for dispersal of their seeds, according to a then-
controversial hypothesis proposed by Janzen and Paul Martin in a 1982 paper in
*Science*. The megafauna swallowed the seeds as part of the sweet, nutrient-rich
guanacaste fruit, which dropped to the ground at a time of year when other food
was scarce. The seeds were then spread by defecation.

As the Pleistocene megafauna were eliminated, the megafauna-dependent
tree populations dwindled. Guanacaste trees largely retreated to the northern
regions of Mesoamerica, Janzen argued, only to reappear in the Pacific lowlands
of Costa Rica during the past 500 years, this time their seeds carried in the digestive tracts of horses ridden by the conquistadors, and further spread by the cattle
and horses of settlers. Some tropical ecologists considered Janzen's Pleistocene
megafauna hypothesis one of his Just-So stories, yet it certainly was intriguing.

When the conquistadors arrived in the early 1500s, coming through Guanacaste on roughly the same route as today's Pan-Am Highway, they found neither gold nor silver. They proceeded to decimate the native population by
waging war, transmitting disease, disrupting crop production, and enslaving
thousands of Indians. In a key bit of historical luck for European settlers who followed, the conquistadors sought their riches elsewhere in Latin America. The
early Spanish colonists concentrated in the Central Valley. According to the traditional, widely held view of Costa Rican history, out of these roots grew an indi-
vidualistic agrarian society based on independence and pacifism, not the
subjugation common elsewhere in the region. More recently, historians have

found that pacts formed among economic and political elites actually shaped the democratic institutions, which basically preserved the self-interest of the elites. Whatever the origins, the Costa Rican national character is often generalized as literate, socially mobile, progressive, Catholic, politically independent, nationalistic, and distinguished by a strong sense of family honor. Like many generalizations about a people, this is part myth, part reality. Also important to this history was a wave of European immigration in the late nineteenth and early twentieth century.

The Republic of Costa Rica, about the size of West Virginia, is a special place, and the Ticos, as citizens of the country like to call themselves, are special people. Although still a developing country, Costa Rica is tenaciously democratic, and its people have the highest level of education and standard of living anywhere in Central America. The winners of a short revolution in 1948 after a disputed presidential election abolished the army and established a constitution with strong checks and balances on political power. Costa Rica's government is similar to that of the United States, with executive, legislative, and judicial branches. There is only one house in the legislative branch, known as the Legislative Assembly. Elections for president and assembly deputies are held every four years, and these officials can serve only one term. A national debt crisis has fueled poverty, and voters have become disillusioned about a perceived lack of choice between the two main political parties, but competitive elections have helped maintain a stable, civilian government in the second half of the twentieth century, in stark contrast with Costa Rica's Central American neighbors, where military regimes and coups d'etat have been the norm.

Education through sixth grade is compulsory and publicly funded, and government estimates put the literacy rate at more than 90 percent, which may be somewhat rosier than reality. Health care is heavily funded by the government. Money for these and other social programs has dwindled somewhat as debt restructuring demanded by international lending agencies forced government cutbacks. Even so, social services have been more available in Costa Rica than elsewhere, partly because the government spends only a few percent of its budget on two "civilian militia" forces, the Civil Guard and the Rural Assistance Guard. These comprise more than 7000 men who very much resemble light infantry, with jungle fatigues, M–16 rifles, and other small arms.

Some of the social values and norms of the Ticos are especially fascinating and pertinent to this story. One of them, known as "decision-making *a la tica*," was well described in *The Costa Ricans*, by Richard Biesanz, Karen Zubris Biesanz, and Mavis Hiltunen Biesanz. Decision-making *a la tica* means "constant bargaining in an effort to avoid conflict, even when the problem is not really resolved." This doesn't arise from cowardice; to be sure, Tico culture is based on a healthy current of machismo. Rather, it arises out of the value Costa Ricans place on peace. As one historian put it, "solving problems *a la tica* means preventing blood running down to the river. And this, in a continent bathed in blood, in which a concrete individual life is worth hardly anything, is something extraordinary."

A related value is the eagerness to *quedar bien*—to get along. This "often wins out over other values, such as keeping one's word," the Biesanzes write. "It is easier to promise to have something ready *ahorita* or *mañana* and thus avoid possible friction at the moment than it is to tell someone that it cannot be done soon, or tomorrow, or perhaps ever." Costa Rican social norms also dictate courtesy, dignity, and refraining from embarrassing another person. Ticos can be easily offended. The authors also cited a general fear of change and innovation and a tendency to "think small and go slow" in government and the private sector. "History has been a succession of small happenings with few crises," they wrote. "Changes are accepted only *poco a poco*, little by little, with a great suspicion of large-scale organized planning."

From the beginnings of Costa Rican national history, Guanacaste Province has been an isolated frontier. In the mid–1500s, the Spanish crown began issuing royal land grants to Guanacaste municipalities and to soldiers who served in the conquest. These parcels, often with ambiguous boundaries, formed a vague patchwork across the region. This lack of specificity set the stage for disputes in later centuries.

Several such holdings were awarded in and around what today is Santa Rosa National Park. Vernon Cruz, a former administrator of the park, documented these early land records in his 1975 report, *Administración del Parque Nacional Santa Rosa*. In 1663, the Queen of Spain granted to Juan Martín de Villa Faña the pieces that today are Naranjo—the part of Santa Rosa that fronts on the Pacific—and Potrero Grande—which includes the Santa Elena Peninsula. In 1721, the queen granted title to Sergeant Pedro Ledesma the property called Santa Rosa. "This is the first historical mention of Santa Rosa," writes Cruz. As the decades passed, Santa Rosa and adjoining lands passed to different individuals and down through heirs, at one time even falling into the hands of a priest. The ranch produced mules to help carry cargo, some of it bound for the gold fields of California, from the Río San Juan, which separates Costa Rica and Nicaragua in the east, to the Pacific.

Out of these crude beginnings arose what Edelman describes as a "cattle frontier culture." Elite absentee landlords, many of them residents of Rivas, Nicaragua, often left their haciendas in the hands of the *sabaneros*. Cattle were left to graze in pastures and forests, often mixed with cattle owned by other landlords. The *sabaneros* would round the cattle up and drive them to market.

On the new haciendas and the lands between them, the shade-rich guanacaste trees that had been dispersed across the landscape became refuges from heavy rain and searing sun. Cowboys and travelers used them as rest stops, and settlers sometimes built dwellings nearby.

The Guanacaste frontier, sparsely populated as it was, was home to a wide array of rural people other than *sabaneros*, including peasant farmers, loggers, hunters, small merchants, wage laborers, and miners. They lived a tough life, and many still do. Food was sometimes scarce, and the climate harsh, especially the dry-season heat and the rainy-season floods. Malaria afflicted the population well into the twentieth century. Out of this grew the culture of machismo which, although not entirely peculiar to Guanacaste, had a unique Guanacastecan flavor. In the words of one resident cited by Edelman, the toughness of these people derived from "their Guanacastecan origin and their having lived among the vigorous trees of its forests and its resolute inhabitants. . . . In that region, a child of ten breaks wild horses, fights bulls—and goes to school. That is the secret of why Guanacaste with frequency produces men of firm and indomitable character. Accustomed to subduing horses, they easily acquire the desire to subdue men."

Hacienda owners generally took a benevolent, paternalistic attitude in dealing with *sabaneros* and other workers, but there were conflicts. The landowners often complained that these same people shot cattle, burned pastures, stole fences, and hunted and fished on hacienda land.

Part of this seemingly unruly behavior can be understood by grasping the notion of squatters' rights, the somewhat pejorative term for a common land-occupation concept that later became important in the development of Santa Rosa National Park. Peasants frequently poured onto remote areas on the ranches and farms owned by wealthy families, initiating settlements of their own. In the 1880s, Costa Rica provided for squatters' rights, giving peasants the right to title if they occupied land they believed to be state land, even if the land was private. The catch was that the landlord must also fail to challenge the squatters for three months. Later national laws added to the rights of the landless to gain title to unused land if they cultivated it. On the Guanacaste frontier, landlords strove to control the problem, yet they also had to be careful not to anger squatters and other peasants, who on occasion attacked and sometimes killed landlords.

Rustling persisted well into the late twentieth century. Edelman recounted the 1972 case of a rancher near Quebrada Grande who tracked some of his cattle and the rustlers who stole them. The trail was a common one followed in the Guanacaste rustling trade of the past—through the mountains to Nicaragua. More often in recent decades, cattle disappear aboard trucks without leaving a trace. Authorities sometimes set up checkpoints along Guanacaste roads to examine cattle trucks for proper documentation. Rustlers stole feral cattle from the forests of Santa Rosa as recently as the early 1980s.

Guanacaste's history of shady characters also includes the land speculator. Examples abound. A large wave of speculation moved through the northern edge of Guanacaste in the 1880s and '90s when it became clear that the United States planned to build a canal across the Central American isthmus. One of the proposed routes would have run across Lake Nicaragua, whose long southern shore comes within three miles of Guanacaste's border.

Over the decades, the holdings of various landowners shrank, grew, and shifted, through inheritance, marriage, consolidation, and sale. By the middle of the twentieth century, twenty-two properties, each averaging about 58 square miles, together made up nearly half the land in the province. Many of these were held by families who had owned them for decades or centuries, some even since the original colonial land title.

Into this mix came an odd combination of property-seekers who, for a variety of financial, political, or social reasons, bid for pieces of Guanacaste Province. Graham Greene could have written a novel about them. They included Nicaraguan dictators, exiles from Cuba and Nicaragua, and U.S. operatives, including some from President Ronald Reagan's administration. Another class of buyer was dubbed colloquially the "crazy gringo," a North American whom locals thought wasted money buying land on which to try farming methods that wouldn't work in the tropics. The crazy gringos bid for haciendas such as El Hacha and Santa María, which would later become pieces of the Guanacaste Conservation Area or lie near it. Landowners concealed their identities behind corporate names. Some even fragmented ownership of a large piece of property by dividing it into smaller sections, each owned by a different foreign corporation.

Guanacaste Province was a kind of bridge between Costa Rican and Nicaraguan politics and culture. Its people both enjoyed and suffered the consequences. A few years following Costa Rica's independence from Spain in 1821, Guanacastecans voted to join Costa Rica rather than remain a province of war-torn Nicaragua, to which they had many ties. They gained the relative peace of democratic Costa Rica, but the area also became a place where several conflicts played out.

A sense of detachment ran deep in Costa Rica's northwest, as many residents felt that the central government discriminated against them, ignoring the province when it came to constructing roads, railroads, and other elements of the national infrastructure. This aura of neglect, and concern about the Sandinista government in Nicaragua, may have been partly why wealthy Guanacaste landowners in the 1980s formed, as Edelman puts it, a "shadowy autonomy movement widely believed to have a small paramilitary arm. Some merely sought greater economic and political autonomy within Costa Rica, while others spoke seriously of securing U.S. links for Guanacaste similar to those of Puerto Rico."

The guanacaste tree that towers in front of the Casona likely sprouted sometime in the middle of the eighteenth century. By that time, Hacienda Santa Rosa already had been one of the largest cattle ranches in Guanacaste for a century, routinely maintaining more than a thousand head of cattle, as well as hundreds of horses, mules, and other animals. A century after it sprouted, the guanacaste tree was tall and strong when bullets whizzed, bayonets flashed, and shells exploded around it as a Costa Rican militia routed a band of mercenaries sent from Nicaragua by the notorious North American adventurer William Walker. Among other results, the battle, which lasted for only 15 minutes, gave Santa Rosa the stature of a national shrine.

A Tennessee native (and University of Pennsylvania-trained physician), Walker was of a genre of nineteenth-century North Americans who hired bands of mercenaries and attempted to set up colonies in countries with which the United States was otherwise at peace. The mercenaries and their leaders were known as filibusterers or, in Latin America, *filibusteros*. They were crazy gringos of another sort. Their aim, in part, was to gain territory for the U.S. slave states.

Walker fought his way to power in Nicaragua, and, anticipating an attack, Costa Rica mustered an army. In March of 1856, Walker sent a band of 240 *filibusteros* into Costa Rica to scout. While resting at the Casona, they were surprised by at least 500 Ticos under the command of General don José Joaquín Mora in the early afternoon on March 20.

Some of the *filibusteros* fled, some fanned out in front of the Casona, and some took up positions in the corral in front. The Costa Ricans, armed with artillery, rifles, pistols, and machetes, made several tactical maneuvers to outflank the foreigners. As Colonel Lorenzo Salazar's troops filed down to the west of the corral, the Ticos heard the *filibusteros* shout, "The greasers are coming!" Salazar's men split into four columns and attacked in the corral directly and on the southern flank. Others snuck over the hill to the east of the Casona and attacked from behind. Another squad fired cannons set up in front of a line of trees to the northwest of the Casona.

In the face of this spirited assault, the remaining *filibusteros* one by one took refuge in the main house. One of the Costa Rican officers, Captain Manuel Quirós, charged into the Casona's courtyard and, falling in a hail of bullets, called out a final order to his men to attack the house, "Entren ustedes!" Enter they did. Many of the *filibusteros* were killed inside the house, while others made a desperate charge out and escaped to the northeast. The remainder, including the wounded, put up a white flag in surrender. They were marched to Liberia, immediately court-martialed, put up against a wall, and shot. Historian Carlos Meléndez, who published an account of the battle a century later, concluded that 26 *filibusteros* and 19 Ticos were killed during the short, intense battle.

The incident is now a Costa Rican legend. In a 1987 interview with *La Nación*, Timoteo Gallo Hernandez, an elderly Guanacaste man (he credited his longevity to "a youthful encounter with God and a steady diet of seafood"), recalled his grandfather's description of the battle of Santa Rosa and the Ticos who fought it. "Although their arms were nothing like those of the Yanquis, they were very brave," Gallo said. "And for that reason, they won the war."

The fight with Walker momentarily melted away the parochial conflicts that had troubled Costa Ricans of different regions, uniting them for the first time with a sense of nationalism. Among many Ticos who explained to me the significance of these events was Pedro León, a biologist at the University of Costa Rica and one of the leaders of the Guanacaste project. "We like to say, 'This is the one battle we've had with the U.S., and we won it,'" León told me. "It's a joke, of course, because it wasn't the U.S. Army. It was a wild man who tried to take over Central American countries at a time when taking over countries was viewed as appropriate."

The Casona also figured prominently in a battle in 1919, when government troops delivered the first defeat to a rebel force in a furious fight. A larger engagement at Santa Rosa in 1955 pitted another Tico militia, many of them high school students, against an invasion from Nicaragua headed by Costa Ricans defeated in the 1948 War of National Liberation, also known as the 1948 Revolution. This attack, which was supported by Nicaraguan dictator Anastasio Somoza García, was turned back, partly by the volunteers and partly by threats from the Organization of American States. Fighting was heavy around the Casona, and the rusty hulks of two burned-out armored vehicles, known as *tanquetas*, still sit along a back road into Santa Rosa about two miles from the park entrance.

The Somoza influence in Guanacaste Province neither began nor ended with that defeat. Somoza, the leader of Nicaragua from 1937 until he was assassinated in 1956 in Panama, smuggled cattle into Guanacaste and fattened them on a friend's ranch. One son, Luis Somoza Debayle, president of Nicaragua from 1956 to 1963, acquired Santa Rosa in 1966 when it was bought by a company he owned. Another son, Anastasio Somoza Debayle, who ran Nicaragua from 1963 until overthrown by the Sandinistas in 1979, became a partner in companies that in the early 1960s bought Hacienda Murciélago on the Santa Elena peninsula. He eventually added eleven adjacent farms, thousands of head of cattle, two airstrips, roads, buildings, and a dock. According to one estimate, the Somoza brothers' empire in Guanacaste at one time covered about 223,000 acres, or 350 square miles.

The aim of the Somozas in buying large chunks of Guanacaste appeared to be both financial and strategic. Anti-Somoza forces, beginning with small groups of Nicaraguan exiles in the 1940s, had based much of their training in the province and had even launched invasions from the region.

As Europeans settled the Guanacaste landscape, they cleared forests and created pasture. Logging accelerated in the late nineteenth century with the boom in the market for tropical hardwoods in Europe, North America, and Peru. Work crews moved first through the easily accessible forests in the Guanacaste lowlands, leaving the tougher work on the volcano slopes for future generations. The province's elite made a fortune in timber exports beginning in the late 1800s, and the value of wood triggered boundary disputes and unauthorized logging on both private and government land.

Another major ecological change that swept across the province around the turn of the century and after was fueled by an invasion of exotic pasture grasses. Several varieties of African and Australian grasses were introduced by ranchers. Among them were the African savanna grasses pará, guinea grass, and jaragua. Ranchers believed that cattle grew better on them than on native grasses. As part of what is called the Africanization of the neotropics, the exotic grasses quickly invaded open areas, aided by annual burning by landowners. Jaragua, with its ability to support an intense dry-season fire, aggressively took over the landscape. One enthusiastic observer quoted by Edelman, noting jaragua's ability to survive drought and still feed cattle, called it "the salvation of Guanacaste."

From the 1950s through the 1970s, a beef export boom brought a new intensity to ranching operations in the province. It also brought more devastation to Guanacaste's forests. Joseph Tosi, Jr., a Massachusetts native, remembered a particularly dramatic example. It occurred in a 5000-acre forest on the banks of the Río Tempisque near the town of Filadelfia, about 30 miles as the crow flies south of Santa Rosa. Tosi, a land-use geologist and forest ecologist who has worked in Costa Rica since the early 1950s, first saw the site in 1952. "It was loaded with wildlife," he said, shaking his head during a 1996 interview at the Tropical Science Center, an agricultural research and consulting organization where he worked in the San José suburb of San Pedro. "Two or three of us went up there and did a forest inventory for the owner. We found guanacaste trees greater than 100 inches in diameter, straight and tall." When the scientists traveled through the same spot in 1960, the forest was gone. "They took the cream and burned the rest," he said. "It was typical of the culture at the time."

A map showing the shrinkage of closed canopy forest in Costa Rica from 1940 to 1983 reveals a stunning eradication, especially on the Pacific side, where dry forest removal was far greater than deforestation of Atlantic-side rain forest. Like a puddle of water evaporating in the noonday sun into a few isolated drops, the once vast forests of Guanacaste had been relegated on the map to small patches atop volcanoes. The islands of forest in Santa Rosa were too small to register.

Costa Rican and other biologists were well aware of the accelerating transformation of natural areas into farms and ranches. They expressed concern that great forests were wasting away; that mangroves, the nurseries for fish and shrimp, were dying from pollution; that archaeological and historical treasures were being vandalized. They began to publicize the benefits of preserving natural areas for scientific research, education, tourism, and recreation. Such areas also served as gene banks—storehouses of genetic information that might one day be in high demand. They noted that foreign scientists spent money in Costa Rica when they came to conduct research, and they pointed out the importance of the scientific knowledge gained, not just for foreign scientists, but for Costa Ricans as well. Part of this scientific contribution, as one of them put it, was "cataloging the outdoor classroom."

Meanwhile, anti-Somoza feelings continued to surge in Costa Rica, and at least in part to defuse it, Luis Somoza offered the San José government the Casona and 62 acres around it in 1966, shortly after his company had acquired Santa Rosa.

That same year, Kenton Miller entered the Santa Rosa scene. Miller, then in his mid-twenties, was a forestry officer with the Food and Agricultural Organization of the United Nations, based in Turrialba, at the Interamerican Institute of Agricultural Sciences. The Chicago native was finishing his doctorate at the State University of New York at Syracuse, and he had been brought to Turrialba to establish a graduate program in what biologists called "wildland management." The term encompassed not just national parks and wilderness areas, but also areas whose resources were extracted on a controlled basis, like a U.S. national forest. Miller's course took graduate students all over the "back country" of Costa Rica, as he put it in an interview thirty years later.

In the midst of one such course, Miller was invited by the government tourism institute, Instituto Costarricense de Turismo (ICT), to examine Hacienda Santa Rosa, which the institute was considering acquiring and shielding with reserve status. He and his students happily complied, and they found the site to be a virtually uncharted wildland.

"We literally went in with aerial photos and a pocket stereoscope and navigated our way across the savanna," he said. "There were no roads, only cattle tracks from the Pan-Am Highway to the hacienda. We went back and forth, and with the help of the locals living there, we got down to the beach." Along the way they saw dozens of free-wandering cattle in pastures and woods. Much of the forest had been cleared and planted with pasture grass. Most of the rest had been selectively logged and allowed to grow back, at least partially. When Miller and his students reached the beach, they reveled at the sight of sea turtles emerging from the water and laying eggs in the sand. Except for a little evidence of hunting, fishing, egg poaching, and sorghum farming by the locals, the lowland areas near the beach were "essentially untouched—there wasn't a footprint of a human anywhere." Although these forests had been altered by humans for centuries, they still appeared magnificent compared with those in southern Guanacaste Province.

The group lived in the Casona while completing the study, which concluded with a recommendation far grander than the tourism institute originally envisioned. "Rather than a 1,000-hectare [4-square-mile] historic monument, we proposed an 11,000–hectare [42-square-mile] national park that would contain the historic site but would also pick up what at that moment looked like that last dry forest remnant in northwestern Costa Rica," Miller said. "It turns out it was basically the last one in the whole of Central America."

In July 1966, however, the government, refusing Luis Somoza's donation offer, opted instead to expropriate and pay for the Casona and nearly 4 square miles around it, declaring the setting a national monument and placing it in the custody of the ICT. Still, tourism officials liked Miller's proposal. They contracted with his group to create a management plan for the proposed larger version of the park.

Miller, his graduate students, and a small group of Peace Corps volunteers returned to the hacienda, hiked the whole area again over the next year or so, and in 1968 submitted a report, "*Proyecto Parque Nacional Histórico Santa Rosa*," the first plan for the formation and management of a national park in Costa Rica. The plan proposed allowing most of the hacienda to return to wildland, except for a small zone around the Casona where a few cattle, pastures, and other trappings of the cowboy culture so revered in Guanacaste would be preserved. Such historical exhibits had their precedent in the Great Smoky Mountains National Park, in the eastern United States. The plan also recommended creating campgrounds and picnic areas and building a good road off the Pan-Am Highway so visitors could easily get to the Casona, the beach, and other areas.

The rising anti-Somoza passion among Costa Ricans had created widespread political support for making the historic site at Santa Rosa a national park, and since the government liked Miller's plan, ICT officials started negotiating with the Somoza family to acquire the land. As part of that process, Miller and several Costa Rican officials met in early 1968 with Anastasio Somoza Debayle. (Somoza's brother, Luis, had died the year before.)

Miller recalls the blazingly hot morning when he and the Costa Ricans stood in the shade of the Casona's veranda and watched a Mercedes carrying the ruthless right-wing dictator speed across the savanna, kicking up a huge cloud of dust. The car pulled up, and several Nicaraguans got out, each carrying a submachine gun. Somoza stepped out with one, too. After a few moments, the others put their guns back in the car and walked over to the veranda, but before Somoza joined the group, he strolled to a tree in front of them and hung his submachine gun on a branch, a gesture clearly meant for all to see.

This was the first time Somoza himself had met with the Costa Rican planning team. They sat on the veranda drinking rum, then, in Spanish, began to discuss a map of the proposed park boundaries. Suddenly, Somoza looked up at the North American biologist and said, in perfect American English, "What the hell are you trying to do, Miller, create another Yellowstone?"

It wasn't an angry question, really. Just imperious. There was a sense, too, that Somoza knew he was losing an opportunity to make money on tourism. "You know, there's only one thing wrong with this plan, Miller," the dictator added. "It's your idea instead of mine."

Somoza later offered to donate the land for the park, but Costa Rican officials insisted on paying.

"It was a point of honor," Miller said.

In recounting that story, Miller was quick to point out that the plan for Santa Rosa National Park built upon ideas being floated at that time by Joe Tosi and another North American biologist who taught and conducted research in Costa Rica, Leslie Holdridge.

For instance, as part of a 1967 report on Guanacaste climate, physiography, and soil conditions, Tosi assessed the "highest sustainable land-use potential" of the region for forestry, livestock grazing, and cultivation. North of Liberia, he found much of the land on the steeper volcano slopes and on the Santa Rosa Plateau that ran west of Volcán Orosí unsuitable for anything but forests. Even forestry would yield low return on the soils of the plateau, which were shallow, stony, and nearly infertile. "Very similar land-use potentials would be expected for the rugged hills of the Santa Elena Peninsula," which was just west of the study area, Tosi wrote. In short, much of the land in that area, from the volcanoes to the ocean, should be set aside as national parks for recreation and tourism, and he was especially concerned that the land on the volcanoes be publicly administered for watershed protection, wildlife management, and scenic conservation.

Tosi also raised what he called the "tantalizing possibility" of creating a national game park on the extensive Santa Rosa plains and the Santa Elena Peninsula. Native species would be included, but Tosi also suggested bringing in African herbivores, perhaps rhinos, hippos, and antelope. "The climates and vegetation are very similar to those of East Africa, which has been able to capitalize on this resource through tourism," he wrote. "Many of the large browse animals from that continent would probably prosper." Holdridge had suggested that international wildlife organizations might well back such a project.

In retrospect, the idea of bringing in large, exotic animals to northwestern Guanacaste Province sounds a bit strange, given today's conservation biology credo of restoring a wildland to its natural state. Yet consider that at that time, the mid–1960s, Costa Rica was a desperately poor country, with a stagnant economy, exploring its options for industrialization, modernized agriculture, and other methods to achieve economic growth. Guanacaste Province, essentially a wild and uncharted land, was thought of as a potential resource for raw materials and large commercial agriculture. When the biologists brought up the idea of African animals, they were merely suggesting a form of economic development that they believed would have a relatively light touch on the land, what today is called ecotourism. Besides, rhinos and hippos weren't all that different from the Pleistocene megafauna that once roamed Guanacaste.

The African game never arrived, but Miller, Tosi, and Holdridge had formed a vision. They were the first to see that a large piece of Guanacaste Province—from the tops of the volcanoes, down across Santa Rosa and Santa Elena, and to the Pacific—ought to be conserved. "It would have been unique," Miller told me, "but there was no way in hell, because the land ownership pattern was all subdivided among holding companies owned by Somoza and front-ended by Costa Rican names. There was no way to figure out who owned what. The Costa Ricans decided, 'Let's get what we can and get a single chunk under one ownership.'"

What a dream it was, though.

"It is a great place to be. It is the place where I thought I was going to grow old and die."

That's how Alvaro Ugalde began to answer when I asked him in a 1995 interview about his early years at Santa Rosa. Ugalde, a tough-talking, tough-looking fireplug of a man, was appointed superintendent of the new national park in 1970. The birth of this park was a milestone in the history of Costa Rican conservation, and it was clear from the way he talked about it that Ugalde was in love with the place and the time.

Santa Rosa was one of Costa Rica's first national parks. In 1970 and 1971, the Legislative Assembly passed laws and President José Figueres issued executive decrees establishing national parks in various parts of the country, including Volcán Poás, Santa Rosa, Tortuguero, and Cahuita.

If ever there was a place and time for Ugalde's kind of toughness, it was Santa Rosa in the early 1970s. The nearly 40-square-mile terrestrial part of the park (expanded by about 3 square miles in 1977) consisted of old pastures, coastal habitat, and forest. The forest was almost entirely secondary successional forest, growing back after having been at some point in the recent past either logged or burned. Ugalde faced several striking barriers—he called them "threats"—to achieving the kind of natural area that Miller and the others envisioned. The barriers included squatters, cattle, fires, hunters, and troublesome neighbors, not to mention the Park Service's chronic lack of money. To overcome them, Ugalde had little more than his determination, some knowledge gained while studying national parks management for a few months in the United States, and Miller's management plan.

During his first year at Santa Rosa, Ugalde lived in the broken-down Casona, sleeping in a tiny, second-story room with a leaky roof. As tourists began to filter through the park, he learned how to hang his laundry in a concealed place. He was joined by volunteers from the U.S. Peace Corps, and members of the Costa Rican National Youth Movement worked in the park during school breaks. They set about building picnic areas, trails, campgrounds, and other facilities. Ugalde chuckled as he recalled the day he took a vehicle and designated the road around the new administration area:

"It's very clear in my mind. I said, 'I'm going to mark the streets.' It was at the end of the rainy season, and the grass was above the windshield, of course. So I put it in four-wheel-drive and just kept looking at the grass as it fell in front of the car, making that loop that is still there today. Now they have all those new buildings and labs—around that loop that I made with that four-wheel-drive some 25 years ago."

The first challenge Ugalde faced was removing squatters. This came at a time when conservationists were literally risking their lives to establish national parks. The politics of doing so was tough enough, but big landowners and squatters constantly resisted. "People were getting shot at to set these up," Miller once told me. The squatters, a group of families who lived behind Naranjo beach, had already threatened Ugalde and later did the same to some of the new park employees. Ugalde was well aware of Guanacaste's history of squatter resistance, but the group in Naranjo was getting ready to cut and burn more of the old secondary successional forest there, having already cleared more than half of it.

The episode was the first test of government resolve to protect the new park. "It was a tough one, to tell you the truth, because all I had were a couple of horses, and myself, and supposedly a decree calling it a park," Ugalde recalled. "The government had paid for that land, but I had nothing else." He paused and thought for a few moments. "Of course," he added, as if remembering a source of spiritual strength, "I had a machete."

Ugalde controlled the upland area of savanna and forest, and the squatters controlled Naranjo. One day he rode down to their settlement. "I just got off the horse and asked for a cup of coffee. They kind of grumpily gave me a cup and I said, 'Well, let's talk.'" He chuckled again, recalling what he told them next, saying it in a very firm voice. "'We'd better talk about this, because you are in trouble.'"

There was no shooting; in fact, Ugalde doesn't remember ever being shot at. But he told them that the law would not allow them to stay. He pledged to find a way for them to continue farming elsewhere, saying that if he failed he would resign. They agreed. He and Mario Boza, then the Park Service director, persuaded the Lands and Colonization Institute to relocate the families to an agrarian reform project elsewhere in the province, and the problem was solved.

Another challenge literally staring Ugalde in the face when he arrived at Santa Rosa was cattle. The park was encircled by cattle ranches, and one way or another, at least a thousand had found their way into the park, trampling and eating seedlings and bruising the fragile soil. (No one truly understood at the time that the cattle also were keeping the grass down to a level where it provided less fuel for fires.) Some ranchers let their herds roam freely around the park.

Others had rented sections of jaragua from ICT after it bought the land from Somoza but before Ugalde arrived.

Ugalde and his helpers insisted that the cattle be removed. The ranchers, who had strong political connections, laughed at them. "There was no respect for us as an authority," he recalled. "We were young kids. I was 20, 22, something like that. These guys were powerful people." Among the connections the ranchers had was Daniel Oduber, president of the Legislative Assembly and a Guanacaste rancher himself. Oduber eventually would become the nation's president and one of the most important supporters of the national park system and Santa Rosa, but in 1970 he backed Pedro Abreu, owner of Hacienda Rosa María on Santa Rosa's southern boundary and, by some accounts, a malevolent neighbor to the park.

Between the time ICT bought Santa Rosa and the law passed establishing the park, Abreu had moved his fences around roughly 150 acres of the park. Ugalde and Boza took him to court to get the land back and have the cattle removed. They waged a public relations campaign, instigating newspaper stories that warned of Santa Rosa's imminent demise. After a three-year court battle, the Rural Guard killed an estimated 1000 cattle. The meat was sent to hospitals and schools. "That's the only way it stopped," Ugalde said. "He realized we were serious about it."

Ugalde combatted fire and poachers, too, and he and Boza skirmished in the political arena as well. "We learned how to deal with Congress, not to say how to fight Congress, because Santa Rosa was at stake just a few months after its birth," Ugalde told me. Their campaign, waged largely against Oduber in those days, left them a bit fearful of the influential politician, concerned that he would hold their actions against them. Eventually, though, Ugalde and Oduber became friends. Oduber offered water from his ranch to the south when the park's supply ran dry. When Oduber dropped by, Ugalde would show him their progress and talk about their needs. Among other political acts in support of the park, on May 4, 1977, Oduber issued an executive decree officially expanding Santa Rosa to include many portions of its major drainage basins, although not the headwaters of its two most important seasonal rivers, Quebrada Guapote and the Río Poza Salada.

Santa Rosa caught the eye of international conservation groups early in its life. On November 10, 1970, it received its first donation, a small grant from the Sierra Club Foundation. The following October, the World Wildlife Fund donated $5000 to construct a biological research center at the park, and the Organization for Tropical Studies contributed too. Ugalde and the volunteers built the first road from the Pan-Am Highway into the administration area. With guidance by historic restoration experts from the Organization of American States, they continued restoring the Casona and developed a visitors' center and interpretive exhibit as part of a public education program. They created latrines and picnic and camping areas, and they brought in water service and electric lines. They improved the nature trail, the Sendero Indio Desnudo.

In the midst of Santa Rosa's early struggles, Ugalde and Boza staged an event in the shadow of the Casona's guanacaste tree on March 20, 1971, at which Figueres signed the decree officially establishing the park. Against a backdrop of colorfully dressed schoolchildren, as well as Oduber and other politicians, First Lady Karen Olsen de Figueres cut a ribbon inaugurating the park. It was the 115th anniversary of the battle against Walker's *filibusteros*, and the first of many such ceremonies to come.

Ugalde gives special credit to several Peace Corps volunteers who worked at his side in these early years. Tex Hawkins was the first to assist him, and others included Douglas Boucher, Steve Cornelius, Curt and Marge Freese, and Keith Leber.

Boucher, now a botanist at Hood College, in Frederick, Maryland, remembers the surprise he and others in his group felt when they arrived at Santa Rosa in mid–1971 to find that the park basically "remained to be established," as he put it. All in their early twenties, their primary task was to explore the park and document its organisms. Boucher had just finished his bachelor's degree in biology at Yale, and his only field work with plants had been a summer botany project in Rock Creek Park, in Washington, D.C. "It turned out that that was the only plant experience in the whole group, so Mario [Boza] said, 'You're our botanist,'" Boucher told me. He spent his first few months at Santa Rosa roaming the park, collecting all kinds of plants, and taking them to the National Herbarium in San José for help with identification. He wrote four separate reports for the Park Service on the plants and habitats of Santa Rosa, as well as a fire-control plan. Cornelius concentrated his efforts on the rich diversity of wildlife down by the beach, including the sea turtles that lay eggs at Playa Naranjo. Freese surveyed monkeys, and Leber focused on birds.

As Boucher, Cornelius, Freese, and Leber wandered the mosaic of forest and pasture, they realized the complex of habitats it contained. There were patches of moist old-growth forest with relatively mild alteration, like the Bosque Húmedo, hidden among the secondary successional dry forest islands. More evergreen forests lined the river banks. Even the grassland wasn't unidimensional. In addition to jaragua pastures with scattered guanacaste trees, some areas, known as *sonsocuite* savanna, usually punctuated with calabash trees, became wet and mucky in the rainy season. When the dry season came, the soil would crack deeply as it dried out. Some pastures had extremely rocky soil with a thin cover of grass.

There was a small trace of resentment about the North Americans, exemplified by a comment left in the Casona's guestbook at the time, saying that it

was shameful that park personnel, referring to "*los gringos*," were living in the historic buildings. It was not a typical response, but it was one that stuck in the
minds of some of the volunteers. It also expressed a sentiment that lingered
through the years, to lesser or greater degrees, as other gringos came and went,
doing their work in biology and conservation.

Even so, the Peace Corps volunteers retained a strong spirit and a dedication
to helping Ugalde build the new park. "These were the early years of the environmental movement," Boucher recalled. "Earth Day had just happened. The
idea of national parks was rare at that time in Latin America, and we had a real
feeling that we were helping something really important to get started. We were
working in an organization with two visionaries, Mario and Alvaro. If it hadn't
been for them, this would all be cattle pasture now."

In March 1972, President José Figueres ordered the Civil Guard to Santa Rosa.
The troops dug trenches, erected bunkers, and held military exercises. Figueres,
leader of the victorious forces in the 1948 revolution, feared a coup, possibly
backed by Somoza. Ugalde thought the rumor and reaction were ridiculous. "I
remember having trouble with them, and I kept saying, 'Why the hell in Santa
Rosa?' But they were there, filling their sacks full of sand. I was furious, because,
you know, I was the clear authority of that park." Nothing happened, and the
Civil Guard eventually left.

When Ugalde opposed the Ministry of Agriculture's plan to cut jaragua in
Santa Rosa and give it to local ranchers hit by the Guanacaste drought of 1972,
the minister transferred him to Volcán Poás National Park. "It was a very sad
moment to me," Ugalde said, "because of all those dreams of growing old in
Santa Rosa that I had as a very naive kid."

Vernon Cruz, a young agronomist, and other Santa Rosa superintendents
after him, continued the campaign against cattle, hunters, and fire. Among other
notions about park-building that Cruz advanced was the idea, as he put it, of
"the possibilities for re-establishing the forest."

Ugalde became Park Service director two years later and led an international
fund-raising campaign that brought several million dollars into the system. In
time, more pieces of land eventually became linked to Santa Rosa, including an
inholding called Argelia, which fronts on the southern section of Playa Naranjo;
Murciélago, the former hacienda to the north of Santa Rosa over the Santa Elena
mountains, which was expropriated from the Somoza family in 1980; and
Rincón de la Vieja, a volcano with mud baths and hot springs about 15 miles
southeast of Santa Rosa.

"Those three chunks—Santa Rosa, Murciélago, and Rincón—were the
nucleuses where the so-called 'project' later took off, to join together all of them
in what we have today," Ugalde told me. "We were pushing for expansion very
early in the history of that park. We knew it was not enough."

Men in uniform returned to Santa Rosa for several months after President Rodrigo Carazo took office in 1978, when Costa Rica became deeply involved in the Sandinista effort to topple the Somoza regime in Nicaragua. The Carazo administration gave asylum to political refugees from the civil war, allowed Sandinista officials to shelter their government in Costa Rica, and helped Sandinista troops live and train in the north and northwest. The rebels launched many of their attacks into southern Nicaragua from Costa Rica. Santa Rosa became a training camp for both the Costa Rican Civil Guard and Sandinistas, although some sources told me that Sandinistas didn't train there. According to others, the troops, working with Venezuelan military advisers, bivouacked in the administration area and took target practice with rifles and mortars, shooting at trees and fences. The soldiers told scientists to stay out of certain places, but they sometimes inadvertently lobbed shells into the administration area. The Sandinistas based themselves at El Pelón de la Altura, an hacienda just east of Santa Rosa, and, in the weeks leading up to their victory, on the Somoza family's remaining piece of property in Murciélago. The Santa Rosa occupation concerned Ugalde most. As director of the Park Service in San José at the time, he worried that the destructive activities might escalate, and he campaigned to get the soldiers out of the park.

Although the feeling was not universal, especially in Guanacaste, there was widespread anti-Somoza sentiment in Costa Rica. Many felt that the Nicaraguan strongman represented a threat to Costa Rican national security. Some feared a Nicaraguan invasion, or at least that Somoza would seek refuge on his Guanacaste ranch if the Sandinistas ousted him from Nicaragua. Others charged that Somoza used Murciélago's coast to harbor Nicaraguan fishermen who, protected by the general's gunboats, illegally fished the Golfo de Papagayo and other Costa Rican waters.

In September Carazo, in part responding to this anti-Somoza feeling, expropriated Murciélago, which included a house and small airport. The ranch was split in three. Most eventually became Sector Murciélago of Santa Rosa National Park. A few small farmers were allowed to stay on another part of the property with good soil, and the Civil Guard kept a training base on a separate small piece.

Soon, the anti-Sandinista contra rebels were moving through remote patches of Guanacaste Province in their attempt to establish a southern front inside Nicaragua. President Monge allowed U.S. operatives to construct the Santa Elena airstrip and what was billed as a Voice of America radio tower in Hacienda Poco Sol, a property on the east side of the Pan-Am Highway near Santa Rosa. Many Costa Ricans believed it to be a U.S. communications and intelligence-gathering facility. Monge also permitted U.S. Special Forces to train contra rebels and Costa Rican Civil Guards at the Murciélago base.

The mid–1980s was a time of great turmoil in Costa Rica. Some critics wondered whether the republic could maintain its political and social stability in the face of growing economic hardship, brought on mainly by its mounting international debt and declining prices for coffee, bananas, beef, and other commodities. Civil wars in Nicaragua and El Salvador threatened to spill across its borders. In some ways, they already had, with the guerrilla activity and thousands of refugees entering from other Central American countries.

Costa Rica did not consider itself aligned with any nation. However, because of the strained economy, it accepted more than $1 billion in economic aid from Washington in the 1980s. U.S. leaders hoped to keep Costa Rica stable and to enlist its support against what Uncle Sam viewed as Communist infiltration of Central America.

Costa Rica stood out as a peaceful beacon in a region bloodied by civil war. The country was known in the tourism literature, not entirely without reason, as the "Switzerland of Central America." The government's official stance was one of "permanent, active, and unarmed neutrality," as Monge had put it in 1983. Oscar Arias, his successor, had continued Costa Rica's bent against military solutions to the region's conflicts, maintaining a policy that peace could be obtained only by social justice and economic development.

Costa Rica had been invaded in the past, and some thought it could happen again. The invasions had all come through what is now Guanacaste Province, and San José had become increasingly concerned about the leftward tilt of the Sandinista government.

Even so, the nation maintained its sterling international reputation among conservationists and scientists. Conservationists loved it because of the remarkable biological diversity and richness contained within its parks and preserves, a system totaling 2026 square miles, or more than 10 percent of the country. The country had long been a favorite spot for scientists who wanted to study tropical forests under some of the best conditions anywhere—in a nation less permeated by the political, military, and social turmoil common to many developing states, and where the forests were relatively easily accessible. Costa Rica was, as environment writer Chris Wille once put it, "ecologically avant-garde."

In August of 1996, a child's drawing hung on a wall inside a building at Santa Rosa. It had been sketched by a first grader from a local school using colored pencils, and someone had taped it up on the wall with half a dozen or so other drawings. This one showed the fighting in front of the Casona in 1856. The *filibusteros* were shooting from behind the stone foundation of the old house. Two Ticos who had charged across the open ground in front had fallen, bleeding. Another stood nearby, shooting a rifle. A fourth fired from behind the guanacaste tree.

Now, ten years after I had first seen it, the guanacaste's branches seemed wrinkled, twisted, and senescent, the thinning canopy of leaves reminiscent of an old, balding man. The branches, as they had been long before any crazy gringos set foot here, were still adorned by birds, lizards, and the white flowers of epiphytes—plants that grow on other plants while getting moisture and nutrients from the air. Probably sown out of the dropping of a cow or horse, this gnarled yet magnificent veteran had, in its better than two centuries of life, survived repeated waves of logging, fire, drought, and war that rolled across the Santa Rosa landscape. It remained an apt symbol of the toughness of Guanacaste Province, and some of the paradoxical dimensions of the Costa Rican soul— noble and peace-seeking on one hand, and on the other hand strong enough to withstand a fight when necessary.

"Many Costa Ricans struggle with the idea of the role of the military," León, the University of Costa Rica biologist, told me. "They take pride in the fact that they have no military, saying that for nearly fifty years they've gotten along quite well without one, and, in fact, it's been to their benefit. On the other hand, they also take pride in their victories on the field of battle, especially in 1856 at Santa Rosa." Meléndez, the historian, once wrote that "Santa Rosa is a place and a symbol . . . of liberty and national sovereignty." Even a schoolchild's drawing reflects it.

Today, Santa Rosa and its guanacaste trees also have become patriotic symbols of ecological restoration. A sign in front of the old tree gave standard information: It's the national tree, lent its name to the province, and so on. But the sign also noted that the guanacaste tree is contributing to the natural regeneration of the dry tropical forest, concluding, in Spanish, in large letters and quotation marks, as if the words emanated from deep within the reddish-brown heartwood of the tree itself: "I have seen Costa Ricans defend this ground . . . against the *filibusteros*, just like the noble park guards in the battle for conservation."

# · 5 ·

# The Fires of Guanacaste

Alvaro Ugalde's first encounter with fire at Santa Rosa came during the dry season in early 1970. He was working at the Casona when workers came up and told him a fire was approaching. He climbed the hill behind the historic building and saw clouds of smoke billowing up to the north, not far from where the entrance road runs today. Ugalde confesses that his reaction was naive.

"'Well, let's jump in the car," he told them. "Let's go stop that fire."

"No, no, Alvaro," the others said, laughing like the worldly parents of an idealistic teenager. "No, no, no."

Ugalde shot back: "Come on you cowards, let's go."

So they piled in the car and took off. They got out where a patch of forest broke into savanna. They saw a ten-foot-high wall of flame rushing at them through the jaragua, as if exploding out of a lake of gasoline.

"The only thing I could do was just turn around and get the hell out of there," Ugalde remembered. "Some of the *peónes* who knew better said, 'It'll stop here, where there is forest.' I couldn't believe them, but it stopped there."

The most formidable enemy of the Guanacaste dry forests is also its most exotic enemy, since fire is not a natural phenomenon in the region. Lightning plays no role in starting dry-season fires there, unlike forests in the United States and Australia. Storms are absent during the dry season. In the rainy season, lightning is accompanied by rain. An ungrazed pasture full of years of dry grass can burn in the rainy season, but then again, the pasture is hardly natural tinder.

Instead, fires get set as a regular part of Costa Rican life. During the dry season, ranchers and farmers burn the dry, brown cover off their fields to stop the advance of forest into their pastures and trigger growth of grass shoots, which cattle love.

However, misjudgments and carelessness occur in setting these fires, and therein lies the threat. On hot, sunny days with winds commonly blowing at twenty or thirty miles an hour, and swirling gusts double that, even if the ranchers exercise care—and many don't—fires frequently get out of control. Flames often rush into the park, either communicated directly through adjoining jaragua fields or leaping as burning, wind-carried embers cross roads or trails that might separate the two properties, igniting parched forests and grasslands far from the place where the fire was started. Fences are simply incinerated.

The majority of fires that invaded Santa Rosa were started on nearby ranches and farms. Hunters also set fires inside and outside the park to open meadows and concentrate game. Sometimes people accidentally set fires in the park, but they are also set intentionally by people who just like to watch the tall swards of jaragua burn. The grass along the Pan-Am Highway is a common spot for someone to stop his car, get out, and set a match. It happens a few times every year. "It's not clear at all what the rationale is," Janzen once told me. "I don't think it's actually directed against the park. I think it's directed against a big stand of grass. Maybe somebody says, 'I'll put a match to this and see what happens.' It's spectacular."

Fire has deep roots in Guanacaste history. Indians may have burned small patches in the initial stages of forest removal, and the white settlers, of course, used fire to clear pastures and fields. Peasants often set fires as an expression of resistance to the policies of landlords.

Since the annual burn was part of tradition and highly useful to the ranchers, residents often view fire as inevitable. Some believe that unless the Park Service has tried to establish special relations with the locals, the park officials carry the responsibility for dealing with a pasture fire once it jumps into the forest. "Fires, unlike cattle, are not viewed as belonging to anybody, and nobody retrieves them when they wander into your park," Janzen wrote in an essay, "The Eternal External Threat," published in 1986.

Once fire enters Santa Rosa, it finds plenty of good fuel, mostly in the form of jaragua. Fire spreads through fields of the tinder-dry grass as if soaked in gas. The conflagration incinerates seedlings and saplings hidden beneath the grass; it torches young trees that might have outcompeted the jaragua within a few years and created a new patch of forest.

Although the rush through jaragua is swift and violent, the forest fires of Guanacaste are generally quite a bit more tame than the dramatic fires of the western United States and Australia, where a single blaze can leave hundreds of square miles of mature forest in Dresden-like condition. In Guanacaste, fire rarely destroys an entire block of forest. Rather, it cuts away at the margin between the grassland and trees. After racing across a jaragua patch, no matter how vast, it usually subsides shortly after entering the forest and begins to creep through the litter layer below established trees. With nightfall, a rise in humidity and fall in air temperature may help extinguish it.

Sometimes wind can thrust an intense fire tens of yards into the trees. The later in the dry season, the deeper it cuts. The following year, a new stand of jaragua will grow in place of the burned trees. The net effect is an advance of the jaragua downwind. If fire visits in subsequent years and the wind is right, the jaragua continues to push back the forest, like waves from an expanding ocean clearing the beach a few inches at a time. The process can continue until all that's left is pasture.

This gradual assault proceeds even without the incessant push of powerful waves of fire. Fire sometimes creeps through the forest-floor litter, leaving behind a form of damage that at first appears insignificant. The heat of these litter fires kills small sections of a tree's cambium, the formative cells beneath its bark. The tree can heal that spot in five to twenty years. However, if a fire should strike again during that time, it can penetrate to the tree core and kill it. "A light litter fire, then, cuts down large living trees as surely as if done with a chainsaw," Janzen has noted. During the next rainy season, with the tree's crown no longer blocking sunlight from the forest floor, grasses and herbaceous plants grow rapidly there, providing fuel for fires in subsequent dry seasons. This area becomes a brushy pasture that also begins to expand at the expense of the forest, eating away from within. Old logs and standing dead trees can burn for days, providing a source of cinders that can be blown far and wide to ignite other fires behind the lines.

To the casual observer, one of the most striking differences between fire fighting in Guanacaste forests and fire fighting in the U.S. West, besides the relative magnitudes of the blazes, was technology. By the end of the 1980s, U.S. fire fighters had access to digital images of a forest fire from a plane flying above it. They also could use portable personal computers to compare those images with maps showing locations, roads, land cover, and water sources. They could talk via long-range radio, parachute out of planes in remote areas, use bulldozers and other heavy equipment, and sometimes rely on aircraft to dump water and chemicals directly on a fire.

In contrast, Costa Rican forest fire fighting at the time was a low-technology affair—brooms, a few fifty-five-gallon drums of water transported by truck, and a bright blue plastic backpack of water connected to a hand pump and spray nozzle. "Aerial reconnaissance" meant haphazard looks from atop the nearest hill. Communication was face to face; there were no radios. Messages in the field were sent by runner and driver. Two groups of the same party leaving park headquarters minutes apart to fight a newly discovered fire often misinterpreted directions to the field rendezvous point and spent an hour cruising the countryside, trying to find each other. Getting to a fire was difficult, especially if it was burning in a canyon or beyond the crest of a hill. A crew could drive to where they thought the fire was, but sometimes it wasn't there. Even in broad daylight, crews could become lost on the way to a fire on the sinuous network of rocky dirt roads in and around the park.

If the fire was newborn and small, the crew sometimes pounded it into submission with brooms soaked in buckets of water. Often the blaze was huge, so park guards, scientists, and other volunteers fought it for several days and nights with few breaks and little sleep. Choices about how best to fight the fire were made depending on wind direction, topography, and vegetation. A load of water or fire-suppressing chemical dropped from an airplane or helicopter was the stuff of dreams.

Ugalde and his charges routinely fought fires that burned right past the Casona. He remembered one fire that came rushing over the hill behind the house, whipped by a ferocious dry-season wind. "There were flames on the roof, with the jaragua flying, burning on top of the historic building," he said. "We almost lost that house." The fire continued southwest, sneaking along trails and other jaragua-covered open areas, all the way to the beach.

Protecting the Casona was a constant worry. Farm workers employed by the Park Service kept the fields near the house trimmed with machetes as close to the ground as possible. Even so, fire repeatedly charged up the hill behind the Casona before crews beat it out with brooms and shovels.

The Peace Corps volunteers had no fire-fighting experience, so they learned on the job. In his first year at Santa Rosa, during the 1971–72 fire season, Doug Boucher watched two large fires threaten the Casona and another burn through the nature trail forest behind it. The latter fire approached across a savanna, and the volunteers figured it would stop when it burned a few yards beyond the edge of the forest. "But it kept on going into the forest, and we realized it was not stopping," Boucher said. "We went out there with brooms, trying to beat it out. It was mostly low flames in the litter layer, but it was tough to fight. The wind was blowing smoke into our faces. Everyone out there was coughing, choking. It went on like that for three or four hours before we got it out." But it didn't reach the Casona.

Boucher was assigned to study the park's fire-control problem. Each time he heard of a fire, he hopped on his motorcycle and dashed off to the front line. "I would try to see where it started, which way it was going, and how fast," he said. "We realized very soon that there was no way to stop the fire in the savanna. The flames would get five to ten meters [fifteen to thirty feet] high, and you get this enormous amount of noise—crackling and thudding. All these insects would come out ahead of it. Sometimes when the wind was strong, the flames moved very rapidly horizontally."

The burned field would cool quickly. Five minutes after the fire swept through, Boucher would be in there, walking the ash-covered terrain, carefully examining the remains. He earned a reputation for returning to the Casona coated with ash. "Many's the time he would appear late in the afternoon, blackened by smoke and soot," recalled fellow Peace Corps volunteer Keith Leber, now an editor at the University of Hawaii Press. The young park guards, tickled by Boucher's almost religious devotion to studying fire, dubbed him "El Zopilote," after the black vultures that commonly gathered at a fire scene looking for the spoils.

Later, Boucher set up sampling plots to gauge how much damage occurred, what grew back, and how it regenerated. Even though it was "a scene of complete destruction," the grass would re-sprout within a few days, its reddish tips slowly turning green. Boucher learned, too, from Misael Martínez, who had been a *sabanero* at Santa Rosa when the ranch was owned by Somoza. Martínez, who then worked for the Park Service, had the uncanny ability to tell the Peace Corps volunteer how far a fire would jump under certain conditions and why a fire break wouldn't work if it wasn't fifty or, in many places, a hundred yards wide.

Boucher eventually wrote a fire-control plan that called, basically, for controlled burning. "We had zero money for any investment in tools or labor to fight fires," he explained. "We had to come up with some sort of way to deal with them that would essentially cost nothing." After discussions with Ugalde, Martínez, and others, Boucher concluded that firebreaks along the boundary wouldn't work. "Sparks from the fires would go very large distances," Boucher said. "Even if we cleared a substantial stretch along our fenceline, they would still go right over it."

Eventually the conservationists settled on a program of controlled burns that rotated, year by year, among a particular spatial arrangement of jaragua pastures.

In essence, the pattern kept the pasture to be burned surrounded by previously burned pastures. The burns were conducted early in the dry season, as soon as the susceptible grass was dry enough to burn well, but not so dry as to virtually explode. "We suspected that even if we did this every year, it probably would just stabilize the situation," Boucher said.

In fact, for nearly two decades, the extreme wind and the multitude of rogue fires from outside the park overwhelmed the preparations and adaptations that park personnel had decided on. Also, when cattle were finally removed from the park in 1978, jaragua that had been grazed short in many areas now grew large. That gave grass fires the upper hand, and savanna made huge inroads into the forest. In many areas, it overran small fragments of forest that had been slowly growing back on rocky outcrops.

Add to that the humble fire-fighting capability, the perennial lack of money, and the extreme mental and physical fatigue that seeped in during roughly five months of fire fighting each year, and it is easy to understand why the motivation of some of the Park Service workers might wane in the course of a fire-fighting campaign.

In retrospect, one of the more interesting ideas in Boucher's report on fire control was the notion that the forest might one day be restored. Since the pastures were an artificial system—the introduced jaragua grass—reestablishment of a vast forest was the logical aim of fire prevention. "I remember having debates when working on the plan as to whether the eventual goal should be to totally eliminate the savanna or almost totally," Boucher said. "The question was, 'Do we want one hundred percent dry forest, or ninety-five percent and save some of the nice views?'"

The plan called for maintaining some savanna in front of the Casona, mainly to preserve a semblance of the scene at the time of the 1856 battle, and a large savanna near the entrance to the park, so that visitors could turn around half a mile or so in and look back at the panorama of volcanoes. "Aside from that, the feeling was that the dry forest was a very endangered ecosystem, and we had one of the biggest chunks of actual or potential dry forest in Costa Rica and all of Central America. So the initial impetus was, let's stop the fires from taking more of the forest, not to mention the fences. But the guidance I got was, let's think in terms of something that will allow the forest to grow back."

It would be easy to overdramatize the dangers of fire fighting in Guanacaste and the shocking moments that people often encounter. No one is aware of any deaths or serious injuries resulting from fires in and around Santa Rosa. However, even though relatively tamer than U.S. forest fires, the fires of Guanacaste can still be extremely unpleasant. They start unpredictably, move fast, shift unexpectedly, confront crews with walls of flame and smoke up to 200 feet high, suddenly surround them, choke them with smoke, and draw oxygen out of their lungs.

Battling on the front line of a rapidly advancing fire in the tropical dry forest can be as hazardous, confusing, and physically and emotionally taxing as at any fire. Exhaustion is common, and the tropical heat can leave the body unusually vulnerable. Frank Joyce told me that while he helped to fight a steadily advancing fire in Santa Rosa for several days in a row, a wind change stoked a wall of flame and shoved it toward him. He turned to run but fell in his tracks as cramps, brought on by dehydration and loss of electrolytes, immobilized his leg muscles. He quickly massaged them enough to drag himself out of danger. "It was close," he said.

Ugalde remembered several times when fire lunged at, and even surrounded, him and his crew. Typically, they would concentrate their efforts on swatting down one line of fire, then another line would rush at them from another direction. "All of a sudden it becomes very hard to breathe," he told me. "You say to yourself, 'Well, how do I get out of here?' You cannot see on one side, and then you hear this terrible noise coming towards you."

One afternoon in the 1988 dry season, I rode into a trap with a Canadian college student. The leader of one fire-fighting crew had assigned us to find and bring back a group that had gone to a distant front of the fire. We sped down a road toward it in a little Suzuki jeep, stopping a few yards short of a crest where wind-whipped smoke rose. We jumped out of the jeep and walked forward, shouting for anyone who might hear. We were answered only by the crackling and whooshing of the fire to the front and—suddenly—on both flanks. Almost as if on cue, thick clouds of smoke washed over us from one side, and we realized the fire was encircling us. We exchanged looks of concern but spoke calmly, as if to keep from alerting our would-be ambusher that we were on to him.

"I guess they're not here," said the student.

"No. I guess not."

In an instant, we jumped back in the jeep, and he executed the fastest, most wonderful dust-kicking three-point reverse turn I'd ever seen. Years later, watching *Star Wars* and other movies in which space pilots peeled away from an exploding enemy starship, I recalled this escape.

Despite bursts of adrenaline and the constant misery of heat, thirst, minor burns, and smoke inhalation, fire fighting is mostly an exercise in tedium. There is constant worry about how hard the wind is blowing and what the terrain is like. Then there is the hard job of keeping track of the main fire, its offshoots, and the *contrafuegos*. Fire crews must anticipate where these might spread.

Often, during the peak of the fire season, crews will be out until well after midnight fighting a fire. They roll out of bed early to check the battleground and to troubleshoot. "You've got to go back to that fire in the morning to see that it's in fact really out, because the stumps and logs keep burning," Janzen once told me. "Given that you think you really do have it out, then you have to keep checking it all day long, because as the day dries out and the wind blows across it, little sparks begin to fan and burn in the logs, and the fires start, and the wind picks up embers from that and blows them on to an unburned area, and you get a fire all over again."

Crews sometimes forgot who the enemy was as emotions rose and fell and mental and physical fatigue set in. Disagreements flared over strategy and tactics, and fire fighters vented anger that more properly should have been directed at the person who caused the fire, except that this was rarely established. As a fire burned on and on over several days, the Santa Rosa fire fighters could be worn down to the point of demoralization. In the back of their minds was the realization that no matter how hard they worked to save a piece of forest, it could be burned anyway, at that time, later that dry season, or the next year.

Sadly, it often happened.

The one fire I helped fight in Guanacaste was a case in point. It was the February 1988 blaze on Finca Rosa María, against which Julio Díaz had set the *contrafuego*. Eventually, the main fire overwhelmed all efforts to control it. Weary park guards, scientists, and volunteers set more *contrafuegos*, babysat them, and repeatedly returned to various fronts of the battle over several days. The Civil Guard was called in to help. From the Monument to the Heroes at night, the glow of the fire could be seen as it approached Santa Rosa, periodically brightening with gusts of wind. The fire roared into Santa Rosa and cut through the *Cañón del Tigre* before, days later, finally subsiding.

## · 6 ·

# The Living Dead

On the misty slope of Volcán Cacao, beyond the barbed-wire fence that lined the road, the enormous tree stood alone amid a vast cattle pasture. The tree, a member of the genus *Nectandra*, a kind of small-fruited avocado, was easily 100 feet tall. Its narrow trunk rose high and naked before splitting into branches that bore leaves. This broad green crown radiated brightly, like an emerald, seemingly in defiance of the waves of gray fog that passed down through it and around it from the top of the mountain. One moment the *Nectandra* was shrouded in fog, disappearing from view; the next, it emerged, ghostlike. It was easy to imagine that the tree itself was moving, not the mist. And it was easy to imagine the other giants that once stood with this tree to make up a majestic blanket of cloud forest. In this patch of phantom forest, all but this one had been cut and burned to make way for the green gauze of grass that now rolled halfway up the volcano. It was an eerie sight, made all the more so by the blackened hulks of dead trees lying in the pasture. Although the forest was physically obliterated, its spirit seemed to linger.

I've observed this scene and countless others like it in Guanacaste Province, and I was seeing yet another as Ian Gauld and I drove eastward one afternoon up into the Cordillera de Guanacaste between Cacao and Volcán Rincón de La Vieja, in June of 1986. It suddenly dawned on me that such scenes could completely cover the landscape of northwestern Costa Rica if the Guanacaste project failed.

We were on our way to a night of insect collecting on Finca San Gabriel, just over the continental divide on the Atlantic side of Costa Rica. Gauld, an entomologist at the Natural History Museum, in London, was at the wheel of Janzen's Land Cruiser. Then thirty-nine, he was a friendly man with bushy brown hair and a beard, and as we bounced along the rocky road, he would now and then ask me to hold the steering wheel while he reached into the glove compartment for a Marlboro and a match.

We exchanged waves with Ticos walking along the side of the road or riding on horseback. We ambled through the town of Quebrada Grande and, higher up, through a makeshift village of long, turquoise, barracks-like buildings. This was a camp assembled by the Costa Rican government to house refugees from the war in El Salvador. Every so often a herd of cows blocked the road; we simply sat and waited, the engine running and the windshield wipers working against the mist. We swooped down hillsides, across bridges, and up the opposite inclines. Crossing the rickety bridges was a white-knuckle affair for me, although it didn't seem to bother Gauld, who accelerated as we approached to enable the Land Cruiser to make it up the other slope. A typical bridge was made of two-by-six planks laid across two beams. The planks rattled as the Land Cruiser dashed across, and they sometimes flipped up behind us. Planks were missing from a couple of the bridges, and sometimes there were large gaps between the bridge and the road.

All the way up the slope, we passed pastures littered with fallen trees and ones standing alone like the *Nectandra*. The upright trees appeared to be as alive as could be, but actually, they were the "living dead."

The living dead is a term coined by Janzen to describe an organism caught in the onslaught of habitat destruction, but not slaughtered, at least not directly. A living dead species is in a kind of no-man's-land between life and death, on its way toward extinction. It has become part of what biologists call the agro-ecosystem—the homogenized landscape of formerly diverse natural habitat that has been transformed, completely or nearly so, into a farm or pasture. Nearby might be a small wildland island or two, surrounded by "biological desert."

The mist-shrouded *Nectandra* was a good example of a living dead individual. As the term implies, it was both alive and dead at the same time.

Already probably a century old, the tree was physiologically alive in the sense that it tapped the energy of the sun, grew, and periodically produced new flowers, leaves, and seeds. It might have lived several decades more as part of its normal, individual lifespan. However, it was reproductively dead in the sense that it occupied a habitat in which any seeds it shed would be trampled by cattle, burned by pasture fires, outcompeted by grasses, or desiccated by the piercing sun of the dry season. Technically, depending on the species, a living dead tree still might contribute to the genetic recombination of its kind by spreading its pollen on the wind or on the backs of bees or other pollinating animals. Or its seeds, possibly pollinated by trees from forest stands elsewhere, might be carried off by fruit-eating birds and bats, to be deposited perhaps in a habitat more receptive to germination. In this sense, the living dead tree still served as a reservoir, however minuscule, of genetic diversity, helping to exchange genes among forest fragments. But the key point is that any offspring it tried to sow in its inhospitable neighborhood would die. Thus, generally speaking, its next generation could not germinate and grow, could not be born and mature. Since it could not achieve what its ancestors achieved, could not fully continue the age-old chain of self-restoration in the forest, could not play its natural role in the existential mission of its species to survive, it was essentially as useless and irrelevant biologically as if it lay on the floor of the biological desert with the others.

The living dead, Janzen wrote in his 1986 essay, "The Future of Tropical Ecology," are "standing, living and breathing, but as dead as is the litter, since they have no reproductive future."

Moths, birds, butterflies, jaguars, herbaceous plants—these and other wild organisms of Santa Rosa and nearby parts of Guanacaste Province at the time were potential members of the living dead, albeit less conspicuous than the many solitary *Nectandras*, mahoganies, guanacaste trees, and others that rose above the landscape. As forests were cleared, large expanses with complex mosaics of habitats shrank. The smaller and less diverse these islands became, the harder it was for organisms to maintain healthy breeding populations. Gradually, species slipped into living dead status. Animals and plants that depended on them for food or seed dispersal soon shared the same fading fate. A tree species whose individuals could live for 400 years might persist in the island through many human generations, while an insect species might disappear overnight.

Since the low-grade agroecosystem often had enough remnants of forest plants and animals to give the appearance that nature was still present, the message of impending extinction was often difficult to grasp, much less communicate. By continuing to exist, existence of living dead species, Janzen wrote, "allow a farmer to clear a forest, yet show trees to his daughter—a cruel deception. They allow the tropical farmer to live on his ranch, yet not be consigned to the most boring habitat on earth."

The living dead showed, as he put it, "what once was." They represented the past of the tropical forest, but they also were a harbinger of its bleak future.

It was late afternoon by the time Gauld pulled the Land Cruiser off the road near a thick stand of forest and set the brake. The view was obscured by large, low clouds that whipped by just above our heads. The wind was cold and damp, the Atlantic rain forest thicker than the Santa Rosa dry forest.

He strung up three collecting sheets between trees, on a gate, and off the side of the Land Cruiser. These were white tarps about eight feet wide and six feet tall, and from the top of each sheet he hung two long tubes—one a black light and the other a white fluorescent light. He connected the lights to twelve-volt car batteries on the ground.

The object of this setup was to draw insects out of the landscape—for as far as they could see the light—and onto the sheets. "The black light gets most of them," he explained. "The white light is mostly for us." Once on the sheet, he could pick off insects of interest, drop them into a killing jar, and a few minutes later transfer them into a morgue that would preserve them.

As he adjusted the third sheet, a Costa Rican farm worker with a machete in his hand popped out through the thick forest, startling both of us. The Tico smiled and said hello. He looked at the sheet for a moment, smiled some more, and walked on, shaking his head. Gauld told me that locals often stop at collecting sites to find out what's happening, and when researchers try to explain, the Ticos often seem befuddled or amused. "Los gringos locos," the locals called the scientists. Crazy gringos.

Night fell shortly after six, and Gauld started the lights. "It's a waiting game, mostly," he said.

The black light produced a lavender shimmer on the sheet, and what followed was reminiscent of a screen at a drive-in movie theater. First in the cast of characters to arrive were dozens of kinds of moths. They were mainly dull and gray, but they came in all sizes, from smaller than a fingernail to larger than a 3x5 index card. Tiny wasps flew in, as did metallic green beetles. They took their places all across the sheet.

Gauld recalled his best night of black-light collecting, in the forests of Borneo. "I've seen sheets like this where you need to stand this far out," he said, positioned about six feet from our sheet, "because there is this sort of sphere of living things flying around in a great mass. Things are landing on the sheet three layers deep, and there is just a heaving mass of things on the ground. It's almost like a whirlwind. You put on something with a hood, and tighten the hood up around you, and you dash in and grab a few things and dash out again."

We talked about Gauld's specialty, the Hymenoptera, an order of insects that comprises the bees, ants, wasps, and their relatives. His main interests were the parasitic and predatory wasps. These are not the social wasps that most of us know, the kind that build large nests and live as a colony. Rather, their strategy for survival relies, generally, on the ability of the female wasp to fly around, find a particular kind of insect or other arthropod, and lay an egg in it. The larva that hatches out then consumes "that piece of meat," as Gauld put it. A variation on this theme is for the wasp to paralyze the arthropod with a sting and bring it back to a nest chamber for the larva to consume there.

This basic strategy leads to all sorts of tactics. Since the prey needs to feed a larva that will eventually grow to the size of the adult wasp, the food item actually has to be *bigger* than the adult wasp. That presents special challenges to Mom, the hunter. "Large arthropods tend to be rare, or very well concealed, or extremely dangerous," Gauld said, chuckling.

Among the most dramatic examples of wasps that take on dangerous prey are several species of the genus of *Pepsis*, also known as the tarantula hawks. These huge, brightly colored wasps, common in Guanacaste forests, attack mygalomorph spiders, also known as tarantulas. "I can't imagine a more hazardous object to attack than a mygalomorph," he said. "They are wicked spiders. There was one near the light the other night, and it took a big sphingid [moth] and just killed it like that." And he clapped his hands.

The tarantula hawk overcomes the spider with quickness and paralyzes it with a sting. Then it drags the spider into a burrow and lays an egg on it.

Other wasps attack less hazardous large animals, like moths in the pupal stage. Over the eons, moth species have responded by evolving ways to hide their vulnerable pupal stage, and wasps have evolved ways of finding them. Some wasps bring several small caterpillars back to a nest rather than one big one. Wasps without nests insert an egg in a caterpillar when it's small, and the young wasp waits until the caterpillar grows large before eating it.

A kind of evolutionary chemical warfare between wasps and caterpillars has produced defense systems in some caterpillars that knock out young wasps. To respond, some wasps evolved venoms that override the defense system. In one species, the wasp transmits a virus that defeats the defense. In another, the wasp injects venoms that manipulate the host's development process, persuading the caterpillar to form a pupa earlier than normal. On and on go the variations.

Whether by wasps, flies, or other organisms, parasitism and predation play a crucial role in the balance of life in a tropical forest. These two processes help determine how many caterpillars there are, and that is important because caterpillars eat more leaves in the forest than any other group of organisms.

Sorting out the impact of caterpillars was one goal of the Moths of Costa Rica project, begun in the late 1970s. As part of the study, Janzen and colleagues were attempting to document the effect of caterpillars on Santa Rosa's 680 species of broad-leafed plants, as well as the effect of the plants, parasites, and predators on the caterpillars. As Janzen once told me, they were investigating "the moth fauna—what it is, what it eats, and who eats it." They used black-lighting and other collecting techniques to determine that about 9000 species of moths were native to the country. That alone represented incredible diversity; consider, as an example, that Costa Rica has about 130 species of sphingid moths, while 115 species of that family are found in all of North America. Collecting revealed that at least 3000 species of parasitic wasps lived in the Santa Rosa region.

As part of the study, the researchers trekked the forest to find caterpillars, especially during the growing season. Often, this was done at night, when the artificial light of a flashlight helped the collectors more easily distinguish green caterpillars from green foliage. They ran black lights to capture adult specimens for identification and to obtain eggs. They hung cages in the forest containing virgin female moths, a technique that helped capture, mark, release, and recapture the males—and thus census them. They raised caterpillars and placed them in the forest, watching what they ate and how much. They noted how the plants responded.

To discover predators, the researchers watched birds, spiders, and other creatures and noted which caterpillars they took, and sometimes got information by dissecting dead animals to see what they had eaten. The issue of finding parasites was a bit more intricate. By bringing caterpillars back to the lab and rearing them in plastic bags, Janzen watched what hatched out. It could be a healthy moth, or, if the caterpillar had been parasitized, a wasp, fly, or other parasite could emerge after consuming the caterpillar or pupa. By collecting thousands of caterpillars over many years and keeping careful records, the researchers were gradually uncovering the details of this complex interplay, including rates of parasitism and whether some moth species were more vulnerable than others. So far they had found that close to 85 percent of some species were parasitized, while other species didn't appear to be parasitized at all. It was another important piece of the ecological puzzle that is the tropical forest.

The wind had picked up, and the small number and variety of insects coming to the sheet disappointed Gauld. Tonight's harvest was a poor one, scientifically speaking, because wind tends to make night-flying insects sit tight. It hampers male moths, for instance, from tracing the trail of pheromones emitted by females, while the females, more inclined to eat than mate, have trouble following the fragrant odor trail of their food sources. So Gauld called an end to our watch. He left the lights running, and we climbed into a tent pitched near one of the sheets.

The moths and other insects gathered this night would be brought down to Santa Rosa for processing and study. Such work requires precision, patience, and time, and it might seem esoteric and even a bit strange to the casual observer. But it can pay off. Linnaeus might have been thinking about this kind of study, albeit sans electricity, when he wrote that "the great book of nature . . . gradually unfolds herself to him who, with patience and perseverance, will search into her mysteries."

It paid off in Guanacaste.

Black-lighting and other collecting techniques revealed in 1986 and 1987 that many kinds of insects found atop the cordillera didn't live there, but were actually just passing through. This was a key discovery, and it illustrated why, to restore endangered tropical ecosystems like these forests, knowing the taxonomy and life history of organisms was crucial.

This lesson had its origins in the idea—widely held until the 1980s—that the Mesoamerican dry forests were distinct and separate from the cloud forests on the volcanoes, and even more distinct from the vast rain forests to the east, on the Atlantic side of the isthmus. Biologists also assumed that the main dry-season migration in the dry forest consisted of animals such as birds, lizards, monkeys, beetles, and butterflies moving a few hundred yards to evergreen forests along seasonal watercourses, a few rivers, and water holes. There, they took refuge against extreme heat and aridity, emerging with the onset of the rainy season and reoccupying the newly green dry forest, rejoining the many animals that stayed. Biologists hypothesized that such dry-season "refugia," as the moist areas were termed, contributed to the extreme diversity of animals in the dry forest.

However, this short-distance migration is only one of the methods animals have evolved to survive the ordeal of the dry season. Studies of moths and birds in Guanacaste in the 1980s showed that some species that spent the rainy season in the dry forest migrated before the dry season to the moister forests in the cordillera and to the remnants of semi-evergreen forest in places such as the sheltered ravines of Cerro El Hacha. These refugia were suddenly found to be critical to the survival of many dry-forest species.

Even more surprising, some of the animals, such as birds, moths, and the endangered white-lipped peccary, were found to travel scores of miles over the cordillera to refugia in the Atlantic-side rain forests. This cross-country migration of millions of animals between the Pacific and Atlantic sides was among the most astonishing ecological stories of the tropics.

The key study that uncovered this phenomenon was led by Janzen. It found that with the approach of the dry season, dozens of species of sphingid moths fly east across the cordillera to the rain forest, only to return with the onset of the rainy season months later. Sphingids, a group of large, heavy-bodied moths that include the sphinx and hawk moths, have a rapid wingbeat and flower-feeding behavior that calls to mind hummingbirds. At dawn, dusk, or dark, they hover above blossoms and uncurl their extremely long tongues to drink the nectar.

Sphingids were collected at black lights only a few hours after the first rains of the rainy season at Santa Rosa. They came in by the thousands to a light which only the night before had drawn a few dozen. They could not have emerged from their pupal stage and become flying adults so quickly after a first rain, as was the classical view of this burst, so they must have come from somewhere else. Adults of these same species were common in the Atlantic rain forest, a habitat that also supports the host plants for their caterpillars, as does Santa Rosa. In addition, the adult moths were found in the high-elevation cloud forests of the cordillera, where the host plants did not grow. This and other evidence—including black-lighting on the volcano slopes—indicated that they must be flying through the cloud forests to the dry forest on a journey that was timed to coincide with the onset of the rainy season.

Janzen proposed that the moths did this to take advantage of the burst of greenery and the annual low in carnivore density following the arrival of rain on the west side. The adult moths laid eggs on the new leaves, and after a few weeks, the next generation had eaten its fill, pupated, and emerged as adults. Curiously, most of these moths did not stay in the dry forest, where an ample food supply could easily sustain a second generation months before the dry season set in again. In fact, all but a few of the newly emerged adults headed back across the cordillera to the rain forest.

Janzen suggested that they fled the area to avoid the annual population explosion of caterpillar carnivores, such as birds, spiders, ants, and scorpions, as well as parasitoids, such as ichneumonid wasps and tachinid flies. These resident enemies of sphingid caterpillars would soon be getting back up to significant numbers after a hiatus during the dry season. In essence, having taken advantage of the burst of dry-forest leaf production with the onset of the rainy season, the moths cut and run. In the eastern lowland rain forest, carnivore pressure, though significant, was probably not as much a threat as that looming in the mid-rainy-season dry forest.

This cross-country seasonal migration has since been found in other species of moths, and in butterflies, flies, and peccaries. Assuming the relevant habitat was intact, the phenomenon was believed to occur not just in northern Costa Rica, but in most of Central America, where the volcanic backbone creates a similar border between a wet eastern side and seasonally dry western side in its rain shadow.

This migration story raised several questions. Why don't the moths just stay in the rain forest? How do they know when to head for the dry forest, and how do they navigate? How did this behavior evolve? The answers to these and other puzzles are now the stuff of debates among scientists, but the important point for conservation was this: The dry forest and rain forest ecosystems are not separate unto themselves. They are inextricably linked, depending on each other for survival. To save these species, both kinds of forests must be saved, and in some cases restored.

So, too, must the flyway corridor that linked the two kinds of forests. If a piece of corridor was missing, migrants that needed to cross in order to survive might not be willing or able to make the crossing. The organisms in each ecosystem that depended on the migrants would die as well. For instance, if key pieces of the corridor disappeared and the migrating sphingids slipped into living dead status, or went extinct, the populations of carnivores and parasitic organisms that depended on them would suffer. So, too, would particular species of epiphytes, shrubs, trees, and vines that depended on the moths to pollinate them.

"Obliteration of either wet or dry forest will obliterate these animals," Janzen wrote in the Guanacaste project proposal. "One cannot view Costa Rica's national park system as a series of islands but rather must view it as a network partly connected by migrants." By 1986, the destruction in Guanacaste Province had raised serious questions about how migrating species were responding as the dry forest patches shrunk and the flyway corridor dwindled. How many of the migrants were losing their way and dying?

The Guanacaste project plan addressed these concerns by proposing to buy, maintain, and—where necessary—restore the natural corridors.

Gauld and I were up at four, stepping out of the tent and into a cloud in the pre-dawn darkness. It had rained on and off during the night, and now gray curtains of fog whipped across the Cacao-Rincón gap at ground level. The sheets were covered with insects: black wasps of all sizes, a rainbow of beetles and bugs, bright-green katydids, bees, flies, and leafhoppers. A large wolf spider walked across the top of one sheet with a katydid in its clutches, and in a lower corner crouched a water scorpion, a big, black predator with raptor-like front legs.

Gauld walked carefully toward the black light, looking at the ground for specimens, then moved in quickly to the sheet, snatching insects and placing them in the killing jars. A few minutes later, I moved them to the morgues. "You can collect for days and days and days and not get anything of interest, and then one night you'll strike gold," he said. Nothing struck him as scientific gold right then, but who knew what might be revealed by others back at Santa Rosa.

As I stood in front of the sheet nearest the tent, an insect rain spattered against my poncho. A few beetles and moths found their way beneath the poncho, buzzing and fluttering up my sleeves and around my ears inside the hood. Gauld told how a few weeks earlier, shopping at a supermarket in Liberia, he was handing money to the woman at the checkout counter when a moth crawled out from under his sleeve and flew away. She jumped back but said nothing. No doubt she was thinking, "Los gringos locos."

Just after five, daylight arrived quickly, as if someone had flipped a switch. Gauld unhooked the batteries, shook the remaining animals off the sheets, and packed the equipment into the Land Cruiser. By half past five we were bouncing along the road down the slope, through cow herds, over rickety bridges, past silvery milk canisters left for pickup in front of farm houses.

And past the living dead.

I recalled something Gauld had said the night before when I asked his opinion on the Guanacaste project. He had thought for a moment, then responded, "So many stories of conservation are tales of woe, of fighting a rearguard action. Now Janzen comes along with this incredible contribution: 'Let's grow it. Let's go on the offensive.' It's important not just from the aesthetic point of view. It's important to show that it's possible."

# · 7 ·

# The Battle Plan

A war-like mentality is required for the biodiversity crisis.

—Daniel Janzen

Shortly after Janzen and Hallwachs finished the first draft of the Guanacaste project plan, another "peculiar accident" occurred. In December 1985, Janzen attended a conference called Tropical Forest Conservation, at the National Zoological Park, Smithsonian Institution, in Washington, D.C., where dozens of scientists and conservationists gave talks on their efforts in various parts of the world. These perspectives led to the book *Saving the Tropical Forests* by Judith Gradwohl and Russell Greenberg. During the meeting, Janzen approached the organizer, Michael Robinson, director of the zoo, and asked whether he could have five minutes before it ended to present the idea of regrowing a tropical forest. "He said, 'Sure,'" Janzen recalled. "So I went down and gave this five-minute talk, literally just sliced in there, just ten slides, real quick, and it caused this incredible explosion."

The explosion, which unfolded over the next weeks, months, and years, came from conservationists at the Nature Conservancy, World Wildlife Fund, and other groups who were upset that Janzen seemed to be undercutting their prime fund-raising message. The groups had trumpeted a single theme to potential donors: tropical forests must be bought and saved before they are cut down, because once they're cut, they're gone forever. Along came Janzen spouting what seemed like a direct contradiction: Hey, we can grow it back. "I had no idea what I stuck my foot into," he told me years later.

The theme of the conservation groups was based partly on research in the Amazon, where Robert Goodland and Howard Irwin called the forest "ecologically a desert covered by trees" because the soil that supported it was frequently bankrupt of nutrients. These forests had evolved intricate strategies for capturing nutrients directly from rain and keeping them in the living forest, but scientists found that when cleared and burned, the forests often could not regenerate in the poor soil.

As Janzen ticked off the steps in his plan to restore a forest, he noted that the conventional wisdom about the finality of deforestation had never been tested. The conservation groups had argued that donors could handle only one message, that they would be confused by the different approach and bored by the thought that regrowing a tropical forest might take a thousand years. This would destroy the groups' focused, emotional appeal to get money. Besides, they said, rain forest is different from dry forest.

Conservation groups already were a competitive bunch. Part of each group's role, like any organization, was to perpetuate itself, so it could continue to employ people and produce its "products": conservation and the publicity and fund raising that made it possible. With a finite amount of money and publicity available for conservation, organizations that got the biggest pieces would be the best at doing conservation, or so the argument went. It was no surprise, then, that when a newcomer stepped in with what appeared to be a radical new message and went after some of the publicity and money, there was, as one conservationist put it, a "certain edginess." As provocative as the newcomer might sound, as much excitement as his idea might produce for the conservation world, the idea detracted from the efforts of others. There was a slight loss of publicity for everyone else, a slight loss of money, a slight shift in the balance of power. Not only did conservation groups stiffen at the prospect of a challenge to their dogma from the Penn scientist, some of them began to believe, as Janzen once told *Popular Science*, that he had "completely flipped."

None of this, of course, caused Janzen pause for a single moment. His conversion from strict research scientist into scientist-activist, or, as Pedro León had put it, a "born-again conservationist," produced a man on a Messianic mission. If Janzen was one of those people who never rested, he became more so after the Corcovado episode. U.S. environment writer and conservationist Chris Wille described him as reacting with "the zeal of a scientist who has seen his laboratory in flames." In early 1986, Janzen did what any good proselytizer does: took his plan on the road.

In addition to sensing the first hints of that "certain edginess," Janzen remembered the day at the National Zoo for two other events. One occurred when Robinson asked whether he would like to return to Washington to give a longer talk. That invitation led to the project's first public fund-raising lecture, on February 25 at the zoo, "How to Grow a Tropical National Park: A Case Study in Restoration Ecology."

The other event occurred when Liliana Madrigal, a Costa Rican then working for the Nature Conservancy, approached him. Their conversation, Janzen recalled, went something like this:

"Which conservation organization is going to be your formal sponsor?" Madrigal asked.

"Huh?" said Janzen, who didn't know what she was talking about.

"Well, why don't you let it be the Nature Conservancy?"

"Yeah, well sure. Whatever."

After discussions with Madrigal and others, the Conservancy agreed to become the conduit for donations to the Guanacaste project, despite some initial reservations by some of the group's leaders about the restoration message. It was, for example, the perfect place to handle the $24,000 that had just been sent to Janzen by the Swedish journalist who had confronted him at the Crafoord Prize ceremony in 1984. Her articles had by then raised the money, and suddenly the biologist had become someone who could make use of that money. The Conservancy handled that and many other sums of money for the Guanacaste project, and it helped, along with the Park Service and other organizations, to publish *Guanacaste National Park: Tropical Ecological and Cultural Restoration*, the thin, salmon-colored book that served as the restoration plan.

Within the 103 pages of the book were maps, more than four dozen black-and-white photos, a description of the region, and the proposed park's restoration strategy, organizational infrastructure, use programs, land-acquisition outlook, and budget. The book described the Guanacaste project's three main goals: first, to use existing dry forest fragments as the seed source to restore about 270 square miles of "topographically diverse land" to a "dry forest that is sufficiently large and diverse to maintain into perpetuity all animal and plant species, and their habitats, known to originally occupy the site"; second, to maintain this wildland so that it offered "a menu of material goods," ranging from genetic information to watershed protection to tourism profits; third, to use the wildland to stimulate "a reawakening to the intellectual and cultural offerings of the natural world."

The restoration strategy had evolved from the observations of Janzen and the others before him about how fire, hunting, logging, ranching, and farming had harmed the forest and its various systems. Janzen had looked for what would counter those particular challenges. The main restoration tool was nature itself. The plants, birds, mammals, insects, and other life forms banked in Santa Rosa's forest fragments constituted an aggressive force that would reinvade pastures, degraded forest, and other disrupted habitats. The biology of this process, it was now known, would restore not just animals and plants, but the *interactions among them*. This was reestablishing nature's individual details and the ties that bound them together.

To enable it, the project leaders would have to stop the fires, stop the hunting, and keep some cattle in jaragua fields during the rainy season to help lower the fire risk. Hunting cessation would ease the threat to seed-dispersing animals such as monkeys, coyotes, deer, curassows, tapirs, and peccaries. Although poaching was only a minor problem at Santa Rosa at the time, elsewhere in the broader area of the proposed project, pleasure hunters from San José, Liberia, and La Cruz and local meat hunters were taking a toll on the animals. The white-lipped peccary was thought to be closest to extinction in the area. Since the primary issue wasn't rural hunters desperate for meat to feed their families, Janzen thought the most important step would be to subject the pleasure hunters to "an intensive and personal education campaign" by Guanacaste National Park (GNP) biologists.

Fire-control efforts, including fire-lane maintenance and better fire-fighting plans, would help protect existing forests and developing trees. Mainly, though, most of the fire-fighting problem could be solved with a few radios, some trucks with water tanks and high-pressure pumps, and a fire program director and crew whose sole responsibility was to protect the park. "The past five years of fire control (and lack of it for the past 14 years) at Santa Rosa make it clear that the technology of fire elimination is feasible and straightforward," Janzen wrote. The challenge was "the social problem of insuring that the technology is applied year after year without fail."

Park personnel and volunteers would accelerate natural reforestation by planting trees in selected spots by hand. Gradually, the expanding islands of trees would connect. Within twenty years, the smaller pastures would become mostly trees, while larger areas, like Santa Elena, would require fifty to 200 years to fill with young forest. A hundred to a thousand years would be required for the whole region "to begin to approximate the full structure of pristine dry forest."

As for what Janzen termed the GNP organizational infrastructure, a carefully designed constitution, one specific to the needs of the region, was key. It would be crafted during workshops attended by representatives of more than a dozen Costa Rican institutions. Administrators of the new park would live in or near the park and not be transferred to other parks in the system unless they wanted to be moved. Although the park would be the property of the

Park Service and would obtain some of its funds from that agency, it would rely mostly on investment revenue produced by the Guanacaste National Park Endowment Fund.

Janzen outlined several "use programs," which were aimed at making the park relevant to people who lived outside it—nationally and internationally, to be sure, but especially locally. Later, this would be called "biocultural restoration," and the project's emphasis on it, in addition to ecological restoration, became one of its most noteworthy characteristics. Rather than planting trees, the thrust of biocultural restoration was to plant biological understanding in the local culture by encouraging interaction between the park and its nearly 40,000 neighbors.

The message of taxonomy came through in the proposal. Arguing that "we are woefully ignorant" of what plants and animals inhabit the land that would be GNP, Janzen proposed an inventory of the park's organisms, which would be conducted by taxonomists from around the world as a contribution to the restoration effort. "As researchers, we cannot come to understand what holds GNP together without names for the units in the matrix." Nor could the "biological stories" of Guanacaste be brought to the public, nor the organisms and their stories be connected with scientific information elsewhere in the world.

Janzen estimated that "a minimum start-up endowment" of $3 million would be needed to provide the "minimum" $300,000 annual management budget for the park. In addition, it would take $8.8 million to acquire the 181 square miles of land envisioned, an area that would also include some donations. He noted the uncertain status of much of that land. "What crises are there before the end of 1986? Today, as you read this, the tiny patches of pristine semi-evergreen forest on the sides of Cerro El Hacha are being cut by members of the Colonia who are preparing new fields for corn, rice, and beans. The owner of Finca Jenny could decide at any moment to convert her forest to sawlogs or cashew plantations. The renters of Rosa María's croplands may not wish to abide by the owner's restrictions on pesticide use. A forest guard must be hired and provisioned to patrol the donated portions of Hacienda Orosí. We have no promises or understanding from the owners of Santa Elena."

I had been at Santa Rosa for a little over a week, that June of 1986, and I'd spent time in the field with George Godfrey hunting caterpillars, Frank Joyce watching birds and wasps, Ian Gauld collecting night-flying insects on Cacao, and Colin and Lauren Chapman, a wonderful husband-and-wife team of Canadian biologists researching monkeys and tropical fish. Now, I had my first chance to interview Janzen, to find out more about the man and his mission.

I walked through the open door of his house, past a sign that read, in Spanish, "Dangerous—in a deteriorated condition" and a window with a bullet hole, the record of an errant slug from the pistol of a cowboy killing a horse. I had to duck to avoid plastic bags hanging on lines strung from roof beams. Most of the bags contained caterpillars, pupae, and leaves, part of the Moths of Costa Rica project. Some of them served as tropical storage cabinets, holding food, boxes of color slides, and other items.

Wasps lazily circled amid the cobweb-laden beams. The house was home to bats, a boa constrictor, and the small paca of mate-marking fame, which scurried around the crumbly concrete floor amid the clutter of killing jars, buckets of seed pods, and car batteries used for black-lighting. Tools hung on the wall next to Janzen's desk, which was covered with notebooks, jars of insects, and an assortment of other odds and ends. A nearby bookcase held titles as varied as *Gray's Anatomy, The Population Biology of Plants, Gorky Park,* and *Short Stories of Katherine Mansfield.* Piles of newspapers and cardboard boxes filled with files lay everywhere. In a bucket on a chair behind me, sea crabs danced a raspy jig.

Janzen sat with his back to the front door, which was propped open. He was bare-chested and pale, his ample beard and wispy, gray hair swirling off his head like smoke from a campfire. He hunched slightly over a spreading board, peering down his nose through reading glasses, meticulously opening the wings of newly collected moths, and pinning them to the board.

Hallwachs, a thin, thoughtful woman with long brown hair, was writing at a nearby desk. From time to time during my interview with Janzen, she would get up to look for something in another part of the house, work in the kitchen, or step outside to speak with someone who had walked up. I declined her offer of a cool drink, but there would be many times over the next few years when I would gratefully accept a glass of lemonade, a brownie, or an explanation of some element of the project or the forest in Janzen's absence.

I sat across the desk from Janzen, and for the next hour or so he answered my questions, looking up every so often from pinning moths to emphasize a point. He spoke in tones that were sometimes friendly and matter-of-fact and sometimes laced with annoyance, not, I gradually realized, necessarily directed at me or a silly question, but at his frustration with the threats faced by tropical forests. He withdrew from the interview twice: once to give orders to a student and once to answer a phone call. Each time, he returned to our interview in stride, undistracted by the interruptions.

He covered Guanacaste project details, the destruction of the tropics, and the responsibility of scientists to do something about it. Janzen had long since found out who the local landowners and other major local figures were and made contact with them. He and the Costa Rican conservationists were awaiting action on applications they had submitted to three major foundations for enough money to make the first land purchases and manage their development. First on the list was Cerro El Hacha (cerro is Spanish for "hill"), some fifteen miles to the north of Santa Rosa. The patch consisted of about nineteen square miles, part of it belonging to Luis Roberto Gallegos, considered a troublemaker by the conservationists, and part belonging to more than a dozen small farmers.

Some of the forests on Cerro El Hacha had been cut and burned several hundred years earlier, but now the farmers were clearing and "improving" forested land on their property with crops, and Janzen feared that much of the rest of the forest would be destroyed in 1987 and 1988. The upper slopes were almost completely deforested, and the new pasture burned each year. Old-growth forest still grew in ravines on the lower slopes, but a checkerboard pattern of corn and bean fields had invaded, and broad, barren valleys were eroding out of deforested spots on the hill. Janzen believed that one of the blocks of forest, about 500 acres, was the largest tall old-growth dry forest remaining in all of Mesoamerica. Also, the forest in the ravines was so moist that it fed ever-flowing creeks through the six-month dry season. However, streams in the cleared areas had stopped flowing. Janzen told me that he wanted Cerro El Hacha to be the project's "first actual purchase," but that would depend on the grants.

Meanwhile, he was trying to engineer the buying of Finca Jenny, which did become the first purchase. Two centuries earlier, this small piece of land—a four-square-kilometer farm on the southeastern edge of Santa Rosa—had been part of Hacienda Santa Rosa. It was deemed critical to the project because it supported a forest that joined with Santa Rosa National Park's largest and deepest evergreen canyon forest, in the Quebrada Puercos area. The owners had just indicated that they were willing to sell Finca Jenny for the government-assessed price. Janzen already had in hand the $24,000 sent by the Swedish journalist, "but it's going to cost $113,000. We're scrounging for money."

Janzen would begin a major U.S. fund-raising campaign when he returned to Philadelphia in the fall. It would include public presentations, private meetings, and applications to foundations. He also was laying a political and fund-raising groundwork in Costa Rica. Oscar Arias's new Minister of the Environment, Alvaro Umaña, would arrive July 4 for a tour of the area and a briefing on the project, which Janzen had described to Umaña the day after he took office. "Right now," Janzen said, "I'm spending a lot of time worrying about political connections, getting the right attitudes in the right ministries and right government officials so that when the right things come along, they have their minds set in the right direction." He was making contacts in agencies that deal with maps, land purchases, and legal issues. An article on GNP would appear in Sunday's edition of *La Nación*, Costa Rica's leading newspaper. Although "more of a symbolic act, it will be the first time that a conservation drive of this kind has ever been initiated in any Latin American country."

The shift into political and fund-raising activity had forced him and Hallwachs to set aside some aspects of their ambitious scientific research projects.

"It adds a whole new complication to life," he said. "It most assuredly does. But we have no choice. It's not a matter of whether it's easy or not. That really isn't relevant. When your car breaks down, you don't worry about whether it's easy to fix or hard to fix. You simply fix it."

He began to get a bit angry, or at least frustrated. The pace of his words picked up, and his voice seemed to attack.

"The problem is, I could sit here and stack up an enormous amount of very pretty research over a period of the next thirty years or so that I have left in me, and that's fine. And then the next generation of people will turn around and look at this forest and it will all be cow pasture. So what's better? Spending my time stacking up thirty years of research, or only doing half as much research and ending up saving the place, and leaving it for other people to do vastly more research? Because what I'll get done in my lifetime is still tiny compared to what the potential of this place is."

Janzen found it difficult to adjust to his new task as politico, but mainly just resented the time it took. The mechanics came relatively easily. In a way, it was like learning mathematics, physics, or some other field he'd had only modest exposure to. On the other hand, many of the skills he had developed to become a successful research ecologist were useful now as well. If a political system was just another kind of ecosystem, ecologists had an advantage in politics because of their training in understanding interactions and energy flow in a system.

The forces threatening tropical forests varied from country to country, and generalizations were difficult. In Costa Rica, he said, it boiled down to the simple fact that people who owned the land wanted to make money with it. He compared it to a landlord who wanted to rent out his building. In Guanacaste, the way to make money with land was to convert it into a biological machine that made a product. "If it's cows today, fine," he said. "But if you take cows off the market, it will be rice. You take rice off the market, it will be cotton. You take cotton off the market, it will be mahogany trees. It doesn't matter. It's just simply that the dirt is a producing object. And when it's just sitting there with forest on it, for most owners, it's not producing anything the owner wants. So he cuts it down. There's nothing more than that."

Deforestation in Costa Rica was becoming more conspicuous for two reasons. One, a growing population meant more and more people actually wanted each piece of land, and if the current owner didn't want to cut down the forest on it, someone else would gladly buy it and cut down the trees. Second, a growing affluence among individual landowners meant a growing desire to make money and a growing urge to convert the forests into a sellable product.

In countries like Mexico and El Salvador, where the land was optimum for farming, this process had occurred hundreds of years ago. Countries with a vast, rugged frontier and low population density, like the Amazonian interior of Brazil, had lasted longer. "The reason why the Amazon is there is because up until very recently there hasn't been an obvious thing you could do with it," Janzen explained. "All this hullabaloo about blaming the cow or blaming the 'hamburger connection,' that's looking at the blood pouring out of the wound and saying, 'Oh horrible, horrible.' I'm much more concerned about where the bullet came from."

In Nicaragua, which was in the midst of the contra war in the summer of 1986, large blocks of existing forest, if all put in a park, would make up more than 17 percent of the country. "The reason it's not cut is not because it's a park, but because it's a war zone," Janzen explained. "The timber companies were in there starting when the war broke out. The tractor drivers just won't do it. They're terrified."

Janzen once told a writer for *Smithsonian* magazine that if humans destroy the tropical forests, "we will have committed the greatest criminal act that life on Earth ever could or will sustain." I asked him why it was important to save the forests, and what it would mean if we lost them. Today, this question is almost a cliché, but it still bears a hearing, and Janzen's take on it at the time, and the way he phrased it, is particularly revealing. He began by noting that there are several answers.

"One is by example. By some historical accident, if you still had a county in Illinois or Missouri or Michigan or Wisconsin with the original pristine forest standing, would the United States population cut it down? All the reasons why they would not apply here just as much as they do in Illinois. I should point out, of course, that the reasons are mostly *not* economic ones. Nor would people say, 'Well, I don't want to clear that one county in Illinois because it will affect the drainage basin of the Mississippi River or it will affect the oxygen content of the world.' They would walk out there and they would look at that black oak tree which is this diameter (he made a big circle with his arms) and say, 'My God, that's the last one in the United States. I don't want to cut one of those down.' They would say that for the same reason you don't take a Rembrandt and make newspaper out of it when you have a paper shortage."

The common phrase used to describe that idea is "aesthetic reasons." Janzen objected to those words because for many people it translates as "pretty sunsets." To him, the idea is far more complex. That complexity gets to his second reason for saving the forest, which is actually implicit in his first reason. He started explaining it, as he often explained things, by using an analogy.

"If the only music you had ever heard in your life was advertising ditties, and I said to you, 'There's certainly a lot more to music than advertising ditties, little thirty-second spots on a TV set.' And you said to me, 'Well, what do you mean? Tell me. Describe what else there is.' It seems to me I've got two options. I can play for you a piece of classical music or a piece of complicated jazz or something. Or I can sit down and try to describe it in words. And if I try to describe it in words I'll by and large fail. But if I can play it for you, I have a chance to open your eyes to some really new thing, and really what I've done is said, 'Your brain is capable in the music area of handling and dealing with much more complicated things than advertising ditties.' But if you've never been exposed to it, you don't even know it exists. There's a whole chunk of your brain that's just never turning over. You're not doing anything with it.

"Well, I would argue that the human brain is made up of a series of things like this, that is, it's got a whole lot of ways that it processes complexity, that it uses complexity. That complexity in present-day society is stimulated by music, mathematics, literature, art, religion, parental training, arguing, debating, politics. There's a whole lot of these complex bodies of thought and development of the mind. What I see very clearly is that the relationship between the mind and biology is missing."

Biology has been presented to the public mainly as a way of producing an agricultural or medical product, and the notion of its relationship to the mind has been lost, he explained. The complexity that biology offers—say, if a person is exposed to the complex ecological interactions in a tropical dry forest—has the same potential for stimulating the human mind as does music, literature, and other forms of art. Approached from that perspective, forests have the same value as art museums, libraries, and symphony halls.

"People are stimulated by the sights and sounds, and by thinking about them," Janzen said. "That stimulation in itself, I would argue, makes a person a better understander of other areas of life. A person who sat and watched and understood mutualism and competition in animals and plants, if he goes into the business world, is going to understand competitors and mutualists a lot better than if he'd studied abstract subjects or something which he's only been exposed to in the human sense." When he talked of the forest's capacity for stimulating the human mind, he was talking not just for residents of the developed world, but also for people who live in the tropics. They might one day achieve the material basics of a good life—education, medical care, roads, electricity, and the like. However, if they simultaneously created "habitat about as interesting as central Illinois," they would lose the intellectual richness available through biological complexity, and, since they wouldn't have access to the rich, non-biological cultural offerings of a large metropolitan city, they would miss their main chance to be exposed to complexity.

This idea resonated with a statement in the Corcovado report: "We were struck by the boring and intellectually undeveloped lives led by the miners." Boring and intellectually undeveloped. The observation seemed a bit elitist, with perhaps a certain element of White Man's Burden in the background. Yet this theme did make sense, and the wonder and stimulation provided by studying animals and plants had been part of the naturalist's aesthetic at least since eighteenth-century England. Further, it was William Beebe who, as curator of birds at the New York Zoological Society in 1904, had compared species extinction with the loss of a musical score or sculpture. Years later, a biologist who knew Janzen well told me: "Others are caught up in human misery now, but Janzen sees people a thousand years from now without nature, and they're miserable and don't even know why."

As far as listing reasons for saving tropical nature, Janzen would stop with aesthetics and complexity. However, a "cheap three," as he put it, was the list of products we take for granted in the developed world, products that originated in the tropics. At the time, it seemed to me a surprisingly long list. Starting just with breakfast, it included coffee, tea, citrus fruits, eggs, sugar, cow's milk, bacon, beans, rice, cinnamon, and pepper. The list continued to expand for lunch, dinner, and various other aspects of daily life, not the least of which was medicines. "The list is enormously long, and of course what everyone leaves off the list at the end is humans," Janzen said. "They originated in the tropics. The only reason they persist in northern regions is that we take away the winter." The point was that to eliminate the forest now would be tantamount to assuming that early agricultural societies had already picked out all the useful tropical plants and animals and put them into some kind of domestication, an absurd assumption. Absurd especially in an age when important uses of forest organisms extend not just to the whole organism, but to genes responsible for important traits, say, drought resistance or increased productivity in a particular plant.

Janzen called reason number 3 a "cheap three" because, although it might be the easiest reason to understand, it ultimately was less sound. It was part of an economic rationale, and "you can never fight economics," he said. "Somebody will always come up with a crop that you can grow anywhere that will be more valuable in straight dollar terms than a natural site. Maybe today tourists are important. Maybe tomorrow gene banks are important. But somebody's always going to come up with a crop that will economically outcompete whatever it is that you want to do. The only lasting value in this system is something that involves a class of value assessment that doesn't depend on technology."

He used the analogy of a public library. A library, he said, has value for its information, not for the sum total of what each book costs. "You just recognize that there are many, many, many uses for this stuff in the library, and that's it. You don't try to figure out the actual dollar value, because as soon as you do, somebody will come along and figure out that it's better economics to turn the library into pulp and use it for something else." The analogy applied well to the Guanacaste terrain. If someone had come up with a crop that would easily make a lot of money there, the forests would have been completely removed long ago. That could still happen.

When I asked him what could be done, he shot back: "Contributions." The Guanacaste project needed small amounts of money from many people or large amounts of money from a few people. That would enable purchase of big chunks of land to get them out of the hands of people who might develop them for crops. "Then you simply have to endow that land with enough funds to put into investments which continually generate the annual operating budgets for maintenance," he said. The roughly $11.8 million needed for the land and endowment was a small amount of money, he said, amounting to $4.72 per Costa Rican to set up all of Guanacaste National Park. "Five dollars—that's the price of a movie." Spurred on by the movie idea, he did a back-of-the-envelope calculation of income if every U.S. resident skipped one movie and gave the money to conservation efforts: roughly 250 million people multiplied by $5 is $1.25 billion. "With that, you could solve all conservation problems in the New World forever. And furthermore the machinery and technical expertise is there. We're able to do it." Janzen estimated that a $3 million endowment for an operating budget would cover Guanacaste National Park "as an adequate base—$5 million would be superb."

The fund-raising campaign was in its infancy, but Janzen already had been accused of being crass in asking for money. In response, he pointed out that conservation donations were like tithings for the local church: a certain percentage of a person's income goes to keep the church running. "The exact same process is going to have to start happening here," he said. Residents of the United States had a special responsibility in this regard, since they were well-endowed financially. He explained it this way: "When we ask Costa Ricans to preserve these things, we're asking them to preserve them for the world, not just Costa Rica—so you, and I, and everybody else can come and use them. Well, you're asking a country the size of West Virginia with a national budget the size of two large Texas counties to maintain a world-level resource, and to use ten percent or five percent of their resources to maintain it. I mean, that's just straight economics. If you go to somebody's house for dinner, you take a bottle of wine."

Think of it not as a penalty, but as a maintenance cost, he said. When you go to a university, you pay tuition. In Costa Rica, Brazil, and other tropical nations where wildlands represent a "world-level resource," the world ought to be stepping in and picking up part of the bill. If a neighbor's child sets up a lemonade stand and needs a dollar to buy lemons, and you make $50,000 a year, pulling a dollar out of your pocket to help isn't difficult. How ridiculous to quibble about small sums for conservation when developed nations spent billions of dollars on war. "As someone said, the money that would go to solve the conservation problem wouldn't buy underwear for the contras," Janzen said. "It irritates me."

The interview wandered back into a discussion of the dimensions of deforestation. Although Costa Rica had nearly one-fifth of its land in national parks and reserves, it also had one of the highest rates of deforestation in the Western

Hemisphere. "The first general rule is that the game is over," Janzen said. "To be real generous, let's say less than five percent of the country is there to argue about. All the rest is already either agricultural land or preserves. Basically Costa Rica is at the point where in five more years there isn't going to be any argument at all. You'll only take the areas that are already preserved and you'll try to make them better, and the rest will just be agricultural land."

Forest land with "reserve" status, as distinct from a national park, had an uncertain future. To some people, for instance, a forest reserve simply meant that the government either owned or controlled land with trees to be cut in the future. Already, the government was unable to curb illegal logging in some reserves. A fair amount of political effort would be needed to get those into the park system and thus under more protection. Much of the vast forest on the slopes of Cacao and Orosí, which Janzen hoped to incorporate into the new park, was under reserve status, for example. Much could be lost, however, if some government official issued a permit to cut on the volcanoes. If that happened, Janzen said, "I'd go lie in front of the bulldozer, literally. I'd get violent before I'd let somebody work on that forest. There are some pieces like that which are that important."

He described how the conversion of natural habitat outside parks by farming and logging made parks themselves into biological islands. It was clear that he was speaking not just about a general case, but about Santa Rosa. "So all of a sudden you're realizing that what you've got is this goddamned gorgeous piece of forest sitting clunk like a little island out in the middle of the Pacific somewhere, and you see this coming and you realize, Jesus, if that's the case, all the traditional island effects are going to go to work on this little piece of dirt. Yeah, we've got jaguars in there, and yeah, we've got tapirs in there, but because they need lots of habitat, they were all being maintained at least in part on the bigger expanse of rather poorly used farmland around the park. Now, when you go and make Holland out of that, all of a sudden you've really got an island, and you realize that island isn't big enough to sustain what you wanted it to sustain. And all of a sudden you're going to have to go into Phase Two, or these beautiful little parks and preserves are just going to go ssshh-hhkkkkkk [he made a chopping motion with his right hand], and get carried down the drain."

By Phase Two, Janzen actually meant two things that happened after the initial phase of forming the park. One was a biological Phase Two in which proponents of the island, the park, sought to extend its boundaries and restore the natural habitat. The other was an educational Phase Two in which the park proponents launched a public education campaign that explained the ecological value of the park and its natural resources. "This has to be going really strong or somebody's going to want that piece of dirt for something else," he said. "Let's just say there's a river flowing through it. They're going to want to divert that river to fill irrigation canals, and you're going to have one hell of a fight on your hands if you haven't already educated people so that the idea never looks acceptable in the first place."

This was the biocultural part of the Guanacaste restoration plan.

Janzen had gotten to know the people of Guanacaste during his two decades there, and he had observed the development of the region from a "frontier society to a thoroughly agriculturalized one," as he described it. He had hired locals as field assistants, given lectures to civic groups, and engaged in day-to-day chatting with Costa Ricans from all walks of life. Although upward mobility in Guanacaste was limited and formal schooling was minimal, the population was literate and displayed great curiosity when confronted with learning experiences.

The essence of biocultural restoration was to address this, to make the park into a "living classroom" through programs for grade school, high school, and university students; civic groups; and tourists from Costa Rica and elsewhere. The goal of biocultural restoration was to bring back the deep understanding of natural history held by past generations of Guanacastecans. "The difference is that their great-grandparents spent their time basically pushing nature back, and by now they've pretty much pushed it back," Janzen said. "But in the process they lost it. These people are now just as culturally deprived as if they could no longer read, hear music, or see color." Thus, the park must be developed into what Janzen called a "user-friendly" social institution that contributes on a daily basis, just like a library, hospital, or school.

In addition to the intellectual dimension of biocultural restoration, of course, the park would offer other, more practical assets, and local residents needed to be taught that, too. The park encompassed a watershed that provided drinking and irrigation water and contained seed and genetic stocks valuable to Costa Rica's agroforestry future. The Guanacaste area would benefit from money acquired as a result of GNP's education, research, and tourism programs, and the locals needed to understand that, too. For instance, the region then absorbed some $200,000 a year thanks to foreign researchers, who required everything from supplies to field assistants when they came to Santa Rosa and environs to conduct their studies. Janzen predicted the figure would top $1 million as GNP evolved and attracted more scientists.

To survive, GNP would have to be viewed with favor by a diverse array of political, economic, and cultural factions. "The most practical outcome is that this program will begin to generate an ongoing populace that understands biology," Janzen said. "In twenty to forty years, these children will be running the park, the neighboring towns, the political systems, the irrigation systems. When someone comes along with a decision to be made about conservation, resource management, or anything else, you want that person to understand the biological processes that are behind that decision because he or she knew about them from the time they were in grade school."

Expanding the park's boundaries was important for keeping healthy numbers of wide-ranging species like jaguars, but it also was needed to make the park sufficiently large and organized that scientists, nearby residents, tourists, and other visitors could come there, study it, and experience it—all without bumping into each other. He cited the example of tourists who might want to travel to Playa Nancite to watch the arrival of sea turtles, but their visit would conflict with the activities of scientists already at the beach to study the event. In the scientists' view, hundreds of tourists running around the beach with flashlights would disturb the turtles; thus, tourists must be barred. "So all of a sudden you've got a gorgeous turtle beach and you don't let anybody see it," Janzen said. "It's like having a famous painting in the Rijksmuseum in Holland, and everybody wants to go see it, and they get there and are told, 'Sorry, looking at it destroys it. Therefore, it's closed.' You'll see how many contributions you get for that museum." With a sufficiently large area, the park could be zoned: pristine areas to be left untouched, areas for scientific research, and areas for tourists and busloads of local schoolchildren.

Janzen, like many leading scientists and conservationists I spoke with over the next decade, knew of no tropical forest restoration project comparable to Guanacaste in scale and scope. The big conservation projects of that time were built around saving pieces of forest that already existed, rather than trying to rehabilitate a forest. Janzen expressed his belief that although Guanacaste was unique, in five years it would not be alone. "People are going to be forced to do it," he said. He saw cause for hope in combatting deforestation. "Here we have a real chance. If we were working in Guatemala or someplace where we had to educate an entire population, it would be another matter. But here, they're already educated. There's an audience out there for this stuff, just like there's an audience out there already for music and literature. I think the hope is different in different places. Here, it's 100 percent. It's just a matter of how long it's going to take us to find the money."

To close the interview, he talked about the North American audience, which he said was ready for a more sophisticated picture of tropical nature than the media and conservation groups had been painting. That simplistic view had two parts. One was the image of the struggle to "save the rain forest." This was jingoistic, and inaccurate. It focused on one ecosystem to the exclusion of a wonderful, diverse mosaic of tropical ecosystems, including dry forest. The second part of the simplistic view was that tropical nature was just "pretty green trees with parrots flying through them," as Janzen put it. But of course there was a lot more, and the general public hungered to graduate to a deeper and more complicated view.

Aldo Leopold had sounded this theme in the 1940s, in *A Sand County Almanac*, when he wrote, "In our attempt to make conservation easy, we have made it trivial." When Janzen gave public talks, he adhered to this theme, presenting intricate stories about tropical organisms and the vast array of ecological connections that tied them together. "Audiences just soak it up," he said. "I was told that it's too complicated and a lot of people won't understand it. Bullshit. They can understand it perfectly well."

# PART II

# Advancing through the World of Wounds

# ▪ *8* ▪

# A Tropical
# Christmas Catalogue

A single 8 1/2-by-11-inch piece of paper covered each of the thousand or so chairs set up in row after row of the ballroom of Chicago's Hyatt Regency Hotel. The top of the flyer showed the photocopied image of a jaguar prancing through a tropical forest. Angling across the upper left corner were the words, "Send your love to life on earth," and below it was a description of the Guanacaste project and how to contribute.

As people filed into the ballroom and selected seats, they picked up the flyers. They had come, on this frigid Saturday night, February 14, 1987, to hear the keynote address at the annual meeting of the American Association for the Advancement of Science, the world's largest general scientific organization. As they waited for the speech to begin, they scanned the flyers:

"Western Mesoamerica had a France-worth of tropical dry forest, but today less than 2% remains. Only 0.09% has conservation status. The dry tropical forest in Guanacaste National Park (GNP) in northwestern Costa Rica wants to buy itself, replant itself, and regrow itself."

The flyer went on to describe the "living classroom" that would be restored and said the project needed $5 million by May to purchase key land. The "spring 1987 shopping menu" solicited $300 contributions—the price of one hectare, or 2.47 acres, of GNP land. The money would buy the donor, "forever," all of the items in a typical GNP hectare: 0.001 jaguar, 0.1 adult guanacaste tree, 0.01 muscovy duck, 400 dung beetles, 200 orchids, 100 scorpions, 20 toads, one million ants, five yards of riverbank, a rattlesnake, 0.000029 of a Costa Rican volcano, and twenty-four other items.

More information was available from Janzen at Penn, the flyer said, and it listed the Nature Conservancy/Guanacaste Fund, in Washington, D.C., as the place to send contributions, whether $300 or "any amount."

The flyer, created by Janzen and Hallwachs at Penn, was part of the plan to raise the $11.8 million to buy land around Santa Rosa and establish an endowment for park management. The flyer had first been published as a "Tropical Christmas Catalog" in late 1986 in a Janzen letter to the journals *Biotropica* and *Bulletin of the Ecological Society of America*. Subsequent versions included the "Tropical Easter Catalogue" and "Tropical Fall Clearance Sale." Ironically, the swath of words "Send your love to life on earth" cutting across the corner of the jaguar image blocked out the figure of Thomas Belt aiming his rifle at the big cat. (The illustration had been borrowed from *The Naturalist in Nicaragua*, where it depicted Belt's unexpected encounter around 1870. The idea for using it in the fliers had been suggested by Susan Abrams, an editor at the University of Chicago Press.)

Janzen began the fund-raising campaign in early 1986 by appealing to private donors, conservation groups, and foundations in letters, formal proposals, and a series of public speeches in Canada, Costa Rica, England, and the United States. The flyers went along with him, and he obtained the help of fellow biologists whenever possible. Peter Raven, the keynote speaker that February night, had insisted on having them placed on the chairs, despite resistance from some high-ranking AAAS officials. Raven, director of the Missouri Botanical Garden and one of the leaders among scientists who warned about the perils of tropical deforestation, had become influential in Costa Rican biology and conservation, including helping to create Braulio Carrillo National Park. Now he was helping Janzen with his plan. Raven's speech, "We're Killing Our World," offered the Guanacaste project as a hopeful example during a time of global environmental crisis.

Another classic GNP fund-raising tool hatched by Janzen and Hallwachs was the somewhat whimsical brochure, "How to Grow a National Park." Beneath that headline were the words "Guanacaste National Park," and below that, the small, folded pamphlet sported a glossy overhead photograph of a few bright green guanacaste seedlings growing out of the middle of a dark brown pile of cow dung. "First ten days of a guanacaste tree (*Enterolobium cyclocarpum*)," the caption read. "200 years to go."

The brochure, designed and written in their Philadelphia kitchen and printed by the Nature Conservancy, was eye-catchingly peculiar, perhaps even a bit outlandish. At one level, the cow-dung photo was an inside joke about seed-dispersal mechanisms in the tropical forest and how a forest restored itself. On another level, it showed a fact of biology. At still another level, the brochure, along with the one-page flyers, were early signs of a creative, energetic fund-raiser.

A group of Costa Rican biologists and conservationists had banded together with Janzen to engineer the Guanacaste project. Key among them were Mario Boza, then president of the National Parks Foundation; Rodrigo Gámez, past president of the Neotrópica Foundation and an adviser on natural resources to the new Arias administration; Pedro León, who succeeded Gámez at the foundation; Alvaro Ugalde, past director and now adviser to the Park Service; and Alvaro Umaña, Arias's new Minister of Natural Resources.

By the third quarter of 1986, the group was pushing hard within Guanacaste Province, broader Costa Rica, and internationally to educate the public, win support among key politicians, and raise money. For example, Janzen gave talks about GNP to the Liberia Rotary Club. He, Umaña, and Gámez hosted a visit to Santa Rosa and the project by First Lady Margarita Peñón and her family. Umaña, Gámez, and other politically connected Costa Ricans got "friendly advice" out through various channels to Guanacaste landowners that the project needed land at low prices.

Janzen pushed hardest on raising money. Grant proposals, the Tropical Christmas Catalogue, and other fund-raising efforts had begun to draw large amounts of money. In July of 1986, the John D. and Catherine T. MacArthur Foundation led off with a half-million-dollar grant, followed soon by a $75,000 management grant from the Pew Charitable Trust, $100,000 from the W. Alton Jones Foundation for land purchase, $161,000 for the education program from the Jessie Smith Noyes Foundation, and some $600,000 in private donations from individuals in Australia, Canada, Costa Rica, England, Holland, and Sweden. A few Guanacaste landowners had donated about 15 square miles to GNP, and, combined with purchases, about a third of the land needed to be bought for the project had been secured.

Another fund-raising victory came when the Arias administration decided to match donations one-to-one in non-negotiable colon bonds (the colon is the currency of Costa Rica). "This means," Janzen wrote in a letter to past and potential donors, "that the $8 million needed to complete GNP's land purchase will generate $8 million in endowment colones." In an agreement between Guanacaste project leaders and the administration, three-fourths of the endowment colones would go into a GNP endowment fund, while the remaining one-fourth would go to the Park Service. The interest from the endowment would be directed partly back into the endowment, to continue its growth and to keep pace with inflation, and into park management and development.

"The press has been kind to GNP, as have all of you," Janzen wrote. "The dry forest is beginning to rear its head. Every dollar is photosynthate; there is no fluff and there is no graft." He called attention to the self-imposed deadline of May or June to acquire the "core" of about 62 square miles owned by five businessmen on the east side of the Pan-Am Highway. That piece was under steadily escalating "assault" by fire, ranching, and logging.

Janzen, Gámez, and a few others criss-crossed Guanacaste, keeping in touch with landowners. Some of the owners, especially those on the east side, seemed resistant, even uncooperative. However, as MacArthur and other foundations began to give their support to the project, Janzen's credibility rose with the landowners he had been lobbying. Owners of more than a dozen small farms near Cerro El Hacha agreed to stop clearing forest on their land and hired a lawyer to represent them in the pending land purchase. Boza and others at the National Parks Foundation were working out the legal details of the deal. In a letter to Raven in late September, Janzen described the farmers as "fantastic," adding that one of them with a large piece of land had volunteered the use of his fields to any of the others who were considering cutting forest to plant that year's bean crop. Janzen made it clear in the letter that the idea had come from the farmer, not from him.

Two of the farmers, Edwin Bermudez Rodriguez of Cerro El Hacha and Vilmar Rodriguez Orozco of the nearby Colonia Bolaños, were hired to manage the Cerro El Hacha region. They thus became the first full-time employees of GNP. Their contract called for them to stay in their houses and enlarge them into biological stations, but they could use a small fraction of their fields to supplement their salaries. They mapped the vegetation and boundaries of the region, a step needed before payments could be determined. They were trained by a local Forestry Department warden on how to handle poachers and public relations, and the project purchased a tractor and mower for use in cutting fire-control lanes. The new employees also launched a nursery for guanacaste, mahogany, and other trees.

The Forestry Department had agreed to turn over to GNP the entire Orosí Forest Reserve—when the project was ready to manage it. The reserve was a 50-square-mile area of land on the slopes of Volcán Orosí and Volcán Cacao that included large sections of rain-forest and cloud-forest refugia. Janzen estimated the value of that transfer of state-owned land at about $700,000, adding that another $1.3 million worth of land within the boundaries of the reserve would have to be purchased from private owners.

The department offered to cancel the logging permit it had given earlier to one of the owners within the reserve, but Janzen demurred, fearing it would unnecessarily anger the man. "I would rather purchase the land out from under the owner than create an enemy and then have to buy his land," he wrote to Raven. "In the meantime, by having a forest warden working for GNP, we insure that no cutting will occur. How much of the hunting we can stop depends on our energy levels."

In September, the project purchased what Janzen called "the two most strategic farms" in the Orosí Forest Reserve. These comprised 500 acres amid old-growth forest at 1000 meters elevation just below the top of Cacao. The two owners were hired, made deputy forest wardens, and assigned to enlarge their houses into biological stations. (Since they were squatters, the land was again purchased later from the true owners as part of large tracts.) On one of the homesteads, the "view from the outhouse door" displayed 80 percent of what would become the Guanacaste Conservation Area. It was an inspiring view, one that had reached a key early donor to the project, a U.S. citizen whose anonymity Janzen still protects. Janzen and Hallwachs had brought him there in July on a personal tour of the project, and he soon after decided to donate $240,000 to the project.

This grant enabled purchase of the two farms on Cacao, and the hiring of the former owners. It also funded the purchase, in November, of Finca Jenny, the 1000-acre farm adjacent to Santa Rosa, just to the southeast. In 1986, Finca Jenny belonged to the Gulf Land Company, a firm owned by a San José resident, and was managed by an administrator who lived on the farm with his family. The farm supported a piece of dry forest believed crucial to the survival of Quebrada Puercos, the deep canyon forest.

The Alton Jones grant had given the project enough money to buy La Guitarra, a forested section that served as a bridge between Cerro El Hacha and the Orosí Forest Reserve. However, the Guanacaste project leaders, keeping their cards close to their chests, moved slowly on this deal. One of the Cerro El Hacha landowners also served as manager of the bridge land. If informed that money was available to purchase the bridge, no doubt he would tip the Cerro El Hacha farmers, who, sensing the project's eagerness to buy a complete swath from Cerro El Hacha to Orosí, might then raise their asking price.

The energy that Janzen had told Raven the project needed to interdict hunting, logging, and other destructive elements burst forth in the form of Julio Quirós. In August, the Forestry Department assigned Quirós, a forest and game warden, to the GNP area. A by-the-book man in his early thirties who brought to mind Ugalde's fireplug toughness, Quirós quickly confronted the Wild West atmosphere of Guanacaste, bringing law to areas not previously policed. In his first two months on the job, he confiscated more than $10,000 in poached timber and a $30,000 bulldozer used in illegal logging. He recovered a tiny margay kitten and other protected animals from local residents, and he and other wardens made scores of arrests for illegal deer hunting.

Umaña, Ugalde, and former president Daniel Oduber were trying to convince Arias administration officials to acquire Santa Elena for the Guanacaste project, thus connecting the Santa Rosa–Santa Elena–Murciélago dry forest region and fulfilling a 1978 expropriation order signed by Oduber just before leaving the presidency. This would be both a major conservation event and a rite of purification for Costa Rica, which had been embarrassed by revelations of the secret airstrip. In his letter to Raven, Janzen described the attempt and wrote: "Cross your fingers."

Santa Elena had a shady history even before the Oliver North episode. Once part of Anastasio Somoza's hacienda, the ranch was bought in the early 1970s by five U.S. investors for $375,000 and valued in the expropriation order at just under $2 million. Ugalde remembered it as a constant source of problems for Santa Rosa in the early years of the park. Hunters and fires routinely entered the park from Santa Elena, and many observers have charged that the airstrip, long before North renovated it, served as a conduit for contraband headed to the United States.

Martha Honey reported in her book, *Hostile Acts: U.S. Policy in Costa Rica in the 1980s* (1994), that the U.S. Army and CIA had by 1982 begun a plan to combine an expanded airfield at Santa Elena with the Murciélago base to create a major installation for infantry, naval, and air forces. However, tainted with corruption, the project, code-named "Yellow Fruit," was closed down.

Oduber, pushed by Ugalde, had begun Costa Rica's efforts to obtain Santa Elena and make it part of Santa Rosa. The government didn't carry through with payment for the expropriated land, though, because the owners wanted more than San José was willing to pay. Rodrigo Carazo, the president from 1978 to 1982, tried to reinitiate the effort.

When he became president in 1986, Arias informed the United States that it did not want the airstrip used, but the White House chose to do so anyway and not tell the Costa Rican president. When the munitions-laden C–123 got stuck at Santa Elena on June 9, 1986, U.S. officials rushed to free it. On June 11, it flew back to El Salvador.

As Richard Secord described in his memoirs, North and his operatives set up a "dummy company" to lease the land. Known as both the Udall Research Corporation and Udall Resources, Inc., S.A., the company was "named, appropriately, after Morris Udall, the noted conservationist," Secord wrote. Udall Research used part of the ranch "under the cover of building a game preserve and tourist attraction." The choice of name carried particularly vicious irony. Morris

K. Udall, a congressman from Arizona for three decades, was, indeed, conservation-minded, as was his brother, Stewart L. Udall, the Secretary of the Interior during the administrations of John F. Kennedy and Lyndon B. Johnson. As both were liberal Democrats, and Morris Udall one of the first of his party to oppose Johnson on Vietnam, it's unlikely either brother would have endorsed North's scheme, to put it mildly.

Udall Resources upgraded the airstrip and constructed barracks. Secord and North bought two C-123Ks and other cargo aircraft widely used during the Vietnam War, building a 1980s version of Secord's famous "Air America," a CIA-run airline that operated in Southeast Asia during that war. "It was a bedraggled little air force," Secord wrote of the Reagan-era "Air America" of Central America, "but with the right crews, we hoped it would keep the Contras going until the CIA's Air Branch or U.S. Air Force reentered the picture."

It's important to remember that these were tense times in northern Guanacaste. There had been several border incidents involving shooting and shelling by Sandinista troops and contras, although most of them were well east of the Pan-Am Highway. Sandinistas crossed the border and burned an abandoned contra camp on July 3 near Las Chorreas, the first incursion since May 1985, when Sandinistas killed two Civil Guards near Las Crucitas. A May 1984 bombing at a jungle press conference held by Eden Pastora, one of the contra leaders, killed three journalists, including *Tico Times* reporter Linda Frazier, a U.S. citizen, and wounded many others. In April 1986 someone bombed the U.S. consulate near San José. Right-wing Guanacastecans were pushing the plan to secede. Two dozen Green Berets stationed at Murciélago trained several hundred Ticos to serve, as one Costa Rican official put it, as "a paramilitary police force to face the threat of spillover from Central American conflicts."

When Janzen went to Umaña in his first few weeks in office, the biologist proposed the idea of consolidating Santa Rosa, Santa Elena, and Murciélago. Janzen also told the minister that flights were going in and out of the airstrip. Sources close to Janzen told me that he had become aware of the airstrip in January when, driving back from some field research, he got stuck in the mud near the Santa Rosa-Santa Elena boundary not far from where a bulldozer operator also had gotten stuck. While they helped each other free their vehicles, the conversation between the local Tico bulldozer operator and the U.S. biologist (who, it should be recalled, was a former Army MP) revealed that the operator was helping to construct the airstrip. Janzen suspected something foul.

From then on, the scientist had to walk a very fine line. He kept quiet about the airstrip. He was, in fact, anti-contra, but he knew he had to keep a low profile on that, too. To make any kind of noise would invite reactions that could seriously harm the project. Those reactions could come from individuals like the many Guanacastecans who supported the U.S. anti-Sandinista policy, from potential funding sources like the U.S. Agency for International Development, and even from the CIA, which could have him out of Guanacaste in a wink if he called attention to the airstrip.

"So far Washington is not against you?" I asked him a year after the plane had gotten stuck.

"Oh, I must have a file an inch thick in the embassy," he said.

"Does it affect how the NSF [National Science Foundation] views you?"

"No, because I don't open my mouth about it. If I opened my mouth about it in any official capacity or any public capacity, then they could feel much more justified. They'd love to get rid of me—not NSF, but the embassy and people like that. They view me as definitely a liability. But bear in mind, this is why I kept my mouth shut about the airstrip from its beginning."

So Janzen kept quiet. That is, until Arias came to power, and even then he didn't talk publicly. Umaña waited for the right time to broach the idea to Arias.

In early September, on orders of Arias, about sixty Civil Guard troops quietly raided the airstrip by land and sea. Although several hundred contras were rumored to have been occupying the airstrip, the troops found the barracks empty. They did find a large supply of aviation fuel. They rolled empty oil drums onto the strip to prevent further use, and about two dozen of the guards bivouacked at the site to secure it.

White House contempt for Arias boiled. According to the congressional Iran-contra report, on hearing of the Arias order, Elliot Abrams, the State Department inter-American affairs undersecretary, responded, "We'll have to squeeze his balls [and] get tough with him." Among Reagan administration officials who pressured Arias to cooperate throughout this period were CIA director William Casey, who traveled to Costa Rica, and Reagan himself, who met with Arias in the White House in December. U.S. economic aid to Costa Rica, the second largest per capita recipient of U.S. funds, after Israel, was delayed in 1986 and began to decline in 1987.

The Arias administration kept the airstrip closing quiet, but within a few days the story found its way to enterprising reporters. Two freelance journalists, Lyle Prescott and John McPhaul, became intrigued by rumors floating around San José that large planes had been flying over the Santa Elena peninsula. Meanwhile, a North American retiree living in Liberia called the *Tico Times* to say that flights of big planes in the area were an open secret. So Prescott, McPhaul, and Jake Dyer of the *Tico Times* drove up to Guanacaste, along with several relatives. "When we got to Santa Rosa, people didn't want to talk about it," recalled Prescott, who had first gone to Costa Rica in 1978 as a field assistant at Santa Rosa for Janzen and had worked her way into a journalism career by freelancing nature articles, then finding work covering the Sandinista victory over Somoza. McPhaul remembered that when the reporters told a U.S. scientist at Santa Rosa what they were looking for, the biologist—not Janzen—said, "Well, it's about time."

The reporters and family members drove back south to Liberia, where Prescott, McPhaul, and Dyer chartered a light plane, telling the pilot they wanted to look for turtles laying eggs at Nancite. "When we got to Nancite, we said, 'Let's fly on over the next ridge and see what's there,'" Prescott told me. "So we went over the ridge, and there was the airstrip." It was huge. The next day they went to Civil Guard headquarters in Liberia and were met with denial that such an airstrip existed. But when the reporters said they had flown over the strip the day before, the officials stopped denying it and instead shouted, "*Who* flew you over?"

Back in San José, McPhaul found Security Minister Hernán Garrón, who denied knowledge of the flights and said only small Civil Guard planes were using the airstrip. McPhaul produced photos of the field taken by Julio Laínez of the *Tico Times.* "I pointed out to him that the large tracks obvious in the picture were not made by small airplanes," McPhaul told me. Garrón became angry, but he confirmed that the Civil Guard had occupied the airstrip weeks earlier. A flood of stories by the *Tico Times* and major U.S. newspapers and networks followed.

Later stories revealed the potential impact of the airstrip shutdown on the proposed Guanacaste National Park. McPhaul's article in the October 24 *Tico Times* quoted an unnamed "high-level government official" as saying that press attention to the airstrip had caused Arias to back away from the idea of annexing it to the conservation project, since it might appear that he was "pressured" to incorporate it or was simply doing it to be politically opportunistic. The airstrip revelation had "spoiled the atmosphere," the official said. Janzen refused to comment. The article cited a "National Park Service official" who said Janzen had attempted to work a deal with one of the Santa Elena owners, Joseph Hamilton, to buy the land. Neither Hamilton nor Janzen would comment on that.

The article continued: "According to Guillermo Canessa of the Ministry of Natural Resources, Janzen wants to leave the acquisition of Finca Santa Elena for last since, 'because of the situation in the north there's not likely to be any development there,' whereas nearby areas outside the peninsula could be turned into pasture if not protected soon.'"

When the newspapers revealed the airstrip, Reagan administration officials immediately attempted to conceal their role in planning, constructing, and paying for it. This cover-up later confused the question of who owned the property sought for the Guanacaste project, and it added friction to the process of legally acquiring it. North reported to his boss, John Poindexter, that he had taken steps to "keep USG [U.S. government] fingerprints off this." In what North called a "damage assessment," he wrote that Udall Resources, Inc., would "cease to exist by noon today."

North, Abrams, and others drafted a statement meant to guide administration officials in answering questions from reporters, and Poindexter approved it.

The statement said that the airfield had been offered to Costa Rica by "the owners of the property who had apparently decided to abandon plans for a tourism project." The memo concluded by denying the use of U.S. funds or personnel in the airstrip project and referring further inquiries to the San José government. North, meanwhile, pushed for U.S. action to, as he put it, "punish" Costa Rica. He advised Poindexter that any steps to help Arias should be suppressed. "Those who counsel such a course of action," he wrote in a memo to his boss, "are unaware of the strategic importance of the air facility at Santa Elena and the damage caused by the Arias' [*sic*] government revelations."

When the airstrip became an embarrassment to Arias, Umaña immediately saw an opportunity to tie politics with conservation. He recommended that the president consolidate the land into Guanacaste National Park. Arias took his time, deliberating carefully.

Umaña clearly remembers the moment Arias finally became convinced to do it: six o'clock in the evening of Sunday, October 4, 1986. The two men had attended a seminar at the Instituto Centroamericano de Administracion de Empresas, a U.S.-supported school that trains business students from all over Latin America. Arias asked his environment minister to ride back with him to San José. Arias had his driver take Umaña's car back, while the president drove his own car, a white Jeep Wagoneer. Umaña sat next to him. In the back seat was Muni Figueres, the sister of former President Figueres and Arias's Minister of Exports. It was dark, and the three discussed the Reagan policy in Central America and what it meant to Costa Rica. They drove through a rainstorm— "probably the heaviest downpour that I have witnessed in Costa Rica," Umaña recalled. Figueres and Umaña did most of the talking, while Arias quietly concentrated on both the driving and the nuances of the issues, including Santa Elena. Arias kept wiping the windshield with his handkerchief. "It was a very, very important moment," Umaña told me. "This problem obviously was bothering him a lot, and this was very much President Arias's style, the way he handled things—on a very personal level." As he wiped the condensation from the windshield, Arias began to shift toward declaring Santa Elena part of the national park.

The following day, October 5, the airborne operation to resupply the contras ceased when a C–123 transporting 10,000 pounds of ammunition to southern front guerrillas was shot down by a Sandinista surface-to-air missile. Various arms of the U.S. government launched investigations of the Iran-contra affair.

I didn't get to the Santa Elena airstrip until the dry season of 1988, but its basic elements remained. The airstrip ran east and west for a mile or so on a natural plain between two parallel ridges that framed the Potrero Grande Valley. Looking up from the airstrip, the Santa Elena mountains seemed rugged and barren except for a few lonely trees and small shrubs. The airstrip lay a bit closer to the northern ridge, and along its southern flank ran the Río Potrero Grande. Between the river and the strip rose several rock levees a few feet high, apparently constructed in an attempt to keep the river from spilling onto the field during the rainy season. The smooth, flat runway was mostly made of dirt. Nestled back in a patch of forest to the north was the wooden barracks.

I stood in the middle of the airstrip, buffeted by gusts of hot wind, and tried to challenge my notion that this overgrown cow pasture was a symbol of U.S. penetration of Latin America. It was so parched, so plain, so pathetically dull—nothing like the fiery image of intervention by the original crazy gringo, William Walker, at Santa Rosa 130 years earlier. Yet, in so many ways, Oliver North had just been another twentieth-century William Walker, and the United States had still been trying to control Central America. North's boys had holed up in Santa Elena rather than Santa Rosa. They had used airplanes rather than rifles. But North, like Walker, had sent them down to do what had to be done. He might not have called the Ticos "greasers," but his contempt for Costa Rican sovereignty reflected the same attitude.

I had gone to Santa Elena with a group of ecotourists, all nice folks from Boston, Chicago, New York, and San Francisco. To them the airstrip was just one more curious stop in their venture around the peninsula. We drove from the flat valley of the airstrip down a sloping landscape covered with palms and mahoganies. We heard tales of huge crocodiles lurking in the mangroves and saw parrots, hawks, ospreys, and herons. We crossed the river several times, and, even in the dry season, our four-wheel-drive truck got stuck. After we all got out and pushed, we drove down farther to the Santa Elena mangrove swamp—the most productive in northern Guanacaste, said one of our guides, Sergio Bonio, a former hunter who was now caretaker of the land for its U.S. owners. It was a key breeding ground for ocean fish and an important night roost for macaws, herons, parrots, and other birds.

We ate a picnic lunch where the forest met the beach, a brown sand crescent stretching more than a mile between imposing black cliffs. Afterward, as the others frolicked on the sand, I looked out on the beautiful curling waves. Surfers, who revel in the fast waves and right point break off this site, have since dubbed the area Ollie's Point. I tried to imagine what thoughts would have come to mind had I been a twenty-year-old California surf bum sitting on my board off Ollie's Point in June of 1986, looking back for one of those bodacious hollow tubes of saltwater, and instead seeing a camouflaged C–123K Provider curving majestically in from the west just above the blue breakers, its landing gear down and flaps fully extended, its two giant Pratt & Whitney R–2800 engines rumbling like a freight train over the barely audible high-pitched whistle of the two auxiliary turbojet engines, the noise penetrating the churning water around me with deep reverberations as the upswept tail of the old bird passed overhead on its approach to the muddy, secret airstrip, floating toward the mangrove swamp and disappearing behind the palm-covered rises of the Potrero Grande Valley. It would have been unreal.

Back in the States for the fall 1986 term at Penn, Janzen's immediate goal was to raise that $5 million by the following May. The money was needed to buy the core, the crucial, large piece of land on the east side of the Pan-Am Highway, a piece Janzen described as GNP's "solar plexus." To succeed in the acquisition, the Guanacaste project leaders would have to excel at a game that became the crux of tropical forest restoration for them, a game they learned to play well: the simultaneous balancing of all-out fund-raising, intense bargaining, and a kind of public relations that was sensitive to the values and norms of Costa Rican society.

The land at stake in the core included five haciendas: Centeno, El Hacha (adjacent to but distinct from Cerro El Hacha), Poco Sol, San Josecito, and part of Tempisquito. The money was necessary in such short order because the five owners planned to sell or further develop their parcels but had agreed to keep them off the market until early 1987, when the project leaders would be able to begin negotiating. The owners were willing to sell to the Guanacaste project— if the project could come up with the money. Timing was critical. "If they sell to someone else, we will really be hurting," Janzen wrote to Raven. "We will have to then pay that someone else a lot more money, or if it is sold in little blocks, we may never be able to buy it."

If Janzen failed to raise the money and the landowners proceeded with their plans, the project leaders could play the expropriation card, bringing to bear their connections within the Costa Rican national political structure. However, that would be viewed widely by the owners, as well as some other Ticos, as a heavy-handed way to start what was being touted as a "user-friendly" national park. Besides, three of the owners were prominent Ticos, including Mario Burgos, president of ICE, the national electricity institute. Burgos owned Poco Sol, where the Voice of America radio transmission station had just been built near the Pan-Am Highway.

If, however, Janzen succeeded in raising the money and the five parcels in the core were purchased, more than 60 percent of the proposed land acquisition for GNP would be completed. That would connect two large wings of conserved land by a central corridor, and it could give the project irreversible momentum. He felt that if this block of land could be acquired in one swift, bold move, the remaining areas proposed for GNP could be pieced together later.

Janzen constantly sought help and suggestions on fund-raising from Raven and other prominent scientists, including such widely respected ecologists as Thomas Lovejoy, then at the World Wildlife Fund; Thomas Eisner, a chemical ecologist at Cornell; and Don Stone, a botanist at Duke and executive director of the Organization for Tropical Studies. Another was Murray Gell-Mann, a Nobel Laureate in physics with the California Institute of Technology and a member of the MacArthur Foundation's board of directors. Gell-Mann, long a nature aficionado, played a key role in guiding MacArthur money to tropical conservation programs, including the Guanacaste project. In talks and letters to these and other supporters, Janzen acknowledged that he was making a large request of them, but he believed the consequences for tropical conservation justified it.

Indeed, Janzen himself was going to extraordinary efforts, setting a blazing pace in public talks and fund-raising from coast to coast. He and his colleagues had arranged meetings with several philanthropic foundations and invited staffers for many of them to come to Guanacaste for a tour of the project site. They also catalyzed a loose group of people in England to publicize the Guanacaste project, including officials at the Fauna and Flora Conservation Society, the British Broadcasting Corporation, and the Natural History Museum. In Sweden, the Swedish Society for the Conservation of Nature launched a fund-raising drive, with the help of newspaper articles.

Janzen, an old hand in the traditional and straightforward competitive review system for obtaining National Science Foundation grants, was learning a new kind of fund-raising on the run. "My special frustration stems from all of this being new ground to me," he told Raven. "When the crunch hits in the field, I know where to lean and when to jump; here I feel extremely helpless." He was trying to raise $11.8 million, nearly $5 million of it in only a few months—an astronomical figure when compared with any other effort in Latin American conservation history. The only campaign that came anywhere close was the effort by Ugalde, with the help of the Nature Conservancy and World Wildlife Fund, to raise $5 million for the entire Costa Rican national park system beginning in 1982.

Fiercely independent by nature, Janzen was now dependent on others more than he ever had been. An added frustration was that he really wanted to devote his energy to the entire endangered tropics, not just to the "extremely parochial task" of fund-raising for the Guanacaste restoration, as he wrote Raven. "Costa Rica, and the tropics as a whole, needs much more than GNP, but somehow unless I can demonstrate one functioning system with GNP, we won't ever get to the bigger picture."

In January 1987, the project leaders concluded a series of land purchases in Cerro El Hacha with fourteen farm families. It took more than an ability to bargain over real estate to pull the land sale off and keep everyone smiling. After Janzen first approached the farmers and asked them to stop working the land and clearing the forest on their property, he returned frequently to talk about the idea of the new park.

When the MacArthur Foundation committed in mid-1986 to a $500,000 donation, he started negotiating with Ticos on price and time of possession. That took until December. "I treated them as gently as I could," he told me. Midway through the negotiations, the farmers asked if they could plant another bean crop. Janzen gave his approval, but the crop was planted only on pieces of land that had already been cleared of forest. "It was a real dialogue, and it worked," he said. The farmers knew that the land they had squatted on had deteriorated. They had gotten the benefits of the first ten years of frontier colonization and were happy to liquidate before it became worthless. Prices were finally settled on, and the date of sale was set for January 23, 1987.

One subtle problem remained. The farmers demanded cash for their land. Janzen agreed, but insisted that the transactions be conducted within the walls of a bank in La Cruz, the nearest town, and that the farmers immediately deposit their proceeds in a savings account. He made this demand to protect the farmers from robbery and to protect himself and the Guanacaste project from bad public relations.

As Janzen put it, "if you pay a farmer 600,000 colones for his farm, he walks out of the bank, and a man sticks a gun in his back and takes it away from him, two days later that farmer is on your doorstep saying, 'I want my farm back.' He and society really won't accept the answer, 'No, you go away. I paid you your money.' This is a patronistic society where the person in my position is expected to have a sense of responsibility about the farmer, who is treated by society more like a draft animal than a thinking human being. The robbery would have been viewed as partly my fault, and furthermore, we would have had a very upset person. Now, the very upset person might not have blamed me personally, but he would nevertheless be very upset, which would then mean that the community would see me as generating upset people."

After much debate, the farmers acceded to Janzen's demand. He made arrangements with the bank manager to open the bank on Saturday for the transactions and to open savings accounts on the spot.

On the other end of the deal, the MacArthur money had gone to the Guanacaste fund at the Nature Conservancy, then was sent to the San José bank account of the Neotrópica Foundation, the Costa Rican conservation organization that coordinated GNP land purchases. On January 22, Pedro León and other officials from the foundation went to the bank with suitcases and withdrew the cash required to pay the farmers. At four o'clock the next morning, they left San José for the La Cruz bank. When they arrived five hours later, they placed legal documents and piles of one-thousand-colon bills on a table set up on one side of the lobby. As León and his assistants sat behind the table, the farmers filed in, signed land title transfers and other documents, counted out their colones, paid off debtors who had been invited, carried armfuls of cash across the lobby to the teller windows, opened accounts, and deposited their money—sums as much as a million colones (about $16,000 at that time).

At one point the lobby was crowded with forty-five people. Janzen stood in the background "taking pictures," as he liked to describe his role at these and similar events involving the project. He could make light of his part in that day's activity, but his presence in the lobby was the final assurance that the deals would go through. There was irony behind that assurance. The Cerro El Hacha farmers were members of a group in Costa Rican society that often fell victim to financial predators from the upper echelons of society. As a result, the farmers simply wouldn't trust other Costa Ricans to make land deals like they were completing that day in La Cruz, but they did trust Janzen. He was, to many of the financially marginal people of the Guanacaste countryside, like a person from another planet. They had never dealt with North Americans, much less a scientist-conservationist, and Janzen had gradually won their trust over the months of negotiation. The result of this alien though solid relationship was more land for the Guanacaste project, no robberies, and a lot of happy farmers.

As biologists worked their way around Cerro El Hacha and other areas destined to be added to the project, they discovered animals and plants new to science—a tantalizing taste of things to come if the region could ever be surveyed intensively. For instance, a species of swallowtail butterfly new to Costa Rica was found in Cerro El Hacha, a surprise because Costa Rican butterflies were considered to be quite well known, especially the large, ornate species like swallowtails. They also discovered two species of birds identical to two others in Santa Rosa but with different songs.

In Santa Elena, a species of agave plant that had never been seen south of the Honduras-Nicaragua border was found growing, its bright yellow flowers bursting forth like fire. Hope had begun to blossom on the debilitated land between Santa Rosa and the volcanoes. Janzen, in his letter to Raven, sounded upbeat as he described the return of a jaguar to Cerro El Hacha, a migratory pass of white-lipped peccary through Poco Sol, and what he believed to be a giant anteater on the slope of Orosí. Of the anteater he wrote, with boyish excitement, "I have never seen a live one in 23 years in Costa Rica, but this one passed 30 meters behind me." However, Janzen and the others knew they had only just begun their advance through the world of wounds.

# · 9 ·

# Earthquakes

The emerald green bird fluttered through the brush, and as it pulled up to enter its treehole nest, a fleck of sunlight caught its breast, which seemed to explode in a flash of crimson glory. The bird, a trogon, leaned into the nest, dropped a thick, green sphingid caterpillar into the open mouth of one of the nestlings, and took off into the forest in search of more food.

Moments later, Roberto Espinoza reached into the nest from the top of an aluminum stepladder propped up against the tree and carefully withdrew the caterpillar from the nestling's mouth. Earlier that day, Espinoza had placed a soft collar around the neck of this bird, as well as the necks of each of its siblings. He had put them on gently but firmly, so that they wouldn't harm the young birds but so the nestlings wouldn't be able to swallow the large caterpillars. Now he released the collar, gave the nestling a substitute caterpillar, and put the original into a plastic bag in an ice-filled thermos.

Espinoza, a long-time field assistant to Janzen, had spent the past few weeks with Frank Joyce in a nearby blind. For hours on end, they watched the trogon nest, dashing to it after an adult trogon delivered a caterpillar, removing the fresh morsel for analysis, and dashing back to the blind before the bird returned. Swift action was required because the adult birds might abandon the nest if they thought it had been raided by a predator. Also, the researchers didn't want the collared nestlings to choke on the food. Later, after the caterpillars were weighed, measured, and identified back at Janzen's house, they were brought to the nest and fed to the young birds as substitutes. Two other researchers manned a separate blind set up a few hundred yards away near another nest.

It was July 15, 1987, and the past several weeks at Santa Rosa had been dry. Such dry spells would often linger in the midst of the rainy season for a few days or weeks, sometimes accompanied by a bit of rain, but usually marked by a piercing tropical sun and brisk wind.

Amid the dry spell, the researchers had begun monitoring the nests. They were watching two kinds of trogons, *Trogon elegans*, the "elegant trogon," and *Trogon melanocephalus*, the "black-headed trogon." Trogons are a family of birds long steeped in mystery and legend, many of them worshipped as gods by the Mayas and Aztecs. The family includes the quetzal, which has the same red breast and metallic green feathers as the elegant trogon, but is larger and has longer tail feathers. Trogons were believed by the Aztecs to be the reincarnation of the bravest warriors.

Ornithologists had long felt that trogons fed on caterpillars, grasshoppers, beetles, and fruit, but, surprisingly, no one had done a detailed study of the nestlings' diet. Only in the past few days had Joyce and Janzen found that black-headed trogons apparently preferred caterpillars of saturniid moths and the elegant trogons specialized in sphingids. This discovery raised several questions about bird behavior and physiology, including why the parents had such distinct preferences for what they brought the nestlings and how the young birds were able to eat these noxious caterpillars.

Although they cautioned that this was just a preliminary finding based on observations of only a few nests, Joyce and Janzen seemed excited as they talked about the possible implications. What made the study so revealing was Janzen's knowledge of insects. He was able to put species names on the individual caterpillars that came in, an ability lacked by birders. Traditionally, ornithologists and entomologists didn't work together, so the bird specialists lumped prey items into broad categories, failing to make crucial distinctions among closely related species. The researchers eventually concluded that although the trogons feed their young a wide variety of prey items, when they can find a saturniid or sphingid caterpillar, that seems to be the food of choice. Also, elegant trogons generally prefer smaller caterpillars than do black-headed trogons.

The trogon study was only a small element of the excitement at Santa Rosa that July—and only one of the many activities that demanded Janzen's attention. Others, all associated with the Guanacaste project, forced him to play, simultaneously, the role of politician, fund-raiser, organizer, and public-relations man, in addition to the usual one of scientist. Janzen was struggling with the Park Service bureaucracy, land negotiations, and fund-raising, and the fate of the project within the matrix of local, national, and international politics remained uncertain. Failure on any front meant failure of the whole.

The Neotrópica Foundation was working closely with government agencies and politicians to obtain the lowest possible prices that would be agreed to by local landowners willing to sell. Janzen was the man on the front lines in Guanacaste, but key co-leaders in this work were Rodrigo Gámez, who had become a special adviser on the environment to President Arias, and Pedro León, an amiable, brilliant molecular biologist from the University of Costa Rica who had succeeded Gámez as president of the foundation. All land bought for the Guanacaste project automatically became property of the foundation, which, as the project leaders then envisioned it, would transfer the land to government ownership when the entire acquisition was completed.

At his house, Janzen worked in the eye of the storm, juggling demands for his attention like a traffic cop at rush hour: making and taking phone calls, meeting with visitors of all sorts, writing letters and grant proposals, giving orders to student volunteers, and conferring with Park Service employees and newly hired management personnel for the project. The new employees were on the payroll of the National Parks Foundation or Neotrópica Foundation, to keep them from what the Guanacaste project leaders saw as the inflexibility of government employment.

The telephone just inside the door to the house rang with maddening frequency. Each time, Janzen seemed to sprint toward it. Usually the call would be for him, and depending on the caller, he spoke Spanish or English, rapidly in either language, and he gestured with his arms and raised his voice to emphasize words. When the call was for José Antonio Salazar, the director of Santa Rosa, or someone else at the park office, Janzen would grab a large, battery-operated megaphone, point it down the rocky road toward the office, and broadcast a tin-crisp "Oficina, teléfono! Oficina, teléfono!"

Although he did so coolheadedly, an irritating twang added to his voice by the electronic tool hinted at the tension that hung in the air between the two buildings, and between the people who worked in them.

Among the most difficult struggles facing the Guanacaste project leaders in July 1987 was, oddly, rejection of the project by a significant element of the Park Service bureaucracy. Janzen and his Costa Rican colleagues were locked in an intense fight with several people in the Park Service who found fault with many of the project's means and goals. In essence, they viewed the project as a threat to the Park Service's power base, especially in San José.

To understand the struggle, it's important to understand the state of the Park Service at the time, and to know that Costa Rica was suffering through a crippling economic crisis in the 1980s. This, and new government policies for fiscal constraint, had taken their toll on the Park Service, a unit of the Ministry of Agriculture. The average salary for an employee declined, as did staff size. A constant problem, the demand for more work to be done with fewer resources, was at its worst level yet. The agency lacked money to buy private inholdings in parks, and it had seen some outside donations disappear, at least partially, into the Ministry's bureaucracy. Incredibly, it faced the possibility that the government might eventually open the parks to logging and other traditional forms of economic development.

The Park Service already had received a healthy dose of criticism from Janzen, in the form of the 1985 Corcovado report. The report faulted the agency for "inactivity," and it recommended a host of changes. "Janzen has studied and lived in a variety of Costa Rican national parks for about half of each of the past 13 years," the report said. "It is clear to him that there are several personnel policies associated with the Parks that are in drastic need of revision." Among the report's recommendations were to create a detailed management plan for each park, improve living conditions for park rangers, and establish more continuity in the commitment of park directors by giving them a longer appointment at a park than the traditional two years. Some of these steps were under consideration by the Park Service.

Along came the Guanacaste project, with its direct challenge to the centralized structure of the fifteen-year-old Park Service and several of its longstanding policies. Janzen, Gámez, and others felt a new set of policies was necessary for the project to succeed. Chief among them were a proposed decentralized management structure for the Park Service, a new policy for hiring local residents for Park Service jobs, and proposals for the use of livestock in the park. As the project leaders pushed on these issues and Park Service regulars and some members of the National Parks Foundation pushed back, tension rose, with major periodic flare-ups, all of them threatening the project.

The traditional Park Service view of its role had almost militaristic overtones. This view held that the responsibility of the agency in San José and its park guards in the field was to protect nature from assault by people who lived around the park. When a park was established, all the inhabitants inside the property were to be removed and the area policed by guards to make sure they didn't move back in. Direction came largely from bureaucrats (read: Generals) in San José, whom critics felt were too removed from what was happening in the field and too rigid to see the need for practical change.

By the mid–1980s, several Costa Rican conservationists had begun to see that the future of protecting national parks would be far more complex than building a wall. Corcovado had seen to that. Instead, the Park Service would have to address the social and economic challenges beyond park borders. To respond to these challenges would require decentralizing the San José-based system so that responsibilities were shared more with park directors, who would be given more autonomy in deciding how to manage, hire, and administer the park budget from their vantage points in the field. This view held, for instance, that the Guanacaste project director should be able to hire employees, even without approval by officials in San José, since it was the park director who would be responsible for and interact daily with the new hires.

Of course, the proposal to change tradition was viewed as a threat. It upset the old order. The staff in San José worried about a future in which, if each park director could hire his own people, he could also set the salary. That, in turn, would lead to differences in salary throughout the system, an idea that offended the egalitarian sensibilities of many Costa Ricans. "You need some system whereby you maintain certain norms and standards and salary levels, so you don't create injustices," Pedro León told me. "But somebody also needs to delegate. How do you go about maintaining general guidelines and yet release some of this power? That was not easy."

Opponents of decentralization also criticized the idea of hiring local residents because it meant sidestepping the Park Service's requirements for university education and professional training in park management. Some local residents hired by the Guanacaste project, for instance, hadn't even completed high school. Thus, by hiring local farmers to manage parts of the park like Cerro El Hacha and Cacao, the project was bringing "illiterates" into the ranks. Park Service leaders also viewed locals as the enemy; after all, they were often the hunters who entered parks to poach, or the farmers who set fires that invaded the forest.

In the ensuing debate over the Guanacaste project, Janzen didn't deny that the local farmers lacked education and professional training. But he argued that, in fact, they were "very professional at manipulating vegetation, because that's what a farmer does. And so all you do is change his goals." The goal, put simply, was no longer to get rid of the forest, but now to grow it back. Locals were by their background and experience the best possible candidates for handling the threats to the biology of the tropical forest from fire, hunting, logging, ranching, and farming. It was an eminently practical idea—to recruit farmers as site managers, to apply their knowledge, energy, and spirit to running the park. Although many residents would need training in biology and communication skills, they already knew how to perform such technical aspects of park management as fighting fires, constructing fences, caring for horses, maintaining trails and buildings, herding cattle, identifying and understanding plants and animals in the forest, and dealing with what Janzen called "biotic challenges" of field work—things like snakes, ticks, disease, thirst, and wounds. Hiring Guanacaste residents also would establish direct as well as subtle links between the park and the province's economic and social structures. It would bond the local community much more tightly to the park than if Park Service employees were brought in from outside, a method that Janzen and others believed had been demonstrated a failure over the previous fifteen years.

A good example was that of the two Espinoza brothers, Roberto and Alvaro. They had been born at Santa Rosa; their grandfather had been manager of the former hacienda. Now in their early twenties, they were the perfect field assistants—they knew the area like the backs of their hands, had deep emotional roots in the land, were interested in natural history, had the intellectual ability to learn scientific techniques, were eminently coachable, and needed jobs. Wherever they went in the province, they were ambassadors for the park. Janzen estimated that about fifty area residents would be required for the staff in the early stages of the Guanacaste project.

Controversy surrounded Janzen's biology-based proposal to put cattle on the land. His idea was simple: keep the grass under control by letting cattle eat it down to ground level. "Living mowing machines," he called them. To him, this was a logical and cost-effective way to keep jaragua under wraps until the fire program gelled, meanwhile giving tree seedlings a chance to grow. Cattle and horses also were needed to speed restoration of trees that once depended on the Pleistocene megafauna for seed dispersal. Eventually, when the young forest had enough of a foothold to shade out the pastures, the cattle would be removed. Janzen argued with the Park Service administration that, because cattle had been removed in 1977 and 1978 after ranging freely for centuries, and because no effective fire-control plan had been put in place, fires were destroying more and more forest. In other words, the Park Service bureaucracy had created a situation in which Santa Rosa had less forest as a fifteen-year-old national park than it had as a new national park, less forest even than when it was a cattle ranch.

To many in the Park Service, however, the idea of putting cattle into a park constituted conservation sacrilege, and it would set a precedent that would give them problems in other parks. The traditional concept of running a national park called for removal of all domestic animals and constant vigilance to keep them out. Indeed, some of the longest and most bitter battles fought by Alvaro Ugalde and local ranchers in the early years of Santa Rosa had been over unwanted cattle, cattle that were "trespassing" in Santa Rosa and "destroying" the park. Like anywhere else, dogmas die hard in Costa Rica, and heroic veterans like Ugalde, who was then an adviser to the Park Service, seemed unable to accept the concept of living mowing machines. He later told me he never "irrationally opposed" using cattle to keep jaragua down. "My concerns with cattle," he said, "had a lot to do with the political problems of the early days of our park history in Santa Rosa."

It was understandable that a national park system charged with protecting threatened natural areas had developed a philosophy of throwing "enemies" out and patrolling park land like police at a bank. It also was understandable that, after almost two decades of such a philosophy, if someone came along and suggested a new kind of management, that that would be seen as a threat to tradition. Yet not everyone in the Park Service reacted negatively to the new ideas. Many in San José and in the field were exploring them and thinking hard about what they could mean for management, not just for Santa Rosa and environs, but for other parks in the system as well.

In August 1986, Janzen presented the Guanacaste project plan to about forty-five Park Service officials at a meeting in San José, and they locked horns. "From the very beginning, he clashed with the Park Service," said one high-ranking official present at the meeting who asked not to be identified. "We probably didn't realize that he wanted to be the leader. Later, that became clear. He wanted the Park Service out of Santa Rosa and Guanacaste." No doubt this also reflected, at least in part, the general fear of innovation among some Ticos.

Many in the Park Service had been taken by surprise, and they became deeply resentful of the gringo behind the new ideas. The Park Service already had deep anti-gringo currents running through it, several U.S. biologists told me, and Janzen became a virtual lightning rod for them. It was Janzen's missionary zeal, and the way he spoke out on these issues, that many officials found to be distasteful. That abrasive style was often effective and not unusual in some segments of U.S. political, business, and scientific circles. However, it offended many Costa Ricans, whose cultural norm was one of decision-making *a la tica* and *quedar bien*—getting along through negotiation, compromise, and conflict-avoidance. The important social aim was, put simply, to avoid a fight. Yet here was the stereotypical "in-your-face" gringo telling Ticos how to run their Park Service, as some saw it.

Among the most resentful of Janzen were Luis Mendez, whom Arias appointed Park Service director when Ugalde resigned in April 1986 to become an adviser to the agency; José María Rodríguez, Park Service deputy director; and Santa Rosa director Salazar. In addition to pushing hard for Park Service restructuring, Janzen had been critical of Salazar's management at Santa Rosa. The Penn professor had gotten along well with other administrators. In a 1984 article he wrote for *The Nature Conservancy News*, he painted a heroic picture of the staff. "Santa Rosa's park guards and administrators sleep at the gates, fight the fires, ride hard under the hot sun to deter poachers, and sit for hours in town fretting over legal battles with neighboring landowners," he wrote. "These people safeguard the park and make it possible for us to count beetles, scrutinize caterpillars, and examine tapir dung in peace."

That picture of the Santa Rosa staff was drawn, of course, for a North American conservation audience. Still, it's telling. It came the year before Janzen's Corcovado shift, before he had launched the Guanacaste project, and before he had come to view the current Park Service bureaucracy—and some of its people—as obstacles to restoring the forest. When he did finally reach that view, the man standing in front of him at Santa Rosa was the newly appointed Salazar, whose performance Janzen saw as inadequate for the vision.

It's important to remember that Janzen was the ranking scientist at Santa Rosa, in terms of experience and residence there. Although he enjoyed enthusiastic support from many Costa Ricans, some of the local Park Service employees resented this status. Some had experienced his hard-charging will, some felt threatened by the project, and some had begun to become jealous of his new role as a celebrity.

In March of 1987, journalists from the Natural History Unit of the British Broadcasting Corporation filmed a fire scene at Santa Rosa as part of a documentary on the Guanacaste project. Janzen had negotiated with Salazar to burn a plot of jaragua. Salazar himself had become intrigued by the possibility of protecting the park with fire lanes. For the shoot, they rounded up a huge team of Park Service people, scientists, and students from a U.S.-Swedish tropical ecology course Janzen was hosting. As one researcher put it, "everyone and their brother and their dog was there." During the burn, the cameraman seemed to focus exclusively on Janzen, which ignited envy among the Santa Rosa administrators. In their view, not only was he getting credit for everything, he was becoming a movie star!

To make matters worse, the fire flared up again after Janzen and the students left for a trip to Volcán Cacao, so the park staff had to put it out alone. Salazar had had enough. He wrote to San José requesting that Janzen's research permit be revoked.

A bit later, while Janzen and Hallwachs were back at Penn, Salazar and Ronald Sanchez, his sub-director, changed the Santa Rosa policies on scientific work. In the view of some of the scientists, Salazar and Sanchez chose a time when Janzen was absent to use Park Service regulations to increase their control over Janzen and other researchers. Salazar himself revoked Janzen's license to do research in the park and said the scientist failed to respect park authority. The director also ordered the gate at the park entrance locked from 5 p.m. to 8 a.m., a move that restricted access by researchers and the new GNP staff.

Meanwhile, the fire season picked up, with several fires entering the park. This tried everyone's patience. To some of the scientists, park employees seemed to take fire fighting nonchalantly; for example, breaking for lunch before a fire was known to be extinguished. Once, key park personnel who should have been helping instead attended a fiesta. The scientists, who constantly pitched in, grew increasingly frustrated at losing more and more research time to fighting fires.

Then, in late April, came an event later known among some at Santa Rosa as "Black Saturday." As Sanchez and one of the scientists argued over who would get to use a jeep owned by Janzen but turned over to the Guanacaste project, Joyce stepped in to make peace. He ended up in a shouting match with Sanchez. They called Janzen in the States, and he asked Joyce and Sanchez to come to agreement among themselves. They made a tentative truce, but more damage had been done.

After returning to Costa Rica, Janzen continued doing research. But Sanchez, who technically held authority over scientific investigations in Santa Rosa, issued more regulations that the scientists found excessively restrictive and dictatorial. These ranged from changes in how to place flags marking study sites to a requirement that a quarterly report be written to the park administration for each research project. In late June, just after the trogon discovery, Fernando Cortes, director of investigations for the Park Service in San José, denied permission to Janzen and Joyce to pursue the research. The move came in the midst of a critical period—when the newborn nestlings were being fed. A few days later, at a July 7 meeting in San José, Luis Mendez, the new director of the Park Service, rescinded Cortes's denial.

Mendez and others in San José were so incensed by the resulting criticism from Janzen that they launched an effort to have him declared persona non grata. They wanted the Legislative Assembly to vote on such a declaration, a rare denigration mainly reserved for traitors, drug traffickers, and other heinous criminals. To be declared persona non grata in Costa Rica technically meant that the person had to leave the country and never come back. Conservation leaders quickly stepped in. Ugalde wrote a letter to Mendez pointing out that the Guanacaste project was not "Janzen's project"—it was a project of the National Parks Foundation, the Neotrópica Foundation, and leading scientists and conservationists.

Ugalde advised the Park Service, "You're not fighting with Dan, you're fighting with us. That includes *me*." Not that Ugalde and others didn't often disagree with Janzen. And not that they didn't find him abrasive. They just didn't clash to the extent of kicking him out of the country. The effort to have Janzen declared persona non grata quickly fizzled. "It was nothing but bullshit," Ugalde told me. "To dare to do something like that would have been the most ridiculous thing. First, what had he done wrong, other than speaking out loud of mismanagement of the system, to give his opinion about it? He was bringing money to the country. He was legally here."

Ugalde paused and chuckled, no doubt thinking about Janzen's success as a fund-raiser, which paved the way for foreign contributions to many Costa Rican conservation projects, not just Guanacaste. "It's a nice problem to have with ourselves," he continued. "You have to deal with the persona. You have to deal with me, too. I'm another son of a gun. If the person has good intentions, you have to live with it. If they're evil, yes, you get rid of them. But that has never been in Dan's mind. It's more naiveness than anything else, in my opinion."

At the same time that they credit Janzen with visionary contributions, many Costa Ricans told me that his lack of diplomacy frequently put the project in jeopardy. It might have failed quickly if not for the efforts of another biologist, Rodrigo Gámez.

Gámez was a plant pathologist who, over the course of two decades of biological research and conservation activity, had quietly become one of the leaders in Costa Rican conservation. He was thin and bald, and behind his friendly face was a look of wisdom, poise, and shrewdness. Among other things, he had directed the Center for Cellular and Molecular Biology at the University of Costa Rica, run the university's School of Plant Sciences in the College of Agriculture, won an Organization of American States award for his research on plant viruses, helped create a national research council of Costa Rica, and served on the boards of the Organization for Tropical Studies, the National Parks Foundation, and the research council.

Somehow Gámez always found himself organizing new institutions, bringing people together to work on new problems. His credibility as a scientist and administrator served him well in this regard. He also had excellent contacts within the Costa Rican political-economic power structure, many of whose members were family friends from his hometown, Heredia. They included Arias, Umaña, and Eduardo Lizano, the highly influential president of the Central Bank during that period and beyond. Arias appointed Gámez his personal adviser on natural resources. Gámez, who had introduced Janzen and Hallwachs to Arias when he was a presidential candidate, was put in charge of a major reorganization of the Park Service and natural resources administration under Arias.

Gámez had direct access to Arias, Lizano, and other key decision-makers at crucial times during the Guanacaste project. His credibility derived from dignified, effective work for more than two decades in academia and conservation. And his diplomatic acumen helped him play the role of conciliator, assuaging the angry reactions of many to Janzen's hard-driving direct approach.

I once asked Gámez to describe his and Janzen's personalities and what made them work in this situation. He paused, thought, and answered with typical diplomatic prudence. "That's difficult. Dan is a genius. He is a very bright person. Often geniuses and bright people are perhaps difficult people to get along with. I don't think I am as bright as Dan is, and although I have made some interesting contributions to science, they are humble compared to the contributions that Dan has made. But I seem to have more skills in getting people to work together.

"As a scientist, a thinker, you can be a leader. But in the conception of the big idea, the future, the initiatives—how do you translate that idea into a reality, particularly as a foreigner?"

Gámez emphasized that he and Janzen "shared the dream." He called his dream "more social orientation" and Janzen's vision "a happy combination."

"When you understand his ideas and his intentions, you realize that it's not a matter of what you do but how you do it," he said. Costa Ricans saw Janzen as "a very important intellectual contributor" to Costa Rican conservation and biology, along with people such as Edward Wilson of Harvard and Thomas Lovejoy, now of the Smithsonian Institution.

Gámez noted that the philosophy of expanding national parks had long been supported in Costa Rican conservation. Then Janzen came along recommending the purchase and restoration of *degraded* land. The conservationists immediately liked that concept, too. "The idea of restoration was very attractive because even if it's a myth—and humans live by myths—it would be historic," Pedro León told me. "Just think of it. It's a very impressive notion. There was no resistance. It was just an idea waiting for its time."

It remained to be seen, however, whether the power of myth, the support of the Tico conservation community, and the diplomatic skills of Rodrigo Gámez would be enough to alleviate the friction produced by the fiery genius and make the dream a reality.

Among the many opportune developments in the Guanacaste project was the election of Oscar Arias as president of Costa Rica. Arias was inaugurated in May of 1986, and his administration began to take several steps in support of conservation. Many of those steps directly benefited the Guanacaste project, perhaps most important the appointment of Alvaro Umaña as Minister of Natural Resources. Within days of taking office, Umaña gave Arias a plan for restructuring the national government's conservation efforts. Within months, the Park Service, Forest Service, and Wildlife Service were transferred to the newly created Ministry of Natural Resources, Energy, and Mines. Commonly known as MIRENEM, the agency's leader now held a cabinet-level position, a move that signaled the priority Arias gave to integrating national efforts in conservation.

The same basic structure was maintained over the next decade as the ministry added new responsibilities and changed names. Not only did this structure force parks people and forestry people to work together—two cadres with a history of antipathy—but it also offered a playing field for the continuing struggle between traditionalists of the centralized Park Service and progressives who wanted power decentralized. Until he left office as chief of MIRENEM in 1990, Umaña, advised officially by Gámez and unofficially by Janzen, nurtured several progressive changes. He took on the Guanacaste restoration as his agency's "pilot project," as he phrased it.

Another key development came when Arias, on July 3, 1987, announced that he planned to go to Santa Rosa on July 25 and issue three orders that would bring the Guanacaste project a giant step closer to reality. The announcement came during a speech similar to the State of the Union message in the United States.

First, Arias said he planned to sign a proclamation annexing Santa Elena to Santa Rosa. That would effectively make it part of the national park system. Through annexation, the Park Service could exert a limited amount of regulatory control over the land, although in the end the 1978 expropriation would have to be finished by paying the owners, and tough negotiations with Hamilton's lawyers could be expected. However, this move would, theoretically, complete an entire, coherent portion of national park on the west side of the Pan-Am Highway. It would make a dramatic statement about the status of national park-building and ecological restoration. And it would open the last piece of Costa Rica that had not been thoroughly explored by biologists. That was mainly because Santa Elena was a kind of northern Alaska, out of reach. The prospects intrigued Janzen, for it was the oldest land in Central America and perched on serpentine, an unusual kind of rock. It also was extraordinarily dry. In his own cursory treks, he had noticed that the composition of plant and animal species changed abruptly when crossing into Santa Elena from Santa Rosa.

Second, Arias planned to declare the core area east of the Pan-Am Highway to have "Zona Protectora" status, a legal condition that forbid environmentally destructive activities. That meant, Janzen explained, "it's no longer the frontier that it was, and it's not yet a national park. It's somewhere in between." The land would still belong to the owners of the five large ranches, but its full status as privately owned land would be gone. Now it would be privately owned land on which deforestation, fires, hunting, and fishing were banned. Farms already on the land could be worked, but that was about it. Ideally, the next two steps would be to buy the land and then declare it national park. In essence, Zona Protectora status made the core unattractive to commercial buyers because it could not be developed further, strengthened the project leaders' ability to protect it, and gave them breathing room to find the money to buy it.

The third move by Arias was to announce the use of a revolutionary financial transaction to help the Guanacaste project raise money. The transaction, called a debt-for-nature swap, had been conceived by the Smithsonian's Lovejoy and was considered a breakthrough when first used in Latin America in July of 1987. The debt-for-nature swap suddenly became a unique way of financing conservation projects in developing countries with large external debt. For the Guanacaste project, it meant, put simply, that each dollar donated to the project would go to the Central Bank to pay off Costa Rican national debt held in U.S. banks, purchased at a big discount. In exchange for that debt "service," the Central Bank would issue bonds that multiplied the value of the donation several times, depending on the interest rate, for the next several years, when the bonds matured.

For Janzen, the Arias visit and declarations represented "an enormous relief." The fund-raising struggle was given a boost, and all the land needed for GNP would be "frozen in place." Perhaps most important, the Guanacaste project would receive, in a very public way, a tremendous political endorsement. "The more political support for the project, the faster it's likely to occur," Janzen said.

The same day that Roberto Espinoza reached into the trogon nest from atop the aluminum ladder, Janzen had called Alvaro Ugalde in San José, and the two men had discussed "*el temblor*," the Spanish word for earthquake. Ugalde asked whether Janzen felt the earthquakes that had hit the Central Valley that morning. The quakes had registered 4.2, 4.4, and 4.2 on the Richter scale, and they brought to eleven the number of earthquakes in Costa Rica since May. No, the quakes hadn't been felt in northern Guanacaste Province, and Janzen joked that every time he called Ugalde, San José experienced an earthquake. "That's just the impact of GNP on Costa Rica," Janzen said. The Park Service was certainly feeling a kind of GNP-initiated earthquake, and, in a sense, on July 3, Arias had launched some quakes of his own.

The Arias moves in Guanacaste didn't quite come out of the blue. Ugalde, Umaña, Gámez, Boza, León, Janzen, and others had been pushing for them at least since the inauguration, and, in the case of Ugalde and Santa Elena, since the 1978 expropriation decree. Still, there was a new air of excitement and anticipation at Santa Rosa. Since the Arias news had rumbled upcountry, Janzen had tried to stay near the house as much as possible to deal with the constantly ringing telephone. He spent hours and hours on the line with Ugalde and the others, giving and getting advice on how to handle the upcoming visit, then turning to discuss it with Hallwachs.

It was about his time that I began to realize one of the best-kept secrets of the Guanacaste project: the seminal role of Winnie Hallwachs. It took me years to fully understand it, and even then only after Janzen detailed it for me. In July of 1987, it was easy to see that she was a woman of brilliance, gentility, resilience, personableness, and sophisticated humor. However, I, like many other visitors to Santa Rosa, mistakenly assumed that she simply served as hostess, and perhaps sometimes as a fencing partner for Janzen's thinking. In truth, while Janzen may have been the project spokesman, the leader, the putative intellectual godfather, Hallwachs played a comparable part all along, quietly, right there in front of us all. She contributed to its intellectual development, its implementation strategies and tactics, and its conservation spin-offs.

The reasons for Hallwachs's ostensibly hidden role include her deliberate choice to keep a low profile and the fact that the world rarely gives credit for the kinds of contributions she made. The major reason, however, was that we were all carried away by Janzen's energy, dynamism, forceful attitude, and superb communications skills. As Hallwachs herself explained in a 1998 interview, "Dan is the guy who gets things done. He's the guy who's bold about his opinions. He makes decisions quickly and he goes with them. He writes and he speaks a lot more easily than I do, and all of those things also mean that he is in most cases the leader, and in virtually all cases the voice."

Janzen's style was, indeed, to move fast, and he had a way of listening to the ideas of others, incorporating them, perhaps forgetting where they had come from, and a while later talking about them as if they were his own, without giving credit to others. At least that was one interpretation. This wasn't because Janzen was mean or selfish, however. Rather, it derived chiefly from a spokesman's predicament of trying to communicate ideas rapidly, effectively, and simply, but not having time to give attribution. There was, too, Janzen's single-mindedness of purpose; to him, what mattered was getting the idea across and getting the job done. He was, as it were, going for the goal. Who footnotes their spoken words, anyway? Yet if one were to go back and give credit where due, as he did in later years of the Guanacaste restoration, much of it would fall to Hallwachs. "She was more than just 'the supporting woman,'" one close friend of theirs told me.

Hallwachs had come to Santa Rosa in 1978 as a young, beginning graduate student at Cornell University and soon started her dissertation research, which focused on the biology and ecology of agoutis. Long interested in rodents, she had chosen the agouti because it moved about in the daytime, which meant she could more easily observe its behavior. Hers was a collateral project to one begun by Janzen to examine another small mammal in the same forest—the spiny pocket mouse, a common, night-foraging mouse. The researchers also were trying to understand the role the mouse and agouti played in preying on and dispersing seeds of various kinds, a role that, as it turns out, has a major impact on the diversity of trees in the forest. Understanding this role was critical to understanding how forests regenerate.

Hallwachs's approach differed from the common one for charting animal behavior: catch, mark, and release the animals, noting who was where and when. It had been important to her to take a different tack. She sought the kind of intimacy acquired by spending long hours in the forest, watching events over and over, and carefully noting the subtle shifts of behavior as equally subtle changes occurred in the forest, brought on, for instance, by the progression of the seasons, the fall of a branch, or the death of a single animal. She spent more than five years doing this with agoutis before the Guanacaste project intruded.

The research fascinated her, and the forest revealed to her key aspects of agouti behavior. One of the most practical insights, in terms of the restoration project, was that agoutis can carry seeds of trees like the guapinol for a hundred yards or more from the source. Hallwachs described the intellectual process leading to such discoveries as a kind of patient watching and pondering, during which she gradually woke up to what the animals were doing in each instance. It was a slow revelation of causes and effects. For example, by placing small pieces of wood across an agouti path, and watching how the animals changed their route accordingly, she realized that a kind of eddy had formed on the forest floor where seeds the animals would otherwise detect and eat were now protected. "I came up with this anomalous result, and I went out and looked and looked and looked and looked at that place, and finally it clicked: for a small animal like this, a piece of wood that's lying along the ground that looks completely insignificant to a large animal like us can actually channel their walking behavior. They don't have the behavior of jumping over these things generally, and when they go around it, they leave a little piece of the forest that is not heavily searched, when the areas outside of it are combed over and over and over and over again by them and others. This same little tiny arrangement of pieces of wood kept the agoutis and peccaries out of there." The capacity for such discernment, such intuition, also was the foundation for her contributions to the Guanacaste project, which inevitably pulled her away from the agouti research as the project escalated.

There were many dimensions to Hallwachs's contributions to the conservation project. Foremost was her intellectual role, played out largely as she and Janzen batted around ideas. "We do these things together," Janzen said in a 1996 interview when I asked him to describe her contributions. By then he was making those contributions known emphatically and regularly. "We are two very different people looking at the same event, and our reactions to the event are exceedingly synergistic," Janzen explained. "I rush to conclusions. I see a given picture quickly. I'm very vocal—in some people's vocabulary, 'extroverted.' Winnie is the opposite of each of those. She studies a situation much more carefully before she comes to a conclusion. She's more introverted. She pays much more attention to what's background, behind what's happened, instead of reacting to the event, per se. While I tend to swing the pendulum far in one direction, she swings it in the other direction, and because we are very much together in perceiving the same thing, and we communicate very, very well, her impressions and my impressions become blurred into one, and then out comes an actual exteriorization of this. Since I'm the vocal member, it's then attributed to me. But I would say these ideas and directions and thoughts and actions are easily fifty-fifty attributable."

In the early years, until several Costa Rican conservationists began to serve official positions in the newly formed Guanacaste National Park and later the Guanacaste Conservation Area, the project was basically a mom-and-pop operation, with Janzen and Hallwachs working virtually around the clock, whether in Santa Rosa or Philadelphia. Their team effort developed the project's ideas, slogans, photographs, and documents. However, like players on a great athletic team, they couldn't always understand, much less articulate, what each did to help the other achieve a thought or a task.

Hallwachs also served what might be called a diplomatic-domestic role. This went far beyond the emotional support she gave Janzen amid the destabilizing flow of ups and downs brought by the project. It meant, in many instances, playing the role of "the woman of the house" that many Costa Rican men assumed was her natural role, in the paternalistic, macho way of Tico culture, particularly the culture of rural Guanacaste. She had successfully rejected this role previously, but she made the sacrifice on behalf of conservation.

"When the project got started, it sunk in on me, watching the way I'd watched agoutis, that something that would be very motivating to this small group of people [almost exclusively men] was to provide some of the things that women traditionally do, like cold lemonade," she once told me. "Sometimes it got to be food, and a house that was not an absolute despair to come back into. So I then began to look around and realized that Guanacaste housewives are exceedingly competitive amongst each other about what they put their time into—cleanliness, food, things of that sort. I felt that part of what the Costa Ricans wanted from me was this kind of female contribution to the best of my ability, and so I also figured that's part of the price."

This sacrifice put her into some unanticipated and frustrating binds, such as when journalists came to Santa Rosa on assignment to do the "Guanacaste story." They assumed that Hallwachs was playing the simple domestic role, and portrayed her as such in their articles. In meetings with top Costa Rican bureaucrats in San José, she was put in an equally awkward bind when the time came to serve coffee and food, a job the women in the room were expected to rise and do. On Janzen's advice, she would stay seated at the table, where she had been quietly listening, observing, and analyzing the content, characters, and flow of the meeting. Years later she discovered that many of the people in those meetings never understood who she was or why she was there. Yet there were times when she served as an ally to Janzen during the debates. Only when she saw that he was losing an argument or missing an angle would she walk on stage.

Mainly what Hallwachs did in such meetings was play part of another key role, that of a conduit of information. In this regard, she was both a gatherer and a disseminator of information, a function so essential to the project's success that it is difficult to overemphasize. She gathered information at meetings and when working in the kitchen of the Santa Rosa house as Janzen conversed with others. One advantage was that she could keep track of key objectives and see priorities emerge from the discussions—trends that sometimes the discussants might not have grasped. "If I see things happening that look as if those very, very high priorities are going to fail, then I move in," she said. That was her style—stay back unless absolutely necessary. It's a style derived from her quiet personality, a slight distrust in her voice and ability to persuade, and a realization that a sense of teamwork develops best when *other* people feel the rewards of contributing.

Being up-to-date, even if quietly so, helped her provide important information or make quick decisions when Janzen and other leaders were out on errands, as they often were. She also was a kind of message-runner for the project, serving as a receiver at receptions, meetings, and other functions, official or social. "It slowly began to dawn on me that when various people took me aside and talked to me about this and that, the bulk of them were expecting me to take mental notes and relay this to Dan afterwards so he would get these messages," she told me. Likewise, at such functions, while Janzen spoke with upper-level Costa Rican officials, Hallwachs often would talk with lower-level people, advising them, for instance, on what the two had heard on recent visits with donors or agencies in the United States. She was fully bilingual and had tremendous empathy, which made her eminently approachable. The Ticos even sought her out to register complaints about the way the project was going, or about Janzen himself.

Hallwachs, inevitably, sacrificed part of her young scientific career. At first she tried to work through the distractions of the conservation project, to complete her dissertation research and writing, but the big blocks of uninterrupted time necessary to concentrate on the work disappeared. At the same time she came to realize, like Janzen, that saving the forest for others to work in was more important than completing individual research projects. By the summer of 1986, both Hallwachs and Janzen had begun to live a peripatetic life, having been forced from their normal role as scientists working on their research sixteen hours a day and according to the needs they saw fit, to that of conservationists working on details of a conservation project which had started to call their schedules. Finally, Hallwachs was forced to put her work on hold and devote all her time and energy to the project. It wasn't until 1994 that she returned to the dissertation and completed her Ph.D.

As the time for Arias to visit Santa Rosa approached, Ugalde, Umaña, Janzen, Hallwachs, and others worried about what to say to journalists who would most certainly want to ask about the political implications of annexing Santa Elena. They discussed the urgent need for Janzen to raise sufficient money by July 25 for a downpayment on Santa Elena swiftly following its decree as having national park status. And they knew not to take anything for granted. Although all the project's land would now be protected, only 43 percent was actually owned by the conservationists.

What if Arias, for political or other reasons, changed his plans and decided not to come to Santa Rosa? Stranger things had happened in Latin America. The three weeks between July 3 and 25 was plenty of time for the intricacies of Costa Rican and international politics to foul the gears. Would the politically powerful landowners to the east protest the Zona Protectora status? Would the Reagan administration bully Arias out of the Santa Elena move? Would the debt-purchase scheme suddenly lose favor among any of the many parties required to bring it to bear?

Janzen told me he wasn't as nervous about that as were his Costa Rican partners. These veterans of the conservation wars had been burned by whimsical politicians before. Not that Arias was necessarily whimsical. There were just so many ways things could fall apart.

Amid the flurry, the trogon work proceeded.

Shortly after lunch, Frank Joyce pedaled up to the house on a timeworn Schwinn Typhoon. He carried a small foam cooler, out of which he pulled plastic bags full of juicy green caterpillars taken from the mouths of trogon babes. He and Janzen sat across from each other at the picnic table and identified the species of each. Joyce weighed them, and as he called out the numbers, Janzen wrote them into a notebook and repeated them. They spoke clearly, calmly, methodically, as if nothing else was happening. If you closed your eyes and just listened, it would have been easy to imagine them sitting in a lab back in the States, except for the relentless wash of wind waves over the forest canopy and the occasional call of a bird.

After half an hour, Joyce was back on the bike, heading to the trogon study area with the caterpillars and a battered jug of iced tea for, as he put it, "the comrades in the field."

# · *10* ·

# Little Dances

As it wandered in the dry soil, the tiny ant stepped over the edge of a small, cone-shaped pit and slid down to the bottom. It struggled to climb out but lost its footing in the powdery soil on the slope of the pit. Suddenly, another creature rose out of the bottom of the pit, seized the ant in a pair of gargantuan, sickle-shaped jaws, and, with a few jerks that kicked up a spray of powder, pulled it beneath the soil. Another meal for the ant lion.

Ant lion pits are a common sight in the loose, dry soil under the eaves of buildings at Santa Rosa. The pits range from about the size of a bottle cap down to a small hole of the kind made by jabbing a pencil point into the ground. They are almost perfectly symmetrical funnels, admirable feats of engineering precision. Ant lions, which belong to the genus *Myrmeleon*, are the larval form of an insect that looks like a small dragonfly as an adult. The young, ant-sized predators also are known as doodlebugs, a name earned for the tracks left in the soil as they erratically scoot backward in search of a place to make a pit.

An ant lion sits at the bottom of its death trap, its jaws open, waiting. Meanwhile, it flips loose soil on the slopes every few seconds, greasing the skids. In his 1930 book, *Demons of the Dust*, Harvard entomologist William Morton Wheeler wrote that insects sliding down those skids are "instantly seized, buried, paralyzed and sucked out; but if it is at first awkwardly or inconveniently grasped, it is either repeatedly tossed into the air or against the walls of the pit till the proper hold has been secured, or if it escapes is showered with sand till it is again brought within reach." Other naturalists have described the ant lion as an "insidious hunter" whose prey is "carried away by the unstable sand" and "overwhelm'd by the impetuous storm" of the soil it flings.

Ant lions repeatedly played out this shocking drama in front of Janzen's house, where I sat at the picnic table and watched the Guanacaste project unfold during July 1987. This was a pivotal time in the project's history, and as I observed Janzen toil seemingly ceaselessly on myriad details, I often wondered how closely this Lilliputian drama at my feet corresponded to the one being played out more broadly across the soils of the province, among humans.

The picnic table was shaded and protected from rain by an overhang of the house's corrugated roof, but dry bursts of wind constantly played havoc with the papers, pens, notebooks, and hats on the table. You could hear the gusts approaching several hundred yards away, sometimes rustling the trees so forcefully that the sound built into a rumbling like heavy surf. Two student volunteers who worked at the table frequently adjusted makeshift paperweights while they processed recently captured moths, withdrawing them from glass jars and opening and pinning their wings on spreading boards. At one end of the table near them was a wooden box with a glass front panel containing a small pit viper. The snake had been brought to Janzen by a park guard who found it in his house, and Janzen was keeping it for officials with the Costa Rican snake institute in San José who would soon retrieve it to make antivenin for snake-bite emergencies.

Janzen was involved in a wide array of what I came to think of as "little dances." It was his term, and when he used it he referred specifically to tactical games that neighboring landowners played with him and other negotiators for the project. These dances were often frustrating escapades in which the owners tried to complicate the bargaining and trap them in a raw deal. For example, a landowner might use different maps to trick them into thinking they were buying the whole property when actually they were getting the rights to only part of it. In another little dance, a seller might dodge his contractual obligations to clean up his land by getting a local judge to issue an order denying him access to the land until he paid off a loan to the previous owner. Then he wouldn't pay off the loan, but he would pocket the Guanacaste project down payment.

For me, "little dances" came to symbolize a broader interplay: among the project and landowners; among Janzen and donors, politicians, and others; and among organisms and the landscape. They were all performances on the Guanacaste project stage.

For long periods, Janzen would stay by the phone, which seemed to ring constantly. Umaña called from San José to relay news about the Santa Elena airstrip, and Janzen expressed his fear that Joseph Hamilton might begin to make improvements that could strengthen his legal case for holding the property. He got a call from Sergio Bonio, jaguar hunter and newly appointed by Hamilton to manage Santa Elena. Bonio wanted to drop by to discuss a management plan. Janzen immediately called Alvaro Ugalde to report this call. "The situation is changing day by day," Janzen told me.

Whenever he got the chance, Janzen dialed up prospective donors in the States to tell them about the upcoming Arias visit. The Guanacaste project was still billed as needing $11.8 million, and Janzen had so far made tremendous steps toward that goal with major U.S. philanthropic foundations: MacArthur, Pew, Jones, Noyes. The Swedish Society for the Conservation of Nature had weighed in. But there were disappointments, too. The Rockefeller Foundation had denied a grant request, and on June 15, the Kresge Foundation, of Troy, Michigan, turned down a request for $4 million to purchase the core.

The Kresge episode illustrated Janzen's creativity—and some would say the lengths to which he would go—in making grant proposals. Admittedly, the proposal had been, as he put it in his cover letter to the foundation, "somewhat unconventional." The traditional objects of the foundation's largesse were mostly buildings—physical structures. "Here I seek funding for a 1,721,600,000 ft2 biology building at a cost of $0.0023 per ft2," he wrote in the opening lines of the proposal. "This living classroom is in northwestern Costa Rica." He went on to describe the mission of tropical forest restoration and biodiversity preservation, embedding in the text images like "a living classroom," "living museums," and "the forest that once stood."

The Kresge grant would have enabled the Guanacaste project leaders to play what Janzen called the "matching grants game," going around to other organizations and waving the $4 million as an incentive for matching participation in a success story. It also would have allowed Janzen to shift much of his concentration from fund-raising to other aspects of the project. "If Kresge comes through, we're golden," he told me a few months before.

Foundation officials told Janzen that the rejected proposal was too unusual, and that GNP wasn't, in fact, a building. At first, Janzen was angry. But he recovered quickly, and he redoubled his efforts to find money for the core. Still, the mood among the GNP supporters at Santa Rosa contained, at least, a new element of caution, if not desperation.

A year earlier, honoring a Janzen request, the owners of property in the core had agreed to delay selling until the project came up with the money to buy it. But when the Kresge grant and other funds failed to materialize in the summer of 1987, Janzen had to plead with them to wait longer. Some of the owners responded well; others didn't. The implications were uncertain. Would some of them become tired of waiting and sell to someone else who had no interest in conservation? In the game he was playing for the core, Janzen couldn't tell what would happen. He just knew that if he didn't get that land, the dream of GNP as a coherent whole would go unfulfilled.

Along came the Arias pledge to place the core under Zona Protectora status. This made it unlikely that the owners would sell to anyone but the Guanacaste project. However, they would still have a right to demand a fair market price. That meant two things: Janzen still had to find the money, and a lot of hard bargaining was ahead. As he put it, the price of this land would have a lot to do with "the personalities of the people bargaining for it."

One morning, a short, thin Costa Rican in his twenties puttered up the road on a small motorcycle and stopped in front of the house. Janzen, who had been leaning over one of the volunteers giving instructions, walked out and shook the man's hand, and they stood under the guacimo tree and talked for a few minutes, in Spanish. Janzen treated the man cordially and gave him a friendly handshake as they parted. However, as the scientist walked back to the picnic table, his brow was furrowed. I asked him to explain the meeting.

"This is good for the park, but it's also a reflection of how bad a life these people are leading," Janzen said, the anger welling in his voice as the motorcycle pulled away. The visitor was one of many people, rich and poor, who owned land in and around the area proposed for GNP and who were now contacting Janzen, by telephone and in person. Although Janzen was part of a team of land negotiators, he had long been the most active at getting out and pressing the flesh.

The young man was poor, struggling to make a living for his family on a small piece of land. He wanted to sell it to the Guanacaste project. The problem was, his farm was outside the proposed boundary. As Janzen explained it, the young man had been victimized by a quick-talking salesmen who sold him, and others like him, a small piece of Guanacaste Province's poorer farmland. Janzen called to mind scenes in the movie, *The Tin Men*, just out that year, in which Richard Dreyfuss and Danny DeVito play the roles of fast-buck artists who unscrupulously sell aluminum siding to people who don't need it.

"It's a very funny movie, but it's a very sad movie," Janzen said. "The American audience won't think of it in this way, but the people whom they sell the aluminum siding to have exactly the mindset of the people who own these little ranches here, who get bought and sold and bought and sold by one scam after another. And in a certain sense, they end up starving. This guy's got ten hectares or thirty hectares [twenty-five or seventy-five acres]. He bought this piece of land for himself and it was a stupid move. He never should have bought it in the first place. They all have their dreams and hopes, and they all think they are on the frontier somewhere and are going to make it big—just like the gold miners in Corcovado."

The young man and his fellow small-time risk-takers in Guanacaste were about to take it on the chin, just like the miners had. They had been victimized by the salesman, as well as Costa Rica's economic crisis of the 1980s, which was shattering the dreams of small farmers and large landowners alike. Loan opportunities had dwindled, landowners had gone into debt, land went unused, and its price plummeted. Property owners of all kinds suddenly became interested in selling their land, some desperately so. Bad for the owners, good for the Guanacaste project.

Dan Janzen was no tin man, and he was no ant lion, either. He seemed genuinely bitter about the predicament of the young Tico, and he went back to coaching the student volunteer.

A short time later I asked about the status of the land acquisitions, and he brought out a map. He moved his finger across dark areas representing the land in hand, like Santa Rosa and Cerro El Hacha, and undarkened areas representing pieces yet to be acquired, like Santa Elena and the five core properties: "That piece we got for free and we really should have had to pay for that. This really is a fluke that we got that. This one, there was never any doubt that we were going to get somewhere down the road, but then we're going to have to pay for it. We got this one for free, see. But I still have to come up with the money for this one."

He was now pointing on the side of the map to the east of the Pan-Am Highway, talking about the prospects for land acquisition after the Zona Protectora decree and more fund-raising success. "What it'll mean is this will all get painted black, which means now I'm talking about closing down this piece here, which is all that's left. And in this game of raising money, the further you are along towards your goal, the faster the money comes in. Nobody wants to be the first, and everybody wants to be last. When we've only got one square here, that'll go like that." And he snapped his fingers.

Later that day, an older Tico drove up in a white Land Cruiser. This was Jorgé Baltodano, the owner of Hacienda Tempisquito, one of the five big pieces of the core, each worth at least half a million dollars in Janzen's estimation. Tempisquito had been in the Baltodano family since 1935, and Baltodano was one of the core owners who had agreed not to develop the property while Janzen came up with the money to buy it.

Janzen took a minute to finish another phone call from Ugalde, then walked over to Baltodano, shook hands, and led him back to the picnic table. The two men hadn't seen each other in roughly four months, and a lot had changed. Each pulled out a map of Guanacaste, and Janzen traced on his a dark line that had been drawn on Baltodano's map. This was supposed to be the boundary of his hacienda. Janzen apologized about the Kresge grant falling through, and he explained that he had made a proposal to the Swedish government for a similarly large grant and that a decision would be made by August or September. If that one failed, Janzen pledged to keep trying until he got the money. He brought Baltodano up to date on the Arias visit and other events involving GNP, and he assured the Costa Rican that he would not get left behind or cheated in the fast-moving current of events.

After Baltodano left, Janzen described him to me as a "typical gentleman landowner." Baltodano had been "very, very patient," not just during this meeting but during the entire time since Janzen had first approached him about holding his land for GNP. He and the other four big property owners were all interested in selling to the project, but they didn't all behave with Baltodano's understanding.

"They all want to sell very badly," he explained. "Some of them are bastards and pressure me by doing things like setting fires on their land, and they just make themselves obnoxious, trying to get me to buy them off in a hurry. Others are sitting there calmly, waiting. But I still feel very embarrassed about this because I've been telling them for a year and a half, 'Hold off. Hold off. Don't sell your land to anybody else.' We have a verbal understanding with three of them that they really wouldn't sell it to somebody else. The other two I'm more suspicious about."

Janzen had assured Baltodano that the Zona Protectora declaration would not affect him since he was doing nothing illegal on his land, and that he would be able to continue farming and ranching. "You see," Janzen explained, "Baltodano is a straightforward, honest person, and in Costa Rican politics, declaring a Zona Protectora also is a first step to expropriation. I just wanted to assure him that no, this is not the first step to expropriation. It's just like it was before, and we're doing this as a way of dealing with other, obnoxious people."

He continued: "When people talk about a representative of Guanacaste, they grant you there's two species. One species does not live here but just owns property as an investment. Baltodano is the other kind. He's long lived here. He grew up here and will die here. He will never be a very rich man, but also is not poor, either. He owns a sawmill in Liberia and is a very prominent member of the local community."

So who were the obnoxious landowners? Janzen didn't want to talk about it. I later learned that some had deliberately set fires on GNP land near their own property; some had even sent their workers to steal the park's horses. In one episode, after tense negotiations, project leaders bought about 2500 acres from a particularly ornery owner, only to discover that because of an error in the documentation, the man still claimed part of it. He built a fence around that part, and the project leaders sued, claiming they had paid for it and been tricked.

Another afternoon, Janzen received a visit from José Joaquin Diaz Duarte. The two men stood by the front of Duarte's truck and talked for nearly an hour, barely hiding the fact that they didn't like each other. There was nothing forceful or threatening in the tones of their voices, but nothing friendly either.

Months before, when Janzen and his fellow negotiators approached him about buying his land on the slope of Volcán Cacao, Duarte had said he was going to sell it to a rich American, but he would do the conservationists a favor by selling it to them instead. The Guanacaste negotiators considered the asking price outrageously high, so they had laughed at him. Duarte responded by closing a road on his property, the only road leading up to the cloud forest cap where the project's new Cacao Biological Station stood. He did it at least partly because he feared that allowing anyone to drive on his road would constitute a form of possession, and since possession was 99 percent of the law, at least in the minds of many Costa Ricans, allowing vehicles on it would put him at a bargaining disadvantage.

Now, Duarte demanded that the project buy the property quickly, or he would sell it to a Canadian mining company that wanted its sulfur deposits. What Duarte didn't understand was that the imminent Zona Protectora declaration would prohibit the mining company, or anyone else, from developing the land.

Janzen reacted like a professional poker player, which is to say not at all. Even though he doubted that the mining company would pay Duarte's inflated asking price anyway, he held back from laying out the implications of the Arias declaration and how it already had severely weakened Duarte's position. "Six months ago," Janzen told me, "I would have been honest with him and said, 'You know, that's going to cause problems, because the mining company can't mine, because it's a Zona Protectora.' If the mining people do want to buy it from him, then I don't want him to go screaming to them. That just creates more political pressure between now and the twenty-fifth."

The two men did not shake hands. Duarte climbed back into the truck and drove away.

The cast of characters in the Guanacaste land game was like anywhere else, Janzen told me. "I don't get to choose whom to deal with," he said. "We're just dropping a cookie cutter down and it takes a sample out of society. So you get dregs and you get good guys. When you start a park, you just draw a big, arbitrary circle around it, and you say, 'I've got to talk to everybody who's in that circle.'"

Word spread that there had been a small *arribada* of sea turtles the night before at Playa Nancite, a beach just to the northwest of Playa Naranjo and quite a bit smaller and more secluded. An *arribada* is a synchronized mass nesting of sea turtles, in this case the olive ridley, also known as the Pacific ridley sea turtle. This endangered turtle nests in only a few sites in the tropics, and Nancite was one of only two such sites in Costa Rica, the other being at Playa Ostional, about sixty miles to the south. During an *arribada*, up to 100,000 female turtles emerge from the surf in the dark, dig holes on the beach, lay eggs, and scuttle back into the ocean. Such large *arribadas* usually occur between July and December at Nancite, for periods of one or more days at various intervals. Smaller *arribadas*, consisting of 500 to 10,000 turtles, can occur at any time of year. I had been to Nancite in 1986 with George Godfrey without seeing any turtles, but there was a chance for a wave of them this night, and I asked Janzen for help to get there.

He arranged for me to ride with José Antonio Salazar, the Santa Rosa director, who was driving down that evening to drop off supplies for a park guard posted at Naranjo and to bring a group of five guards to Nancite. We left at five o'clock, and Salazar drove his Land Cruiser swiftly over the savanna and hard over the bumpy, jolting road down the *Cañón del Tigre*. He and his men kept up a steady stream of chatter and laughter. It was a tight fit in the back, and I repeatedly slammed one shoulder into the rear door and the other into the Tico sitting next to me. Despite holding a hand aloft, I also hit my head on the roof several times. This rough ride probably was a macho thing for Salazar, and I imagined him to be a Tico Teddy Roosevelt. Better than walking to the beach in the noonday sun, I reminded myself.

When we reached the road leading to the Naranjo guard station, Salazar stopped the Land Cruiser and ordered two guards to unload a duffel bag and a cardboard box filled with supplies. The Naranjo guard would walk out to this point the next day and pick up the supplies. The guards got back in and shut the door, and Salazar gunned the engine. Then he hesitated, let up on the gas, got out, walked back to the supplies on the ground, pulled out some of the groceries, and placed a rock the size of a basketball inside. It must have weighed twenty or thirty pounds. He then stuffed the groceries back in and zipped up the bag. He climbed into the jeep amid chuckles, and off we went.

The road ended at the foot of a rocky ridge. We left the Land Cruiser and started to climb the rocks shortly after six o'clock, in quickly fading light. We caught up with Chris Dick, a student volunteer from Hampshire College, in Amherst, Massachusetts, who had hiked down from the administration area that afternoon. It was dark by the time we reached the top, and we switched on our headlamps. For almost another hour, we walked along the ridge and wound our way down a trail single-file into a dense, lowland dry forest, blanketed with pulses of insect chirping. One of the guards, the second in the column and two men ahead of me, let out a yelp and stopped. Coiled in the path before him was a snake he thought was a fer-de-lance, the most feared and dangerous snake in Mesoamerica (Janzen believes there are no fer-de-lances in the Santa Rosa dry forest). The guard who led the column had unknowingly stepped over it, and as the others gathered at a safe distance around the snake, they jokingly chided him for failing to notice it. I couldn't blame him, though, for even with several lights focused on it, the dark brown snake blended almost perfectly into the moist dirt. There were two lessons here. One, no matter how hard you watch for snakes, it's easy to miss them. Two, don't count on the point man to see them, either.

When we arrived at Nancite, Dick and I dropped our packs in a bunk room and headed for the beach. We came out of the forest and onto the sand, a warm but fresh ocean breeze pushing into our faces. Several stars shined in the cloudless sky, yet the beach was pitch dark. I could hear the surf but couldn't see a thing on the sand, and I tried to visualize the stretch of beach running to the right and left in front of me. A brief use of our headlamps revealed the oval hulks of a few turtles, and it was easy to see that this night would not bring a large *arribada*.

Female olive ridleys emerge from the surf burdened with their contribution to the next generation, and, full of determination, crawl up the sandy slope, their noses skimming just below the top layer of sand apparently in search of some olfactory cue. When they find a suitably dry spot, they stop and excavate a nest with their flippers. Into this hole, roughly a foot and a half deep, they drop about a hundred glistening white eggs. They scrape sand over the nest and crawl back into the cover of the sea. The digging, laying, and filling takes about an hour.

Close to a third of the eggs laid in an *arribada* are lost when inadvertently unearthed by other turtles in that wave or subsequent ones. Others are lost when natural predators such as coatis, coyotes, jaguars, ocelots, raccoons, and vultures mine the eggs. They're also scooped up by local people for personal use and for a commercial market partly driven by the mistaken notion that the eggs have aphrodisiac powers. In the national park, of course, any kind of human harvesting of the eggs is considered poaching, and despite its remote location, that had long been a problem at Nancite.

The surviving eggs incubate in the sand for fifty days, after which hatchlings emerge, dig their way to the top, and dash for the surf. At this stage, vultures, crabs, frigate birds, and sharks take a toll. Once in natural balance, this system was now on the verge of collapse. The olive ridley's demise, initiated by heavy poaching, had been accelerated by fishermen from Latin American countries who take adult turtles at sea to serve the turtle products market. Many others died in fish and shrimp nets that lacked devices to exclude turtles.

Scientists still don't know why turtles pick a particular night to mass just beyond the surf line and then storm the beach. A pheromone—a chemical signal given off by the turtles in the water—may be involved. Researchers haven't found much evidence to support the local lore that *arribadas* are linked to the last quarter moon, onshore winds, and rising tides.

Dick and I joined a group of strangers and watched a turtle scour out a pit. A long time later, she dropped her eggs into the bottom, one by one or sometimes two or three at a time, and I imagined that she was sending signals, a kind of reptilian Morse code, into the abyss of time and space. I was awed at the timelessness of this moment—it had been preceded by billions of such moments over millions of years. I looked up at the mosaic of stars, their light gathered into irregular clumps, and considered the life of this individual turtle: a journey over thousands of miles of open ocean, a chance encounter with a mate, and an improbable return journey guided by some mysterious navigational intelligence, back to this beach of her birth and the births of countless previous generations. It was a sublime feeling, of wonder at the grandeur of life, of ancestors and possible descendants. It was humbling, too, like looking at the stars and suddenly realizing how vanishingly small our planet is, and how insignificant the humans who inhabit it. Yet, it is us, not the stars nor the turtles, who appear to be on the verge of obliterating for all time such improbable journeys, such *arribadas*, such seemingly eternal patterns of life.

In the morning, before returning to the Santa Rosa administration area, I walked down to the beach again. It was a stunningly beautiful scene, the dawn light slowly revealing the off-white sand of the beach bordered on either end by huge, black cliffs. The turtles were gone, and the beach was littered with thousands of empty white turtle eggshells, as if a freighter carrying pingpong balls had run aground nearby and spilled its cargo. These were remnants of *arribadas* past, as turtles and predators played out their age-old struggle across this important piece of the Guanacaste landscape.

Halfway down the beach stood the remains of a lifeguard stand, apparently constructed long ago, perhaps even before this secluded beach became part of the national park. The weathered gray frame no longer held a seat, and scattered in the mounds of sand around it were pieces of driftwood and scores of eggshells. As naturally as the frame seemed to blend with the landscape of this remote beach, it was still an extraordinary artifact. I chuckled, remembering the bizarre form a similar scene had taken on the year before, on my first hike to Nancite. It was an image you might have expected in a Charles Addams cartoon in the *New Yorker*: four black vultures, perched nonchalantly on the aging lifeguard stand, their backs to the sea.

Janzen bristled at the request, but he did it anyway.

The cameraman and soundman from the BBC Natural History Unit had asked him to walk through the same patch of forest yet again. The shot had to be just so.

The men were part of a team that had visited Janzen and Hallwachs several times at Santa Rosa and in the States over the previous two years. Their work, in cooperation with the U.S. Public Broadcasting System, culminated in the one-hour documentary about Janzen and the Guanacaste project, *Costa Rica: Paradise Reclaimed*, which aired in Britain and the United States in early 1988.

Now, they wanted footage of him walking through the woods, just a couple hundred yards out of the administration area. They had selected the spot for its lighting and foliage; it was a young piece of forest in secondary succession, quite open but with enough foliage to seem like, well, a tropical forest. Janzen testily remarked that he would never have reason to walk through this area, but he followed their directions. One of the men climbed a tree, braced the camera on a branch, and had Janzen walk toward him, once from one direction and twice from another. Each time, the scientist complied with their request, which also meant he had to stop at an acacia tree, pull off an acacia shoot tip, take a bite out of it, and comment that it tasted like lettuce—you just had to be careful it didn't have ants on it.

The sun came out from behind a cloud and the light changed, so the cameraman had Janzen walk twice more from the second direction and once more from the first. The Englishmen needed to paint with light, of course, and as they persuaded Janzen again and again, they were courteous, even patronizing, the way you might expect two people to behave if they really needed something from a feisty person. Each time, Janzen chafed a bit more. He was reacting partly to what he saw as the shifting focus toward his personality, partly to the time it consumed, and partly to the phony show-biz aspects of repeating "scenes" for the camera. He was a scientist, not an actor. It seemed as if he was experiencing an internal struggle as well, one that pitted his need for publicity against the "pure" scientist's traditional aversion to allowing himself, not his work, to become the star. Publicity, after all, was part of fund-raising, and fund-raising would ultimately dictate the success or failure of the Guanacaste project.

As the men had finished filming over the past few weeks, they had, in Janzen's view, become obnoxious with requests for him to climb the same tree half a dozen times, repeat statements that had come off-the-cuff during interviews, and do things for the camera that were completely out of character. He had come to see them as a nuisance, yet he needed them. That bittersweet paradox symbolized the interaction Janzen had with most members of the news media.

Janzen told me he was disappointed by the way journalists treated him and the Guanacaste project. He had assumed that writers and producers who came to Santa Rosa would do a story about the project, but instead most centered their stories on his hard-driving, eccentric qualities. He worried about the impact that would have on his efforts in Costa Rica, where a gringo trying to get cooperation from the political, financial, and scientific elites usually did best by keeping a low profile.

The BBC film was a case in point. It had started as a story about an imaginative restoration ecology effort in the tropics, then slowly shifted its focus onto Janzen. As one scientist at Santa Rosa phrased it, the crew was making the film into "just another crocodile-chasing story about Dan." Janzen's irritation with this shift gradually mounted until one day he blew up at the two men, criticizing them for not concentrating enough on the Guanacaste project.

Other journalists did much the same. Gringo reporters would to fly down to Costa Rica, drive up to Santa Rosa, sit at Janzen's knee, and get the story, in English, of course. It was easy, and Janzen's personality almost automatically drew attention. "There's no question about it," Rodrigo Gámez told me. "Dan is very—you have a word for it—attractive, sexy, captivating."

The problem was that the reporter got a narrow view, one gringo scientist's perspective on the project and its context. He or she went away thinking—and writing—about the scientist's project and his ideas, as if they were Dan Janzen originals. And they often portrayed Janzen as a swashbuckler. One writer described him as "smudged, sooty, sweaty and looking pretty much like an extra in a pirate movie. The only thing missing is the cutlass in his teeth." It was an easy trap to fall into.

The writers and filmmakers didn't care, or at best didn't really understand. They came from a culture that rewarded stories about the Great White Explorer rescuing the natives from disaster, eccentric though he may seem. It was not unlike the travel writing of the eighteenth and nineteenth centuries, when European explorers, writing for the home audience, described white men encountering the uncivilized cultures of Africa, Asia, and Latin America. The twentieth-century stories datelined "Santa Rosa National Park" may have made for interesting reading back home, but they were in many ways as elitist, imperialist, and even racist as the stories of earlier centuries.

The developed world got a warped message, that some kind of wild-eyed American naturalist, a good-hearted Long John Silver of the jungle, was single-handedly saving Costa Rican forests. That message angered some important Costa Ricans involved in the Guanacaste project, especially within the Park Service. These were bright, experienced, and highly educated and motivated people, and some of them strongly resented the image, as Gámez put it, of "the blond guy who comes to the tropics, makes the discovery, and the natives sort of gather around and see what this guy is doing." Blond, of course, referred to skin color. They also resented not getting credit for their vital contributions.

Many Tico conservationists took the long view. One of them, Ugalde, said journalists had "made history different than reality, because they don't have the time to talk to everybody. So, you interview Dan, you write about Dan. That's just the way life is." After all, Janzen was delivering something they badly needed: money. Ugalde said he felt "sad when people make history begin with Dan, but that has nothing to do with the good things he has done. . . . He needed so much publicity to be able to raise $12 million that he ended up becoming the hero, the leader, the idol. And hell, I don't give a damn. I don't care. I wanted the money. I wanted him to be successful."

For the record, Janzen never said it was "his project." Every time I heard a journalist, myself included, identify the Guanacaste project as "his project" or "his idea," he corrected that simplistic notion immediately. In later years, he did it with something that approached vehemence.

Regardless, the news media made him an icon for the Guanacaste project, and in some ways for all of tropical forest conservation. *Outside* magazine may have taken the latter to a biting, though inadvertent, extreme in its April 1990 issue when it presented a "Green Cosmology," a two-page spread featuring a select few world environmental leaders. Among those pictured were Rachel Carson and Edward Abbey in the "Saints" category, Chico Mendes and John Muir in the "Seraphim" category, John Denver and Jerry Garcia among "Fundraisers/Motivators," David Bowie and Elton John among "Musicians," Ted Danson and Tom Cruise among "Beefcake," Morgan Fairchild and Daryl Hannah among "Blonds," John Ritter and Jay Leno among "Chuckleheads," Peter Matthiessen and Bill McKibben among "Writers," and Prince Charles and Mikhail Gorbachev among "Pols." And there, high center on the right-hand page, under "Scientist Types," were Jacques Cousteau and Daniel Janzen. Janzen was listed as a "Rainforest biologist."

# · *II* ·

# The Clifftop and the Volcano

Dan Janzen crashed through the crisp, shoulder-high jaragua, sidestepping shrubs, small seedlings, and saplings on his way to the tall *Ateleia* tree. At the base of the tree, he flipped his camera behind his back, grasped the trunk, and pulled himself up to a network of branches. A minute later, breathing heavily, he was perched precariously a good forty feet above ground, his rump and feet pressed against different branches to form a tenuous tripod. It was an impressive physical feat for anyone, let alone a man of forty-eight with Ménière's disease.

Twelve miles or so behind him, fluffy white clouds poured over Cacao and Orosí, pushed by the Pacific-bound trade winds. Here on the sultry lowlands, the wind whipped the jaragua, and the *Ateleia* branches swayed ominously. Yet, with the grace and confidence of a spider monkey reaching for a fig, he swung the camera back around into his hands, popped off the front of the case, and began shooting photos of the field below.

This was the Clifftop Regeneration Plot. In July of 1987, the Clifftop plot was a seven-acre mosaic with thousands of young trees of many shapes and sizes steadily rising through the jaragua. It was one of several sites in Santa Rosa National Park where, years earlier, Janzen had begun piecing together the ecology of tropical dry forest succession, and restoration.

Five minutes and a roll of film later, Janzen strode through the plot, pointing out the mahoganies, laurels, and rosewoods and describing what the trees had taught him about how tropical forests regenerate. "Ecological research is taking particular pieces of the forest apart and trying to figure them out," he told me. "After you're here for a while, you begin to see that the new stories link up with the ones you already know. It's like putting together a jigsaw puzzle."

The Clifftop plot ran a couple hundred yards along the western side of the park entrance road. The main goal of research at the plot and the other sites was to discover how to restore pasture and forest fragments into the teeming, wall-to-wall forest that once defined the region. That hadn't been the original aim. "The Clifftop Regeneration Plot started as an academic exercise," Janzen explained. "We didn't think we'd be doing it for restoration information." However, when it dawned on him that the fragments of forest in and around Santa Rosa were shrinking into oblivion, he refocused the Clifftop inquiry onto the processes of tropical forest regeneration, traditionally one of the most neglected areas of ecological study. Like an apprentice carpenter learning about the wide range of tools and materials at hand, he began studying the different kinds of houses that nature builds, and how it constructs them.

The Clifftop plot was only one such house, but learning about it was a seminal experience for Janzen. Until 1980, fires annually swept through the pasture that eventually became the plot. That year, he decided to protect it, "just for fun." For comparison purposes, he and volunteers burned a control plot on the opposite, upwind side of the road, which made sure that the fire didn't leap across coming downwind. They had done it each year since.

In effect, the fire ensured the continued dominance of jaragua on the upwind side of the road. But the Clifftop side immediately began growing back into a forest. The group mapped every tree that sprouted, recording growth rates and survival. More than 12,000 young trees occupied the fire-protected part by the 1987 rainy season.

Even more intriguing than the robust secondary succession was the way it occurred. The mother lode for this burst of life, Janzen discovered, was an upwind stand of trees whose seeds rained on the plot, brought by the prevailing dry-season breezes. Eighteen out of the 36 tree species then in the plot were wind-dispersed. That was 18 of the 23 species of wind-dispersed trees in Santa Rosa "big enough to cut timber out of," as he put it. This showed the aggressiveness of those species, since they comprise only a small fraction of the roughly 250 tree species native to Santa Rosa. "The mother of those young mahogany trees is a hundred-fifty meters down that way," Janzen said, waving an arm toward the giants to the east. "They may look like just little shrubs, but they are honest-to-god trees, and they're going to be that diameter when they're full-grown."

In terms of raw numbers, nine in ten of the plot's trees were wind-dispersed; the rest had been brought in by animals and either defecated or spit out. The first of these probably were carried by deer, which use jaragua stands as resting places, and coyotes, which hunt cotton rats in the grass. Next came mice, bats, other small mammals, and birds. Janzen referred to a crested guan, whose genus is *Penelope*. "Yesterday when I came by here, a big *Penelope* was sitting out on one of these trees," he said. "They look like turkeys. They're a good-sized bird. Well, every time a *Penelope* shits, there's another thirty seeds going into the plot." As each animal cast its cargo, the proportion of animal-dispersed species in this stew increased. "This kind of forest is not a replica of what was once here," Janzen said, stepping out of the plot and into his Land Cruiser. "A whole lot of changes will have to subsequently take place, but the beauty of the forest is, it changes."

Later that morning, Janzen, Hallwachs, and I bobbed along in the Land Cruiser winding up the rocky, muddy roads north of the town of Quebrada Grande. We were bound for Volcán Cacao, where dry forest habitat gave way to a strip of mountainous cloud forest near GNP's eastern boundary. The wipers, inaudible beneath the din of the straining chassis and engine, swept a foggy rain from the windshield. Suddenly, the odor of rotten eggs permeated the air. "We're approaching the sulfur mine and the south boundary of GNP," shouted Janzen, bouncing on the seat but firmly ahold of the wheel. "There's the sulfur mine. We just went by a piece of it. Now we're—" The heavy vehicle bucked even more violently and descended a steep hill toward a narrow, flimsy wooden bridge twenty feet above a rushing river, the Río Gongora.

"Now we're skidding down," Hallwachs said. "This is the adventure part."

"Uh huh," Janzen answered. "Here's where we begin to pray. 'Oh great mother four-wheeler, give us traction.'" Although the vehicle accelerated and continued to buck, he maintained control and guided it over the bridge and up the opposite incline. The road leveled out a bit but kept its rugged edge. "Nothing seems to have broken yet." We rose out of the cold mist on the slope of Cacao, and there, beyond the barbed-wire pasture fences, emerged the profiles of living dead *Nectandra* trees.

Even though the road continued, we parked at the house of a friendly landowner, donned backpacks, and shortly before noon set out in a hard rain across a few miles of pasture toward the cloud forest atop the volcano. The owner of the pasture, José Joaquin Diaz Duarte, was the one who refused to let anyone with the project drive across.

One purpose of our trip was to drop supplies at the house for an OTS field course that would soon come through and, since Hallwachs would serve as a guest lecturer, to bone up on the biology of the place. Janzen also wanted to check in on the Cacao Biological Station, the Guanacaste project's newest research facility. The cloud forest around it was part of the former Orosí Forest Reserve, which recently had been turned over to the Guanacaste project by the Forestry Department after the project bought Centeno, the most important part of the strip from the top of the volcano to the Pan-Am Highway. The ascent would be one of their first opportunities to explore the Cacao section of the forest, to begin to sort out its characteristics and workings. Finally, whether it was a conscious purpose or not, the trip served as a peaceful interlude amid the chaos then swirling around their house in Santa Rosa.

The rain eased, and a gusty breeze ran across the pasture. When a gust hit just right, the barbed wire wailed like so many bassoons trying to harmonize. The bright green grass was dotted with lone trees, including a dramatically sculpted *Nectandra*. It towered nearly 100 feet high. I paused to examine it and take photographs. The slender trunk rose at least two-thirds of its length before splitting into two main branches, which rose even higher before bursting into a dense network of small branches and leaves at the very top. This tree was the sole survivor of a vast blanket of canopy trees that once covered this part of the slope. How beautiful, how sad. The living dead. Being near it engendered a kind of religious feeling, probably a combination of wonder at its height and shape, and awe at its dire fate. The wail of the barbed wire added a touch of eeriness.

An hour or so later, we entered the intact cloud forest and began a steeper climb. The forest was quite different than that in Santa Rosa. Branches were covered with epiphytes, and leaves glistened everywhere. Hallwachs cautioned me to examine branches before grabbing them, since they might have arboreal vipers, snakes that inhabit trees.

Janzen and Hallwachs discovered a mysterious blue-green frog and a strange fruit on the ground that looked like a Dairy Queen ice cream swirl, colored pink and cream. Hallwachs collected the frog, as well as a leech that squirmed down my pants leg. Coming around a bend we startled a troop of six peccaries, which grunted and dashed off into the forest. A swarm of Africanized honeybees passed overhead.

We stopped for lunch at a stream running across the path. Janzen scooped his hand into the water and drank, while Hallwachs passed out brown bread, cheese, pieces of cucumber, and homemade chocolate cookies—doing her job for visiting journalists. As we ate, howler monkeys barked their territorial imperative from distant trees. Butterflies of many hues and patterns swooped around us.

After lunch, Janzen launched off perpendicular to the path on a trek down the middle of a creek, trying to figure out why the forest on one side looked different than that on the other. The canopy was similar on both sides, but the understory of one was only small trees, while the other had an intermediate layer. "This is the sort of thing we just don't know enough about yet," he said, splashing through the ankle-deep water. "In ten or twenty years we will, but right now, it's just like walking into Santa Rosa for the first time."

"That's funny," Janzen said. "We're starting to run into Atlantic species."

We were climbing again, this time up a very muddy path with a much steeper slope. With each few meters gained in altitude, the epiphytes and long vines called lianas seemed to spread over more and more of the forest. Lianas as thick as an arm reached from trunk to canopy, hugging the bark as if mere extensions of it, like veins in an old person's hand. Sometimes veins rose out atop veins. Smaller vines dangled in the air, attached to high branches.

We rose into the buffeting trade winds, which tore in from the northeast and blasted across the very top of Cacao. A large tree lay chest-high across the path. "I think we're getting up into the area where big trees get blown over and snapped out," he said, stepping under it.

The path narrowed and suddenly jumped nearly straight up. This was more of a climb than a walk, up a steep, slick trail cut out of dense forest. The only way to progress was to grab branches and exposed roots and pull while grinding into the scattered rocks for footing. We slipped and fell repeatedly.

At the crest, waves of mist pelted a dense matrix of dripping stems, branches, and leaves. It was cold up here, perhaps 50 to 55 degrees Fahrenheit, with wet gusts shooting in at better than 30 miles an hour. Standing on this ridge in the buffeting wind, it was easy to imagine balancing on the wing of a small airplane flying through clouds.

Janzen pointed to a tiny yellow orchid growing on a tree stump.

"That would be worth a million dollars at a Los Angeles flower show," he said, his voice just short of a shout, competing with the gusts.

"Quite a greenhouse, Daniel," Hallwachs answered.

This spot was a kind of four-way continental divide, between the Pacific and Atlantic on the one hand, and South and North America on the other. It also was the summit of Guanacaste's natural water machine: These plants drew water from the mist, and it percolated through the soil and down the slope, filling the watershed and ensuring a fresh, plentiful water supply for several towns and settlements. Up here somewhere was an old radio transmitter. The CIA had put one on Cacao and one on Orosí in the late 1970s.

We talked about the different ways that nature initiates forests in the biological desert below, in the pastures of Santa Rosa and surrounding lowlands. The first way was to grow peninsulas of wind-dispersed trees downwind of a forest stand, as the Clifftop plot had shown. The second was to grow a forest island. In research similar to that in the Clifftop plot, Janzen was studying how such islands developed. It all started with the rise of a single tree in an area protected from fire. This was known as a nuclear tree, often a guanacaste or cenizero, and usually from a seed dropped by an animal crossing a pasture. If soil and other conditions were right and it grew large enough, the tree attracted birds, deer, peccaries, and other creatures that came to perch, sit in the shade, or eat its fruit. Under or near the nuclear tree they dropped seeds they'd eaten elsewhere. These trees grew, and the island expanded into an increasingly larger forest fragment and eventually linked with other islands to complete an intact, closed-canopy forest.

Nuclear trees and their islands flourished in Santa Rosa's pastures until 1978, when the Park Service removed the cattle. "As soon as they took the cows out, the grass grew tall and the fires were very strong, and we lost an enormous number of nuclear trees in the first three years," Janzen lamented. By eliminating fire and by strategic use of cattle, he hoped to reverse that trend.

Janzen's unique knowledge of how to rebuild the forest had been shaped by years of broad-based research. In addition to the Clifftop and nuclear tree work, he assimilated studies of fire by himself, Doug Boucher, and others. He discovered animal migration within Guanacaste and its importance in pollination schemes across the Pacific-side plains, mountains, and Atlantic rain forest. He hypothesized the role of the megafauna in seed dispersal. With Hallwachs and others, he began to sort out how the behavior of the spiny pocket mouse, agouti, peccary, and other small animals literally dictated the composition of tree species in a forest.

For example, Janzen found that the dry forest mouse, *Liomys salvini*, was a major seed predator, avidly collecting and killing seeds of the guanacaste, guapinol, and other preferred tree species. This thinned the seed shadow of each tree—the area covered by its seeds after they fell or were otherwise transported. He described the nighttime forest floor as "a seething monoculture of *Liomys*" that ran about capturing seeds. A mouse foraged as far as 70 yards from its tunnel home and carried off as many as 500 guanacaste seeds in a single night. Sometimes it burrowed through the dung of larger animals for seeds. It brought its prey back to the tunnel, which was sufficiently deep that even if not eaten, a seed could not germinate successfully. On rare occasion, a mouse became a seed disperser, if it temporarily stockpiled seeds in a surface cache and then got eaten by a predator, or if the predator took a mouse carrying seeds.

Hallwachs found that the agoutis, *Dasyprocta punctata*, which were active in daylight, helped guapinol trees survive in the Santa Rosa area. In a process called scatterhoarding, the rodent carried guapinol seeds one at a time away from the ground beneath the source tree, where 99 percent of seeds and seedlings were killed by predators. It buried each seed in a shallow hole as far as

100 yards or more from the source, then covered it with soil, tamped the soil down, and sometimes even covered it with leaves and twigs. Sometimes the agoutis returned to eat the seed, sometimes not. If buried in an area with few seeds and other food, the chances that a seed or seedling would die went down to 50 percent. In other words, the agouti extended a guapinol tree's seed shadow out to safer ground.

There were other complexities to the restorative processes: species richness, seedling establishment, habitat structure, and rates of species buildup, to name a few. Something as simple as the type of nuclear tree could influence the composition of a forest island; different types offered different kinds of fruit, perches, and shade—any of which could affect the kind of animal that visited and, thus, the kind of seeds dropped nearby.

Various experiments in which volunteers planted trees by hand were helping determine the potential for direct human involvement in restoration, and whether trees could regenerate in some of the area's harshest environments, such as deforested habitats whose soil had eroded down to bedrock. "We picked the worst possible pieces of habitat in the park, and we planted trees in those to see what would happen if a dispersal agent actually brought seeds there," Janzen said. This information was important to the Guanacaste project because it guided conservationists on issues such as where, when, and with what species to encourage forest regeneration.

In the mid-1980s, Janzen stood alone with Hallwachs in having pieced together so much of the puzzle of how to grow a tropical forest. Restoring the cloud forest and rain forest on the volcanoes still posed some problems; research was under way on that. Also, Janzen expected that restoring Santa Elena would take much longer than Santa Rosa. Santa Elena had different soil, microclimate, and vegetation, and "a lot of the area has been so thoroughly deforested that we don't know what was there," he told me. "So if you say, 'Well, let's speed it up by going on out and finding some trees to get nucleii started,' now we're going to have to have a big debate about what trees we're going to plant, because it's all going to be guesswork."

However, the mystery of the tropical dry forest appeared solved. The fact that Janzen understood this system before he pushed to restore it—a simple point, but key—made him a strong, persuasive salesman. As Kenton Miller once told me, "he was able to put real scientific knowledge into the discussion."

Janzen's scientific authority also established him as a leader in the discussion of how to manage forest regeneration. There were many questions about just how to engineer it, and such questions often would require judgments more philosophical than scientific. The issues were as complex as the animals, plants, and habitats themselves. Nature would dictate the final form of the forest, but that would take thousands of years. In the meantime, human management decisions would determine its structure and content for centuries. "If the park is left completely to itself, one kind of mosaic will occur," Janzen explained. "If live-stock are used to depress the grass in certain areas, a different kind of mosaic will occur. If nuclear trees are planted far out into pastures or seeded there in the dung of cattle, a third kind of mosaic will occur."

A group consisting primarily of Costa Rican officials and conservationists, but also including Janzen, would determine the ultimate biological makeup of GNP. Their decisions would affect which species thrived and which became extinct in the region, because as habitat changed, the inhabitants changed, too. "When this finally turns back into primary forest, we will lose a number of species," he said, looking down as if he could see through the mist to the Gua-nacaste lowlands. "The species richness is higher right now than it'll ever be, because what you have are the species in the primary forest, the species in the grass, the species in the young second-growth, and all that. When this *finally* turns back into primary forest, we will lose a number of species."

For instance, the black vulture, which relied on its vision to find carrion, thrived in open pastures, where it could easily spot dead animals. When the forest grew back and a closed canopy obscured the vulture's view of the ground, it could well disappear. Arguably, the black vulture didn't belong in the Gua-nacaste forests at all. They were a so-called beachfront species, eating rotten whales, dead fish, and other such things. "But you have your king vultures and your turkey vultures, because they can smell, and they can go down underneath the foliage," Janzen noted. "People say, 'Well, that's kind of bad, because you're going to lose a list of species.' Well, that's true, and you have to decide what you want."

If the black vultures declined in exchange for a higher density of today's rare species, it would be a good trade-off. Perhaps the managers would decide to maintain species associated with disturbed forest by keeping some places in a disturbed state. Restoration management would entail many such value-based decisions.

The more important point, Janzen reminded me, was to save not only a cer-tain number of species, but combinations of species, their habitats, and the inter-actions among them. Furthermore, the number of species in a habitat did not necessarily equate to its value. A habitat with 500 species might not be as impor-tant to save as one with only ten species. Janzen gave the example of a mangrove swamp. Depending on the location, it might only have a few species of trees and fish. But it was a crucial habitat type, partly because it served as a nursery for the fish, some of which were ocean-going species. "If you go by the number of species present, you would never dream of keeping mangrove swamps," he said. "Yet they're an extremely peculiar setup, with extremely peculiar plants, with all kinds of peculiar things that are ecologically important."

Other examples abounded. In the end, he said, management decisions prob-ably would be based on "a collection of local circumstances." Much later, as I thought about our discussion, I realized the Guanacaste project was perhaps the foremost tropical expression of Aldo Leopold's dictum, "To keep every cog and wheel is the first precaution of intelligent tinkering."

On the hike down the side of Cacao, Janzen stopped to examine several plants and turned over a fallen, rotting branch, still looking for clues to how the forest worked. Ahead of us, a tinamou ran along the path. A chicken-like bird, the tinamou lays its glossy blue eggs on the ground, and the father tends the young birds. This one apparently was leading us away from its brood, crouched in the forest nearby.

"He ran ahead, and so we followed *him*, and therefore we didn't see the chicks, which is how they escape," Janzen said. "That's pretty sophisticated behavior."

"Animals *are* sophisticated," Hallwachs said. "It's not easy to survive out here."

As dusk fell and the howler monkeys again sang their guttural harmony, we descended out of the forest and across the pasture toward the Land Cruiser. The mist cleared, and the sparse lights of Quebrada Grande twinkled below like stars. To the west, atop the gentle curve of the Pacific horizon, the sky was painted high and wide with shades of salmon and pink. We crossed the rapidly darkening landscape, past barbed-wire fences singing in the trade winds, past a few wandering cattle and horses, past the ghostlike pillars of the living dead.

We started down in the Land Cruiser, eventually clearing the last rickety wooden bridge and making it to a blacktop road. After a while, Janzen launched into a song.

"Busted flat in Baton Rouge. Waitin' for a train." He seemed in good spirits, and he belted it out with a twang above the whine of the engine. "Feelin' nearly faded as my jeans."

It was *Me and Bobby McGee*. His voice got stronger as he continued, and he sounded pretty good, a bit more like Roger Miller than Kris Kristofferson. Somehow I just couldn't force myself to imagine him strumming on a guitar. Even so, it showed that Janzen *was* capable of enjoying himself, if he allowed himself to.

"Freedom's just another word for nothin' left to lose . . ."

And we turned onto the Pan-Am Highway and scooted north toward Santa Rosa.

# · *I2* ·

# Home Runs

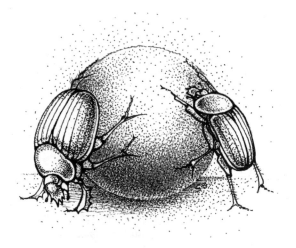

As I strode to the Santa Rosa *comedor* one morning, a movement near my foot caught my eye. I bent down and watched two shiny black beetles roll a dark green object about the size and shape of a malted milk ball across the road. They pushed and rolled the ball in a whirling flurry, quickly enough that they often lost control of their bodies or the ball. Sometimes they fell off the sphere as it meandered down a depression in the road, sometimes they got run over as it lurched ahead or to the side. They were like two lumberjacks in a bizarre barrel-rolling competition.

These were dung beetles, one of the many kinds of insects that feed and breed in the feces of other animals, including the dung of cows and horses at the park. Like other dung beetles, they were nocturnal; these must have been just finishing up a night's work. I had no idea where they were heading with the ball, and, from the looks of it, I'm not sure they did either.

Dung beetles might not be exactly the kind of animal you want to think about as you head for breakfast, but the sight of their seemingly Sisyphean struggle was engagingly funny. *Their* breakfast of choice is scattered widely and randomly through the park. Once found, some species burrow beneath it, pack chunks into underground chambers, and lay an egg in each. Some cut out and shape a chunk into a ball, the better to roll it off to a distant burrow. As they push the ball with their back legs and the ground with their front legs, they can reach speeds up to several inches a second. Placing the dung underground helps keep it moist and protects the offspring from predators and parasites.

To early Christians, these beetles, usually of the scarab beetle family, symbolized heretics, but the ancient Egyptians worshipped them as the Father of the Gods. Because its larvae seemed to rise up magically out of dung, the scarab symbolized regenerative power, resurrection, and reincarnation. Janzen and Hallwachs chose a doubly apt cover image for the Nature Conservancy fundraising pamphlet, the one that showed the photo of guanacaste seedlings bursting forth from a cow pie. Just as the seedlings represented the rebirth of trees in the forest, the pat, which would have come alive at dusk with dung beetles, represented the renaissance of animal life.

The two beetles stayed with their ball as it veered off into a roadside patch of grass and leaf litter, where the going got really tough. One got separated from the ball for a minute and wandered erratically through what must have seemed like a thick forest. It found the ball again and resumed the journey with its partner, and I left them to carry on their little dance.

After breakfast, I returned to my post at Janzen's house and found him back in the eye of the storm. For the next three days, until Arias arrived, on Saturday, July 25, the pace of his phone calls, meetings, writing, and chores seemed to intensify. He hosted visitor after visitor, from nervous landowners to busloads of ecotourists to biologists interested in getting out on the land that would soon fall into the hands of the Guanacaste project. These days were hot and windy, repeatedly punctuated by thunder, lightning, and rain. Sometimes the raindrops assaulted the roof like a million marbles pouring in a thick, steady stream from some giant bucket high in the sky.

Janzen alerted Umaña and Gámez that local politicians were upset that the Arias itinerary contained no time for a meeting with them. He called reporters and invited them to the Arias event. He cranked out a GNP grant proposal for $9.56 million to the Society for the Study of Evolution, a scientific research group that had just announced it had as much as $10 million to give to an institution or field site where evolution studies could be conducted. He exchanged innumerable phone calls with Umaña and others in San José as they considered how to handle issues related to Santa Elena and the airstrip, including what to say to the news media, how to approach a Monday meeting with lawyers for the property, and how to raise a $270,000 downpayment Costa Rica would need to take possession. The list went on and on.

One of the Guanacaste project's characteristics, he told me, was that even when it went well, things got busier, more complicated. "One success leads to a whole batch of next steps, and those next steps actually mean that you have more to do when you're successful, not less. So if you think success means having a sense of relief, or you can begin to relax, it's quite the opposite." When one caller, speaking in Spanish, asked him how things were going, he answered, "*Mas o menos loco.*" More or less crazy.

Wild things were happening in the animal world, too, and they heightened the aura of chaos around Janzen's house. One night, as I sat writing beneath a light bulb in the building where I stayed, a huge cloud of flying termites suddenly rushed in. Hundreds of the half-inch-long insects swarmed around me, part of a periodic sex orgy originating from a termite colony nearby. Males and females paired up, danced, knocked each other's wings off, mated, and wandered out of the building in search of a nest. The place was covered with wings.

Coyotes joined in long, howling choruses near the administration area most evenings. I caught a glimpse of one just before sunset about 200 yards away, and when it saw me it dashed into the forest. One morning, a swarm of Africanized honeybees hovered for several hours across the road from the *comedor* before flying off. Another day, while people sat outside the *comedor* eating supper, a spider monkey jumped on a bird nest directly above a picnic table and spilled the eggs. Another monkey, in a flying leap, snatched a young bird in a different nest. Other monkeys joined in, tearing the bird apart in a frenzy. A wing fell to the ground, and one of the monkeys daringly jumped down to fetch it, then shot up a tree.

A typical few hours in Janzen's day went like this. He was up before six, writing the grant proposal on his Macintosh. Then came several phone calls. Instructions to volunteers. Discussion with a biologist who drove up seeking advice on research at Santa Rosa. A call from his assistant in Philadelphia relaying more information about the grant proposal. At ten o'clock he went to the Bosque Húmedo with the two BBC men, who wanted a final shot of him walking through the forest.

Back at eleven, he met with two researchers from the National Museum in San José. More phone calls, including a long one with León Gonzalez of the National Parks Foundation, the chief formal land negotiator for the Guanacaste project, who sought guidance about the Santa Elena expropriation. Just before two o'clock, while the Mac printed several copies of the proposal, he hopped on the Park Service tractor, which was parked in front of his house, and drove it back to its shed. He had used the tractor the night before to free a vehicle stuck in the mud elsewhere in the park. He washed dishes and monitored the print job. A large boa constrictor curled up in the rafters with three bulges in its otherwise sleek, spotted body, a sign that it had recently eaten three rodents. At 2:40, Janzen grabbed a milk carton from a backroom refrigerator and took half a dozen gulps.

"Lunch," he said.

Janzen actively hunted for $270,000 for Santa Elena, as Hallwachs struggled in her silent way in the background with another set of challenges. Janzen called several scientists in the States, asking them to contact prospective donors on his behalf. He called officials with Conservation International, the Nature Conservancy, and World Wildlife Fund. They were playing what might be called the "politics of exclusivity," each professing to want to help the project acquire Santa Elena, but each demanding full and exclusive credit for doing so. One official with the Nature Conservancy went so far as to tell Janzen that if its arch rival Conservation International was "in," the Conservancy was "out." Janzen interpreted that as a threat.

This wasn't just petty bickering by amateurs. These were skillful politicos, such as William Reilly, president of the World Wildlife Fund, who would soon become director of the U.S. Environmental Protection Agency. Reilly had pressed the flesh on a ten-day tour of Costa Rica a few months earlier, lauding WWF's decade-long string of donations to Costa Rican conservation totaling more than $1.5 million. He visited Janzen in Santa Rosa and, in a *Tico Times* article, praised the Penn professor and the Guanacaste project. "When someone of Janzen's stature talks about treating human beings as part of nature and nature as something that can be improved and restored, that's incredible," Reilly told the newspaper. "There's an exciting future in the rehabilitation of downgraded ecosystems."

One of Janzen's Santa Elena calls, to a man in Boston, revealed much about his style as a solicitor. The man, whose anonymity Janzen protected, already had donated $240,000, enabling the purchase of the two farms on Volcán Cacao, the hiring of their owners, and the acquisition of Finca Jenny.

"A crazy thing is happening down here," Janzen told him. "President Arias is declaring Santa Elena a national park. Why don't you hop on a plane and come down for it?" Apparently the answer was no, because Janzen went on to explain the history of the expropriation effort, the financial situation, and why the new push was happening on such short notice. He described the fate of a grant he once thought possible for the Guanacaste project from the U.S. Agency for International Development, an arm of the State Department. The money had gone to assist Corcovado National Park instead. "With Ollie North's testimony to Congress, there was no way AID could give me that money," he said.

The donor was still noncommittal in the face of Janzen's hints about needing $270,000. So the scientist became more direct. "Would you be willing to go halves with somebody?" he asked. "If I can find someone else to give half—and so far I haven't—would you give the other half?" He said the expropriation offered "a peculiar opportunity" in that it was clearly more than a conservation move; it also was a clear signal to the United States that Costa Rica was in charge of its own destiny. Finding the money was another of the many important steps in the Guanacaste project that, against the odds, continued to get done, Janzen said. Arias was counting on him to raise it, a trust that derived partly from what many Ticos thought of as "the magical atmosphere surrounding this project."

No response. Janzen kept it up, describing his pleas to the conservation organizations. Still nothing. "Can you go to someone and say, 'If you put in $25,000 and I put in $25,000'—to pull together that kind of piece?" Finally, the donor said he would think about it. Janzen suggested he take "a couple of days." He cordially said goodbye and hung up.

Janzen's calls revealed that Reilly might be willing to donate $100,000 to the Santa Elena pot, and perhaps more. The $100,000, Janzen and Umaña reasoned, would be sufficient collateral to take out a loan to make the down payment. Meanwhile, $270,000 in Costa Rican debt could be purchased, and the proceeds would go to the project. However, Reilly needed assurances from officials in San José that raising the Santa Elena money was currently Costa Rica's number one conservation priority.

Once, during a call from Umaña, Janzen stood in his doorway, looking out into the trees. He spoke loudly into the phone, apparently because of a bad connection, but he also was tense, maybe even upset. Umaña was telling him about a meeting the day before between lawyers for the National Parks Foundation and lawyers for Joseph Hamilton, the owner of Santa Elena.

"So you're saying that even though it's declared a national park, we still cannot take possession?" Janzen asked. He listed several problems with that, including the fact that Civil Guards stationed at the airstrip were shooting animals "as a way of living."

A meeting had been scheduled in San José on Monday between both sides, and the two men began to discuss strategy for that meeting when the telephone line produced a burst of heavy static and went dead. Janzen swore and set the receiver down hard. He stared as if trying to melt it, and a few seconds later it rang. He picked it up, but there was nothing at the other end. He hung up slightly less than hard. A few seconds later he picked it up, but now the Santa Rosa office was making a call. He hung up softly, and sighed.

I walked into the kitchen and there was Hallwachs, who had just returned from a few hours of dissertation writing on her Mac. She was mixing a batch of oatmeal cookies and had spread several plates of frozen items wrapped in plastic around the counter. One of the plates contained three or four small snakes, including a rattlesnake and something related to a water moccasin. "It's time to clean out the freezer," she said, responding to the confused look on my face.

While Janzen sat at the Mac writing talking points for the Santa Elena negotiators, Hallwachs and Brad Zlotnick busily cleaned the house, in case the president stopped by during his visit. Zlotnick, of Palo Alto, California, was one of
the volunteers with the Guanacaste project. He had graduated three years before
from Harvard and, while preparing for medical school, had been working at the
Center for Conservation Biology, a group run by Paul Ehrlich at Stanford. Zlotnick had just started as a Stanford medical student, and he was giving several
months of sweat to the project in return for biology credits. Tall, athletic, and
sporting a bushy brown mustache, he was tough but personable, and he always
seemed to have a gleam of curiosity and quiet confidence in his eyes.

Like the other volunteers, Zlotnick was doing this partly for the experience.
He took avid notes about the things he saw and learned—the animals, plants,
ecological lore, and Spanish words and phrases. He enjoyed the rough and
tumble of life in Guanacaste, doing legwork for field experiments, running
errands, and putting up with the biotic challenges of the tropical forest. He and
another volunteer cut the steep trail atop Cacao. He believed in tropical conservation, and he believed in putting his mind and body out into the difficult environment where it was happening. That was true of most of the volunteers, even
though many also got course credits from their universities.

The life of a volunteer was not easy. Janzen may have described it best in a
strangely cautionary form letter he sent to prospective volunteers. After telling
them that help was needed and that they had to pay their own expenses of $15 a
day for room and board, he wrote: "For this, you get to sweat in the sun, sleep
little, be meals to chiggers and ticks, bounce up and down on horses, eat a lot of
smoke and fire, climb trees, babysit for caterpillars, bend over 529 times setting
mouse traps, climb mountains in pouring rain, eat rice and beans and beans and
rice, plant thousands of seeds, and do jillions of other things seven days a week,
from dawn to dark and often after. But there is an awful lot to learn if you keep
your eyes and mind open, and you are doing something that will contribute to
tropical humanity forever." What wasn't mentioned was that a volunteer with the
right stuff also would need a sufficiently hard shell to withstand Janzen's often
rough demeanor.

The day before the Arias visit, after completing the cleaning and other
errands, the volunteers sat at the picnic table, reading the *New York Times, International Herald Tribune*, and other newspapers and commenting on items of
interest. Among the items were John Poindexter's testimony before the Iran-
contra committee, a new wave of fighting in El Salvador, and Umaña's op-ed
piece in the *Wall Street Journal* on the promise of the debt-purchase idea. Zlotnick quoted several passages from a George Will column in the *Tribune* on
rookie home-run "man-child" Mark McGwire and others in a new crop of long-
ball hitters in American baseball. The column seemed to parallel themes then
prevalent in the Guanacaste project as well.

Zlotnick read, quoting Will: "When baseball's best practitioners assembled for the All-Star festivities, not all thoughts were festive. Some brows were furrowed in puzzlement, some eyebrows were elevated by suspicion, and some lips were pursed in disapproval about the fact that baseballs are being hit over fences at a record rate." The volunteers laughed and joked about the home run in the making that was GNP, the various controversies in which it was ensconced, the allies and enemies who would soon be arriving, and the way rumors and little stories surrounded the project. "Explanations should be as simple as possible—but not any simpler," Zlotnick read from Will. They laughed again.

For most journalists, the story of that July in 1987 was not Arias and conservation but rather Arias and the secret airstrip. When reporters called and asked about the airstrip, Janzen's policy was to plead ignorance of the political background and refer them to Umaña or Arias. He was determined to avoid embarrassing the Costa Ricans and jeopardizing the Guanacaste project and its attempt to consolidate Santa Elena. Still, it happened.

The Costa Ricans viewed Janzen as a media expert, and Umaña had leaned on him to get American reporters to attend the Arias ceremony. Janzen, Umaña, and other Tico leaders realized it was unlikely that reporters would fly down for a simple declaration, so Janzen jokingly suggested they have a mock celebration in the middle of the airstrip and plant guanacaste trees. Everyone laughed about it, and the joke spread among a few conservationists in San José, then a few in Washington, and somehow articles appeared in the days leading up to and including the Arias visit on the Associated Press wire, in the *Los Angeles Times*, and in the *New York Times*, all saying that as part of Saturday's ceremony hundreds of local children would plant tree seedlings in the airstrip. The venerable news organizations attributed the information to the Costa Rican Embassy in Washington. Indeed, the conservationists later learned, the embassy's statement announced that children would plant the trees, "symbolically confirming Costa Rica's commitment to conservation and peace."

Preposterous. The tree-planting ceremony was not at all part of the plan for Saturday. Alvaro Ugalde, thinking Janzen had spread the story, called Santa Rosa to demand that the American scientist pipe down. Janzen denied involvement. Like Ugalde, he feared that something would undermine the Santa Elena annexation.

On the same afternoon the AP story crossed the wire, Reilly called to report that World Wildlife Fund (WWF) had come through with $100,000 for Santa Elena and a debt swap. There was one condition: Arias must give WWF prominent public credit in his speech the next day. "This one hundred thousand is a fantastic thing," Janzen told Reilly. About the public acknowledgment, he hedged: "I'll work on Umaña about that. The Arias speech is already written, but we'll see what we can do."

The conservationists were ecstatic about the contribution, but things took an awkward turn three days later when WWF issued a press release from its Washington office. This was, after all, its big coup in the debt-swap competition with the Nature Conservancy and Conservation International, and WWF wanted its fair share of publicity. Notwithstanding the mistake by the embassy, it's important to remember that even though the Arias administration was declaring Santa Elena to be a national park, it was trying to keep the move from becoming an overt slap in Reagan's face. Arias and his advisers felt the WWF press release was just that, overemphasizing the airstrip's links to the Iran-contra scandal.

In the release, Reilly praised Arias for protecting "the largest fragment of tropical dry forest left in Central America," adding that "turning an airstrip into a national park is . . . an important expression of President Arias' commitment to peace in the region." Then he detailed the U.S. government's involvement in the land, including the statement that "Costa Rican officials told us there are six owners of the land, but five of them still remain anonymous. Apparently the owners are embarrassed that they sold the land to a corporation that used it as an airbase to resupply the contras."

"We were very mad about it," one Costa Rican official told me. "We wanted to downplay the political part completely, and WWF did exactly the opposite."

Several journalists arrived the afternoon before the Arias visit. Among them was the husband-and-wife team of Martha Honey and Tony Avirgan, who were working for U.S. television. Honey and Avirgan had long been tracking the airstrip and other angles of the Iran-contra story. They were veteran foreign correspondents, and Avirgan had been seriously injured three years earlier in the bombing at the jungle press conference held by contra leader Eden Pastorain in which three journalists had died.

Janzen led the group on a natural history tour of Santa Rosa, which included a lecture on ecological restoration at the Clifftop Regeneration Plot. But first he held a news conference around the picnic table in front of his house. He stuck mainly to the basics about the Guanacaste project, its origins, and status. He danced around questions on sensitive topics like Santa Elena with the skill of a seasoned politician, which he demonstrated when Honey started questioning him while Avirgan set up the equipment.

"What I'd like to have you do, if you would," she began, "is to cover, you know, the significance of the park, and then describe what's unusual about the ecological management—keeping the farmers in and so on—and the financing of it, this business of reducing the national debt."

"Sure."

"The incorporation of the airstrip."

"What did you say?"

"The secret airstrip, that this is property that you've been wanting for a number of years, and that you're very grateful that the government, that it's going to be nationalized? Then what is the story on the airstrip."

"Well, I don't really know what the story on the airstrip is."

"Well, is it going to be incorporated into the park?"

"Well, the airstrip is in the Hacienda Santa Elena. And Hacienda Santa Elena is being declared a national park. That's all there really is to it."

"But this is property that you had had your . . ."

"The Hacienda Santa Elena, oh yes, is something that we see as an integral part of Guanacaste National Park, from the beginning."

"Is the government buying the hacienda or is it . . ."

"No," Janzen interrupted. "It's declaring the hacienda a national park. O.K.? Then, as is the case with any piece of land in Costa Rica, if you declare it public domain, you then have to go through the usual negotiations to buy it, just like if you were buying it from some rancher. So now you have an argument, a negotiation on your hands between you and the owner as to how much is actually to be paid for it."

"But who's the owner?" Honey asked.

"Joe Hamilton."

"It still is? We're back to him now?"

"Well, he says it is. Who am I to say? You know, we're now dealing with an area that I really don't know anything about. But Joe Hamilton is certainly the person with whom the government will negotiate for the land."

And on it went.

Shortly after six o'clock, a few of us went with Janzen, Hallwachs, and Randall García to Liberia, where Janzen received Guanacaste Province's highest honor, the Hijo Ilustre de Guanacaste (Distinguished Son of Guanacaste) award. García, whom Janzen and Hallwachs considered to be doing an excellent job as director of the Guanacaste project and informal director of all the land to be incorporated, helped explain the event to me. The award was given to twelve people that night for achievements in science, education, music, politics, commerce, religion, and sports. Janzen's award was a clear sign of public support for him and the Guanacaste project.

We filed into the Liberia gymnasium, just off the city's central square. It was hot and sticky inside, and the bleachers were jammed with people. The provincial banner, a large red, green, and white cloth with a guanacaste tree in the center, hung over the stage, which was set up next to a sixty-piece concert band aimed squarely at the bleachers. Among the VIPs on stage to hand out the awards was former president Daniel Oduber, who also received one.

A color guard marched in, the band played, anthems were sung, and speakers extolled the accomplishments of the province in politics, economic development, and other fields. The twelve "sons," including a few women, were called up on stage one by one. The frolicsome acoustics of the gym played havoc with Janzen's hearing problem, and at one point he rose prematurely to get his award, thinking he had heard the word "Pennsylvania." Finally he was called and cited for his efforts to protect Guanacaste Province's natural heritage. He approached the stage, his white shirt open at the collar and the tops of his unlaced field boots flaring out as he shuffled forward. He smiled, shook hands, and silently accepted his plaque.

Afterward, Janzen drove us a short distance to an outdoor restaurant called the Cafe Relax, along the Pan-Am Highway at the edge of Liberia. It was warm and breezy, and the sky was pregnant with the threat of rain. The exuberant conversation around the table, the upbeat music on the radio, and the movement of cars up and down the highway in the night pumped the air full of electricity. Janzen seemed uncharacteristically happy, buoyed perhaps by the award and the impending visit of the president. He bought us all beers, hamburgers, French fries, and milkshakes.

It was good to see him laughing and, well, relaxed. But I remembered what he told me two days before when I had come back to his house from a walk in the woods. He was bent over a tub of wash water, swishing a dirty blanket around. I asked him, partly to be cheery but also to get an update, whether "things look good." He stopped swishing and gave me a grim look, and I knew I had hit a nerve.

"No," he said, forcefully. "Never say that in this business. When you say that, someone stops trying hard, and you fail."

By eight-thirty on Saturday morning, a large crowd had assembled in one of the Casona's stone-walled corrals, near the guanacaste tree. Hundreds of children wore pressed white shirts or traditional costumes of bright red, yellow, and blue. Many of them held banners and Costa Rican flags. Landowners, workers, politicians, park officials, and families sat in rows of chairs or stood on the newly mowed grass. Some stood atop the stone wall.

At about nine o'clock, three helicopters landed in a field just to the west of the corral, and Arias emerged from the middle one. He was short and thin with rich black hair and a blue shirt open at the collar. He walked slowly into the crowd, shaking hands and saying hello, his shoulders slumped, a smile on his face, and distinctive bags under his eyes. At the corral gate, two young children in traditional dress stepped in front of him and led him to a table and lectern at the front.

There he was flanked by Vice President Victoria Garrón de Dorian, Minister of Agriculture Antonio Alvarez, Minister of Natural Resources Umaña, and other politicians, as well as the conservationists Alvaro Ugalde, Mario Boza, Rodrigo Gámez, and Pedro León. Three older students in school uniforms slowly walked in with the Costa Rican flag, and there was complete silence except for the squawks of magpie jays and the flutter of the flag in the breeze. The crowd sang the national anthem.

Alvarez and Umaña each made a brief speech, noting the historical significance of Santa Rosa, the Casona, and Guanacaste Province, as well as the importance of conservation and Costa Rica's national parks. Umaña cited the WWF donation, using a few phrases written by Janzen, who moved through the crowd and around the edges of the corral taking photos.

This moment marked one of the most important peculiar accidents that made the Guanacaste project possible: the presidency of Oscar Arias. Past presidents had helped create national parks, but only after the determined efforts of Boza, Ugalde, and other conservationists to persuade them, against formidable opposition. Arias took presidential support for conservation to another level.

Arias arrived in office with a clear view of what conservation meant and of the effort required to create parks. He was committed to expanding and strengthening the entire national park system, a vision he had begun to acquire in the mid–1970s when, during the Oduber administration, he served as minister of planning. Within his agency, he nurtured a commission on the environment, created by Oduber to provide guidelines on conservation. The commission, whose members included Boza, Ugalde, and León, recommended creation of an institute for natural resources. This institute would represent a major commitment to conservation, including protection of parks and research. "His early experience with this commission was like a seed," León told me. "It incubated for ten years, and in '86 when he took office, one of the first things he did was to create the Ministry of Natural Resources."

Umaña had formally proposed MIRENEM to Arias long before the election. Upon his appointment to the cabinet, Umaña pushed for it. The ministry was established three weeks later, but only after difficult negotiations with the powerful Alvarez, who didn't mind taking the Park Service out of the Agriculture Ministry but wanted to keep the Forestry Department. Arias and his political adviser, John Biehl, "saw the future," as Umaña once told me, and they put Parks and Forestry together in MIRENEM. The Legislative Assembly, for reasons having to do with the standard battles among Costa Rica's political parties, wouldn't approve the new ministry until five days after Arias left office, nearly four years later. As a practical matter that was a minor detail, because the ministry had been funded by the executive branch, but it also meant that Umaña had to have Alvarez co-sign decrees.

The birth of MIRENEM signaled the priority of conservation within the Arias administration, and its pilot project was Guanacaste. "If there had been no MIRENEM, there would have been no Guanacaste National Park, for sure," Umaña noted. The "vision" would have languished without support in the

Agriculture Ministry while the Park Service picked it apart. Even within relatively friendly MIRENEM, the Guanacaste project was considered too expensive, given the concern about how to pay for the sizable number of private inholdings that remained in existing parks. "Arias said the same thing to us," Umaña told me. "'We cannot give you any more money. We do not have money. If you're going to have a project you're going to have to raise the money.' And of course we *did* raise the money."

Arias was introduced to applause. Someone placed a potted guanacaste seedling on the table where he signed the three proclamations—for Zona Protectora, Santa Elena, and debt purchase. Arias stood and talked about Dan Janzen's contributions to Guanacaste and Costa Rica and read from a plaque, addressed from the Costa Rican people to Janzen, recognizing his "tireless and extraordinary work in studying, conserving, and restoring the biological and cultural heritage of Guanacaste." As Janzen shuffled forward in unlaced field boots to accept the plaque, he paused to shoot a photograph of the VIPs at the front. He shook the president's hand, took the plaque, and shuffled back into the crowd.

Arias gave a brief speech, written by Gámez, about the importance of conservation to Costa Rica. He made several references to "Guanacaste National Park." It was, Janzen said later, "the first pure natural resource speech that a Latin American president has ever made." Randall García called it "a home run."

What was left unsaid—about Santa Elena—was equally striking. Arias made no direct mention of the United States and the airstrip, saying only that because of the war in Nicaragua, "people" had tried to involve Guanacaste Province and the government of Costa Rica in "someone else's conflict." He added, "What do we do with this region of conflict? We make it into an area to study nature. We don't send soldiers with instruments of death. We send students, scientists, and naturalists with the instruments they need for their intellectual work." It showed what a measured, mature leader Oscar Arias was, and it illustrated another peculiar accident that benefited the Guanacaste project.

Throughout the first half of the year, as Umaña, Janzen, and others prepared for the official launch of GNP, Arias pushed his policy of closing the Southern Front and negotiating a peace among Central American leaders, without the United States. The Reagan White House regularly criticized the Arias Peace Plan, and the Costa Rican president could have chosen many times in early 1987 to swing back by annexing Santa Elena. Instead, he waited until the annual Guanacaste Day celebration at Santa Rosa, which marked the day Guanacaste Province was annexed to Costa Rica. He billed the Santa Elena annexation as a pure conservation move. But in the strange ways of Latin American politics, in which a leader could do all kinds of things as long as he didn't declare them, it was a distinct message to the Reagan administration about Costa Rican sovereignty and U.S. policy in Central America. As one longtime observer of Costa

Rican politics told me, "It all has meaning, and it will all be proclaimed to have no meaning."

In an interview a decade later, Arias told me that with the Santa Elena annexation, he was trying to teach "a moral lesson to those who tried to deceive the Costa Rican people and the Costa Rican government authorities by helping the contras against the will of the government."

Enthusiastic applause greeted the end of his speech, and Arias wandered through the crowd amid a palpable air of excitement, shaking hands and chatting with anyone who happened to walk up to him. It was an impressive sight, the president of a nation mingling with strangers, seemingly unaccompanied by even a single security agent. (I later learned that agents were there, a few dozen yards away in the forest.) This was close to the border, and tension had been high recently. For the first time in thirty-one years, the July 4 picnic at the U.S. embassy residence in San José had been canceled for security reasons.

Arias seemed relaxed, and I wondered whether I was grafting onto Costa Rica the gringo journalist's view of tight presidential security in the United States. "It's one of the nice things about being a small country," Pedro León told me years later. "Democracy was made for small countries. You have more access to powers and to people. In big countries, it's so awesome."

By ten o'clock Arias was back aboard his helicopter, bound for the city of Alajuela, in the Central Valley, where, much like a U.S. president throwing out a baseball on opening day, he would kick out the ball for the beginning of a big soccer game between Alajuela and a team from Argentina. The next day, he would go to Managua as part of a Central American fly-around to revive his peace plan. Two weeks later, in August 1987, Latin American leaders would accept the plan, which eventually ended the wars in the region. He would address the U.S. Congress in September, by invitation of Congress but against Reagan's wishes. He would be awarded the Nobel Peace Prize on October 13 and serve out his term as president of Costa Rica through early 1990. He then became director of the Arias Foundation for Peace and Human Progress, which he established in 1988 with the prize money. Through the foundation, he worked for demilitarization, equal opportunity for women, and change-oriented philanthropy in Latin America.

About noon, several of the conservationists met privately to debate Guanacaste project policy, funding, and strategy. They included Boza, Gámez, Janzen, León, Ugalde, Umaña, and Geoffrey Barnard of the Nature Conservancy's Washington office. The Conservancy, of course, had backed the Guanacaste project from the beginning. Now, Barnard wanted to be part of the team that negotiated for Santa Elena; some of the others felt there was little to gain from this, except for the Conservancy to be able to brag that it had been in on the negotiations.

They met for more than two hours in the "dance hall," a simple building behind Janzen's house that was little more than a roof, concrete floor, and frame. Janzen used it as a lab. Plastic bags hung from the rafters, and most of the tables were covered with projects. From time to time voices rose, and both Barnard and Umaña were either thrown out of the meeting or left in a huff, returning later.

One of the fights occurred during a long debate on what tactics to use in acquiring Santa Elena. Specifically, that meant figuring out how to deal with Joseph Hamilton and his two Costa Rican lawyers. The group, led by Janzen and Umaña, persuaded Ugalde that he should lead the negotiations, and that the strategy should be to make a down payment of $100,000 on the property as an option toward buying it for the final agreed price, whenever it was settled on.

But bringing Ugalde around to this took a long time. Janzen and Umaña began by pushing for a meeting the following Monday between Hamilton and the Guanacaste project's representatives. Ugalde argued that the group needed to study the situation, confirm who owned the property, get documents and arguments in order, and then perhaps in a few weeks meet with Hamilton. This was Ugalde playing the role of the cautious civil servant. It clashed with Umaña's preference for striking hard when the time was at least close to right.

As the men debated, Janzen erupted angrily when he heard two people walking nearby and thought they were eavesdropping. They turned out to be Hallwachs and Vice President Garrón, who had been among the dignitaries at the ceremony. Hallwachs's duty that afternoon was to entertain the vice president, and the two women had passed the dance hall as part of the tour. Much to the dismay of the conservationists around the table, Janzen bluntly told them both to leave. The vice president!

Janzen argued that Ugalde should represent Costa Rica in meetings with Hamilton. Ugalde was, after all, a tough bargainer. He also was a biologist and conservationist with a full understanding of the issues, and he had a historical connection with the longtime effort to make Santa Elena part of the park system. But Ugalde didn't want the job. He argued that he didn't have the legal power, as the president of the National Parks Foundation, to represent Costa Rica. The others pointed out that he would go into the talks not just as the president of the foundation, but also as Alvaro Ugalde, the famous Costa Rican conservationist and ex-director of the Park Service.

Next, Ugalde worried about raising the money to pay whatever came to be the selling price. Others argued that they would find the money. Umaña became increasingly irritated by what he saw as Ugalde dragging his feet. The minister had blamed Ugalde's hesitancy during the Corcovado crisis for a problem Umaña inherited—finding $3 million in settlement money for the evicted miners. For his part, Ugalde became furious because Umaña insisted the Santa Elena money be raised solely by the National Parks Foundation; Ugalde insisted the government help. When Ugalde suggested that the conservationists seek a multiyear option on Santa Elena, to pay some soon and pass the issue on to the next administration, Umaña stormed out of the meeting.

And so it went. Ugalde finally relented, and the group hatched its negotiating strategy, an immediate, strong demand for the land at a low, but reasonable, price. The meeting concluded after the conservationists discussed how to set up a GNP management committee. Janzen immediately typed a letter for Boza to sign and take to WWF in Washington, where the Costa Rican planned to fly the next day. It was a long letter of gratitude to Reilly, thanking him for the $100,000, reminding him of his commitment to find more, and offering WWF the opportunity to take on the entire Santa Elena purchase.

That night, long after the others had left for San José, I asked Janzen whether the Guanacaste project had gotten to the point he had hoped it would by July 25, 1987. It was drizzling, and he stood in the doorway of his house munching a leftover sandwich. No, he said, he had wanted it to be further along. He had wanted the big properties on the east side of the Pan-Am Highway in hand, mainly. The fact that Santa Elena hadn't come along wasn't a major concern. He had expected that piece to be "the last one on the list."

He hoped to have "the money in the bank" by June of the next year. By then, he hoped that other Costa Ricans would have joined García and Quirós in the GNP administration, that the administration would have coalesced into a group that could find its own way, that the Park Service would be shaped into a supportive structure, and that he, Janzen, would be reduced to the role of mere adviser.

"A year from now, basically, I say to Randall, 'Buddy, you're totally on your own, and I'm around now as somebody you can ask opinions of, but basically you're on your own.' That's already starting now, and it's going to happen more and more. There are going to be all kinds of failures in the system that are going to drive some people nuts, but if they're professional administrators of parks, they will want to stick with it.

"I don't see that for myself. That's not my goal. My goal is to get this thing started up—basically self-running—and then walk away from it."

# · *13* ·

# The Green Magician

At 356 West Fifty-eighth Street, in a second-floor conference room of WNET, the public television station in New York City, Dan Janzen and Winnie Hallwachs sipped wine and nibbled cheese with the powers that be. It was Monday evening, December 14, 1987, and the VIPs had gathered for the screening of the BBC-PBS film about the Guanacaste project. George Page, the venerable voice of the *Nature* documentaries on American public television, was the emcee, and among the few dozen other people were representatives of conservation groups, philanthropic foundations, and the news media, including a writer for the *New Yorker*'s "Talk of the Town" section.

About six o'clock, Page gave a brief introduction to the film and pointed in the audience to Janzen, whom he noted was wearing the "native dress" of New York—in other words, a business suit. Geoff Barnard of the Nature

Conservancy, which presented the film in association with WNET, said a few words about tropical deforestation, concluding, "With Dan Janzen working in Guanacaste, there's hope."

The film, *Costa Rica: Paradise Reclaimed*, ran about fifty minutes. It opened with narrator Page, standing in the woods, saying that Janzen was conducting an "unprecedented experiment" in trying to restore the tropical dry forest. It cut to Janzen climbing a tree, taking photos, and climbing down. "Janzen wants to grow his paradise here," Page said as an aerial shot panned the landscape west of the cordillera, "across this patchwork of ranches and farms." Janzen, cast as the eccentric Great White Explorer, then rescued a coral snake from a well, fought a fire, captured a boa constrictor to show a group of schoolchildren, climbed another tree, walked through several parts of the forest, led a group of visitors in discussions in the woods, measured the circumference of guapinol trees, sat in his house pinning moths, typed on the Mac while a frog slurped up beetles in the foreground and Hallwachs typed in the background, collected moths at a black light, held a margay kitten brought to him by Julio Quirós, and shook Arias's hand at the Casona ceremony.

Meanwhile, Page described the restoration plan. There were plenty of visual vignettes about birds, monkeys, rodents, and leaf-munching caterpillars, and several story points centered around the biology of fig, guanacaste, and guapinol trees. A few local farmers and Guanacaste project workers appeared in various scenes, and Hallwachs was featured in one sequence, sitting on the forest floor photographing agoutis. However, about the only other time Janzen wasn't sloshing around in his unlaced boots with wild shocks of gray hair rising from his head like puffs of smoke, or with a rag wrapped around his skull like a buccaneer, was when he strode down a New York street in "native dress," albeit with a worn rucksack strapped over his shoulders, to address a roomful of suits about the project. At one point, in the forest, he explained the biology of the ant-acacia system while standing behind an acacia, peering over his reading glasses at the camera, brushing off the stinging ants, and rubbing his fingers, hands, and wrists as he talked of thorns, nectar-producing glands, and Beltian bodies. The Great White Explorer didn't even wince. That much hadn't changed since the 1973 film, *Ants and Acacias: An Ecological Study*. He even pulled a leaf from the tree and ate it, and compared its taste to lettuce. Another time he popped a piece of guapinol fruit into his mouth, noting, between munches, that "they're really dry." In front of the black light, insects crawled on his face, and he coughed on moth dust as he explained some basics of insect collecting.

The closing scene was perhaps the most peculiar. It was another aerial shot of Janzen, who was acrobatically propped near the top of the *Ateleia* tree in the Clifftop plot, legs spread on the branches like a contorted iguana. He waved his arms and shouted directions to Hallwachs in the field below, where she stood as a reference point for photographs he was taking, or at least pretending to take.

As the shot panned behind Janzen and moved southwest across Santa Rosa to the Pacific, Janzen shouted excitedly, "Keep going . . . Whoa, whoa . . . Right there . . . Yeah, that's better. Hey, I'm going to fall out of the tree!" Closing the narration, Page again referred to "Janzen's paradise," which was "growing from a few small seeds."

What fun to be a video producer with material like this. The film did an admirable job of doing what many nature films do: tell the story of a unique conservation project, profile an interesting person in the middle of it, and expose viewers to some fascinating bits of natural history. If the filmmakers had engineered it so that Janzen stole the show, so what? For anyone wondering whether such a film should be natural history entertainment or natural history education, well, the answer was clearly "both." Janzen's role had simply made the Guanacaste project story more appealing and more energetic, even though it was a fascinating and important story in its own right.

On one level, the WNET gathering, and the screening of the film, which would be viewed on public television stations across the United States a few weeks hence, represented a culmination in the international recognition for the Guanacaste project. It was the kind of recognition that Dan Janzen, the conservationist, treasured. Such a film was likely to gain more public attention for the Guanacaste project and encourage donors. On another level, the event confirmed the worst fears of Dan Janzen, the scientist. It focused as much or more on him than on the Guanacaste project. Even for someone who knew tropical field biology, who had an ecological education, the film unequivocally portrayed him as an eccentric, zealous genius in the far-flung jungle, heroically saving Mother Nature and the natives as he probed its wonders. In that sense, although there were no caimans in the film, it was just more made-to-order crocodile-chasing for the gringo armchair audience.

In an interview years later in his office at Penn, I mentioned to Janzen that when I asked any of the Costa Ricans involved in the Guanacaste project to discuss the effort, they always began by talking about Janzen. "Very human," he said, agreeing that it was a phenomenon of media exposure. This was what happened when a personality became the story, as it did in *Paradise Reclaimed*.

During the final phase of the BBC production, when Janzen and Hallwachs traveled to England as advisers on the film, they realized the BBC team would exert total editorial control. Janzen argued over several points in the documentary, becoming especially angry that the team kept promoting his role to the exclusion of everyone else's. He recalled the meetings in a sarcastic tone: "Dan Janzen's project, Dan Janzen's project, Dan Janzen's project, Dan Janzen's project, and on and on. I learned not to believe those people at all."

Tens of millions of people saw the film in the United States, but Janzen did what he could to keep it out of Costa Rica. He claimed he "prohibited" its distribution there and, at least as of 1999, it had never been publicly shown. "I really got obnoxious with the BBC," he told me. "Several clandestine copies have shown up in Costa Rica and I've piled on the Costa Ricans that have them. So it's been seen on home videos and things like that."

His great fear was deification. "The minute you start deifying somebody in that society—how shall I put it?—the resentment and jealousy levels by a big piece of society become enormous," he said. The part of society that deifies the "hero" pays attention to him and neglects the project. "That, then, makes the people who are actually carrying this thing out on the ground feel woefully neglected."

The problem was that he needed media exposure in the developed world to raise money. "There ain't no way I can do this without being visible," as he put it. Janzen wasn't aware of the down side of visibility in Costa Rica when he started the project, but it didn't take long to figure out. "We had to lower our profile as low as it could be lowered."

When I asked him whether the risk had been worth taking, he said, "I still don't know. That's a really tough call."

Managing your profile was one media lesson Janzen learned. Another was the frequent powerlessness of someone like him to communicate the truth as he saw it to the public. He developed a cynical view of the media, believing that reporters, with few exceptions, basically wrote what they wanted regardless of the facts, twisting stories depending on personal beliefs or those of their editors.

For example, on July 4, 1986, Umaña and Gámez, in a move to convince the president about the merits of the Guanacaste project, brought Margarita Peñón, Arias's wife, to Santa Rosa for a show-and-tell. A couple of weeks later, back at Penn, Janzen got a frantic call from someone in Costa Rica saying that the president's wife was extremely upset about articles appearing in U.S. newspapers saying that Arias knew about the airstrip because an American biologist had shown it to his wife on the Fourth of July. Janzen got on the phone and ascertained that no such article or television story existed, even though it had been widely reported in Costa Rica that it did. Janzen wrote a letter explaining that to Peñón, and that any such story could not have come from him. He later learned that the rumor had begun at a dinner party of journalists in San José, where someone confused the fact that Janzen, Umaña, and Gámez had flown over the airstrip with the fact that Peñón had visited Santa Rosa the same day. The story might have been created within Costa Rica to discredit Arias.

The point was that Janzen was put in the middle of a false story involving the First Lady. To Janzen, then quite naive about the news media, it showed how loose the truth could be, and how careful he must be about "appearances" when talking with journalists. It also showed how reporters can become part of "little dances."

Janzen said he was "dumbfounded" to discover that stories often originated because of the wishes of powerful interests outside newsrooms. He cited early articles about the Guanacaste project that resulted when friends of his lobbied editors to do "a propaganda favor for the project." No doubt the editors involved believed they had simply been tipped off to a good story. Janzen was further surprised to hear from donors who, upon learning that such articles were impending, asked him to make sure the reporter gave top billing to that donor.

Janzen found that reporters, not so much in the United States or Europe as in Latin America, almost as a rule made mistakes or interpreted statements in wildly eccentric ways. As a result, he changed his method of giving complete answers to questions. "It just sort of dawned on me that if I don't want to make these mistakes, then I have to simply start editing my thoughts and only telling parts of the story," he told me. With some reporters in Costa Rica, before he delivered an answer, he would think to himself, "What is this reporter going to do with this sentence?" In other words, Dan Janzen had learned to act like a politician.

These realizations about the media seemed to disappoint him, like a kid learning the truth about Santa Claus. "I grew up with this idealistic childhood view that newspapers at least strive toward describing things more or less as they were going on," he said. "I was never taught that when you sit down and read newspaper reports of a historic event that what you're reading is something that was manipulated from the very start."

After the film screening, Janzen was asked to stand and make comments. "I'm not used to being cast in this role," he said, tactfully. Instead of giving a talk, he asked for questions. To one about whether squads of humans were replanting trees, Janzen segued into a discussion of how unique it was to even need to consider the makeup of the restored forest. "What the park is going to be for the next three thousand years—that's a question that never before needed to be addressed."

Someone asked how conservationists could reach audiences to "raise funds for saving the rain forest." Janzen said they should focus on specific projects and explain the complexities of each. As he talked, I recalled something he'd said to me in July about North American audiences. Basically, he had become bored with them, as well as others in the developed world. "They're jaded," he said. "I'm so tired of being viewed as a television set. I'm just entertainment in most places in the States." When he gave a public talk, it struck him that it didn't matter whether he put a lot of effort into making it high in quality. "It doesn't cause them to think about their own lives more, doesn't cause them to think about other people's lives more, doesn't cause them to think about nature more. It was just an enjoyable hour. And then basically they just switch off and go on and do whatever else they were going to do the next hour."

European audiences were more interesting and more interested. He became inspired by the curiosity he found in Swedish audiences. The best of all were audiences in the developing world. In Liberia, for example, he could see a fraction of the audience change, actually see their mental horizons expand, as they soaked up new ideas. Perhaps this was a consequence of it all happening in their own backyard, a backyard that North American audiences seemed unable to grasp was theirs, too.

Afterward, Janzen mingled a bit more with the foundation people, looking like he was simply going through the motions with diplomatic discretion, waiting for the right time to leave. They seemed to enjoy basking in the presence of the "star."

He had just been named to the Global 500 Roll of Honour of the United Nations Environment Programme, and *Outside* magazine had named him "Man of the Year." Coincident with the film presentation, the December 1987 issue of *BBC Wildlife* magazine contained an article on the project by the distinguished nature writer Jonathan Maslow. The article itself was an excellent, straightforward, nonlaudatory treatment of the restoration effort and the research that Janzen had done throughout his career to make it possible. A sidebar even contained information on how to make donations to the project. But the headline writers took the usual liberties, for entertainment's sake. On the cover, they called Janzen "The Green Magician of Costa Rica." They labeled the story "Doctor Dry-forest," and the index referred to him as the "miracle-man of Mesoamerica."

Later, Janzen and Hallwachs talked about recent developments in the Guanacaste project's most formative year as we walked a couple of dozen blocks down Eighth Avenue to Penn Station, past vent people shivering in the cold night, and waited on a wooden bench in the station for their train to Philadelphia. Two events seemed foremost on Janzen's mind: one, the upcoming Seventeenth General Assembly of the International Union for Conservation of Nature and Natural Resources, commonly known as IUCN, to be held in Costa Rica, and two, the resolution—just that afternoon—of some fast-moving, opportunistic negotiations for a key piece of land.

The IUCN, whose director general at the time was Kenton Miller, was set to meet in San José from January 29 through February 10, 1988, bringing about 900 representatives of conservation organizations from around the world. Janzen would give a twenty-minute talk on biocultural restoration. Among many tours offered the delegates, IUCN would sponsor two to Santa Rosa. On February 2, Janzen was scheduled to give a tour of the Guanacaste project to Prince Philip, Duke of Edinburgh, one of the world's leading supporters of conservation, who was attending the IUCN meeting. "We can't have any reporters," Janzen told me. A few days after that, he would host some "money people," and a few days later, several officials of the World Bank. And who knew what other VIPs would wander through?

The land deal was complicated, but boiled down to this: Cecil D. Hylton, a millionaire who lived in the Washington, D.C., area and owned many parcels of property in Guanacaste, wanted to donate one of them to some institution by the end of the year to get a tax deduction.

Hylton was one of Guanancaste Province's major absentee landowners, and a bit of a mysterious character. During the 1980s, as the Reagan administration escalated its anti-Sandinista policy, he had bought property in Guanacaste, including Hacienda Orosí and Hacienda El Hacha. The land touched the foot of Cerro El Hacha and ran from the Pan-Am Highway to the base of Volcán Orosí. Much of it was only about 12 miles from the Nicaraguan border. Several people interviewed by Marc Edelman, author of *The Logic of the Latifundio*, told him that El Hacha and other Hylton properties in Guanacaste served as bases for Nicaraguan contras, although Edelman points out that it's possible that on such vast, rugged terrain an owner, especially an absentee one, might never know such a group operated there. Besides, contra activity was concentrated more to the east in Costa Rica, albeit some on ranches owned by other U.S. citizens. Still, one Guanacastecan cited anonymously by Edelman said he found it intriguing that Hylton showed up in the northwest acquiring land just after the Sandinistas took power and the Reagan administration launched its campaign against that government. One of Hylton's companies, Hylton Enterprises Virginia, Inc., bought another huge tract along the Costa Rica-Panama border a few months before the 1989 U.S. invasion of Panama. In the mid–1980s, Hylton donated more than a dozen properties near Cerro El Hacha, worth several million dollars, to the Assemblies of God denomination. The church was headed by Jimmy Swaggart, the North American fundamentalist minister who led Protestant evangelical rallies throughout Central America.

In November of 1987, Gustavo Echeverri, the man who managed all of Hylton's properties in Guanacaste Province, went to Santa Rosa to help one of Echeverri's assistants sell his small farm in the Zona Protectora, an offer that Janzen jumped on. Echeverri also told Janzen that Hylton wanted to donate 28 square miles of land in a different area—to the south of Santa Rosa, outside the proposed GNP boundaries. The land was a cattle ranch known as Horizontes. The deal had to be done quietly, Hylton insisted, and it would have to be completed on December 9, when Hylton would be in San José. To be legal, the donation would have to be received by the Nature Conservancy representative in Costa Rica, Robert Wells.

Unfortunately, Wells's power of attorney had lapsed, so Janzen, Umaña, and others did what they could to push his renewal process through an agonizing series of hurdles within the Tico bureaucracy. Hylton met with Wells in San José on December 10, a day late, and when he found that the deal couldn't go through right then because the renewal process hadn't been completed, he became angry and threatened to donate the land to Swaggart instead. After some persuasion, he agreed to wait until 5 p.m. on Monday, December 14, the evening of the film screening in New York. At 2 p.m. that day, a Costa Rican law student who was Wells's assistant finally pushed the renewal through the bureaucracy, and a short time later, with Hylton present, Wells signed the acceptance document for the donation. The Horizontes deal was done.

Throughout the process Janzen peppered everyone involved with phone calls, except Hylton, and did any legwork he could think of. He wrote the acceptance document that Wells signed. He worked through Echeverri to keep Hylton on board when things looked bleak. He commiserated with Umaña and others about how a low-level bureaucrat could hold the fate of a million-dollar land deal for conservation in his hands. He was even on the phone to San José before the screening that afternoon, checking progress. It was a lesson he had learned as a child in the North Woods: A hunter is successful only if he pursues his quarry persistently and unerringly until it lies at his feet.

On December 10, Janzen sent a memo to more than 600 contributors to the Guanacaste project, giving them a report on the project's status. Their money had been spent with "miserly care," but he asked with "calm urgency" for more contributions. "The nestling is poised for professional flight." He mentioned the soon-to-be-aired *Paradise Reclaimed*, suggesting that they invite friends to view it and "make a fund-raiser of it." He listed thirteen foundations that had made grants and nine countries from which private and government donations had come.

The Guanacaste project had by then received $8.2 million in donated land and cash, just $3.5 million short of its goal. Including Santa Rosa, 53 percent of the project's land had been bought or was otherwise in hand, and down payments made on another 19 percent. The remaining land consisted of small pieces costing from $10,000 to $50,000 and four large pieces ranging from $300,000 to $1 million each. A new ranger station and biological station were under construction on Guanacaste project land on the northeastern flank of Volcán Orosí, as were a new dormitory, laboratory hall, and administration building at Santa Rosa. If the $3.5 million still needed for land purchase could be raised, it would generate at least another $13 million in endowment to run the park, based on the debt-purchase mechanism.

Six resident managers from local haciendas that were purchased to become part of the project had been hired as site managers, including at Cerro El Hacha and Cacao, both of which had been used extensively in 1987 by researchers, volunteers, and workers in reforestation experiments. Poaching had been "reduced to zero," and a fire-control program had been put in place in the area outside Santa Rosa. Planning had begun for planting about 250 acres of native dry-forest trees when the rain started in May.

As if to head off any impression that he was dominating the leadership of the project, high up in the memo Janzen placed the following paragraph: "GNP's Costa Rican authorship is multiple and expanding: Espinoza, Madrigal, Gámez, Boza, Umaña, Varela, Arias, Coronado, Quirós, Mejia, Ugalde, Bermudez, Gonzalez, Rodriguez, Oduber, Echeverri, Brenes, Solórzano, García, Lara, Eras, Traña, Araya, Murillo, Bassey, Marín, Burgos, Baltodano, Mendez, Vargas, and growing." An interim executive committee made up of officials from Costa Rican agencies and foundations had begun to produce "detailed philosophical and managerial guidelines for GNP," and a management plan was expected by the end of 1988. The GNP staff was all Costa Rican.

Although Janzen had written in the memo that "all the owners are eager to sell and our only delay is funding," that was not quite accurate in the case of Santa Elena, the single biggest unacquired piece. Several of the big landowners on the east side of the highway were "really on my back about finding money," he told me later. However, he wouldn't talk about Santa Elena. I eventually learned that Hamilton, through the lawyers that represented him, didn't really seem all that interested in reaching agreement.

Meanwhile, from Ithaca, New York, the U.S. rain-forest mafia weighed in with a letter to the editor of the *New York Times*, published in October. Thomas Eisner of Cornell, Edward O. Wilson of Harvard, and Peter Raven of the Missouri Botanical Garden asked for help in saving "a prime tract of forest," as they called it. "The land, the so-called Santa Elena Hacienda, which until recently was the site of the Central Intelligence Agency-sponsored airstrip for the support of the insurgents in Nicaragua, comprises 40,000 acres midway between two sections of Costa Rica's principal national park, the Santa Rosa Park." Purchasing Santa Elena would consolidate a large natural area and bring under protection an area of great biological and geological importance, they said, adding that the World Wildlife Fund had pledged $100,000 and was trying to raise the remaining money needed. "Our purpose is to call the Santa Elena Project to broader attention, in hopes of generating individual contributions. At a price of $4 an acre, the venture may well be the conservation bargain of the century."

Despite its risks, Janzen's media campaign was paying off. Many of the heavies who attended the IUCN meeting in Costa Rica made the five-hour drive up to Santa Rosa to get an on-site briefing on the Guanacaste project. Among them were high-ranking officials of several philanthropic foundations, conservation groups, and the World Bank. As the years progressed, donations from the foundations represented during those visits became crucial in land purchases.

Although it made little difference in raising money for the project, the visit of Prince Philip showed that its message had reached far and wide—and high, if you were the kind of person who believed in highnesses. The prince, who had broad knowledge of national parks around the world, had heard about the Guanacaste project, and when the Arias administration suggested he visit, he accepted. The purpose was merely educational, but Janzen cooked up something special.

The visit was not publicly announced; it was the prince who insisted on having no journalists. Security officers from both Costa Rica and Britain accompanied him. The prince flew his own plane from San José to Liberia with a few passengers, including Gámez and Umaña. At the Liberia airport, they were met by Janzen and others from the Guanacaste project and Park Service. Janzen drove his Land Cruiser, the lead vehicle in the caravan, with Prince Philip in the passenger seat. Lord Buxton of Alsa, a long-time conservation adviser to the prince, rode with them. The prince engaged the others in a conversation about the history of Guanacaste Province and the political conflict between Costa Rica and Nicaragua. They also discussed the Guanacaste project and the ecology of tropical dry forests, comparing northwestern Costa Rica with African dry forest regions that the prince knew well. "He had done his homework," recalled Gámez, another of the passengers.

Janzen led a tour of Santa Rosa, walking through the forest and giving his standard lecture about tropical ecology. The caravan stopped at the Clifftop plot, where park guards had stationed themselves all around the tall jaragua field on the side of the road opposite the regenerating forest. Janzen led the prince up to a corner of the field, the jaragua gleaming in the sunlight.

"Now, what I want you to do is set this on fire," Janzen said.

"Aaaahhhh!" The prince hesitated, not knowing whether Janzen was serious. Here was one of the world's most prominent conservationists, about to put a match to a Costa Rican national park. What was going on?

What Prince Philip didn't know was that it was time for the jaragua field's annual burn as part of the continuing experiment at Clifftop. Janzen told him repeatedly that he was serious about the request. Finally, with several dozen people watching, the prince reached into his pocket, pulled out a gold cigarette lighter, flicked it on, and placed it into the jaragua. Umaña, mugging with the prince for photographs, had his back to the rapidly expanding fire. "Umaña thought this was the greatest piece of showmanship, of him and Prince Philip burning up the park," Janzen recalled. "He got so excited about it, he forgot to pay attention to the fire." Umaña didn't actually catch on fire, but the prince and others found the close call quite amusing. As usual, park guards and volunteers contained the controlled burn within the field.

The strange beauty behind the idea to have the prince set the fire lay in the fact that fire was still an extremely controversial issue in early 1988. Some conservationists, including some in the Park Service, did not think fire could be controlled. And some still did not understand that controlling fires was the key issue in conserving and restoring the Guanacaste forests. In essence, this world-famous conservationist was appearing to cast *his* vote on the issue. "It's one of those crazy things that occurs to you," Janzen told me, explaining how he got the idea to involve British royalty. "We had to do it anyway, so why don't we have Prince Philip set the match? Because, you know, he's this great conservationist, and here he is, setting fire to a Costa Rican national park."

The group drove to Cerro El Hacha to view some land newly purchased by the project. The convoy left the blacktop road and drove along a dirt road, which, since it was the dry season, kicked up huge clouds of dust. Gámez remembered how Janzen's Land Cruiser pulled up at a barbed wire gate typical of Latin American ranches, in which a pole holding several strands of barbed wire across the road was stuck at the top and bottom in wire loops of a fence post. A few moments passed.

"What are we waiting for?" Prince Philip asked.

"For you to get out and open the gate," Janzen shot back. Gámez saw a look of horror on the face of Lord Buxton, who was maintaining an air of formality.

"Oh yes, I'm sorry," said the prince, who put on no such airs. "I forgot I'm supposed to do this."

He jumped out of the jeep, pulled back the gate, stood amid clouds of dust churned up by the passing vehicles, reattached the gate, and jumped back in the jeep. As the column wound its way over the countryside, he repeated the performance several times, refusing the entreaties of the security agents to let them do it instead. The Costa Ricans were struck by the prince's relaxed demeanor as he peppered them with questions, made jokes, and did his part to keep the tour going. The tour continued until late in the afternoon, when the entourage returned to the Liberia airport for the flight back to San José. The visit left a lasting image in the minds of those who worked on the Guanacaste project. As one of them put it, "It was the day that Prince Philip burned the park."

If the visit of Prince Philip was memorable for its symbolism and royal cachet, the appearance a few days later by two Swedes became memorable for igniting the Guanacaste project's grandest financial achievement.

One of them was Gudren Hubendick, a representative of the Swedish International Development Authority, or SIDA, the Swedish equivalent of the U.S. Agency for International Development. She had been brought to Guanacaste by Max Segnestam, director of the Swedish Society for the Conservation of Nature. The Guanacaste project had, over the past two years, received a large amount of publicity in Sweden. That generated $260,000 in private donations. Students in field ecology courses from Swedish universities had come to Santa Rosa each March for three years to participate in the course Janzen ran for Penn students. He was hopeful that some of the students would return to work with research scientists or as volunteers. Janzen had sent a proposal to SIDA in May of 1987 for $150,000 to buy land, mainly small inholdings in the Orosí Forest Reserve.

The same proposal requested $649,000 for what he called "infrastructure development," that is, education, management, research, and reforestation. The Arias Nobel in October 1987, although awarded by Norway, had heightened interest in Sweden in many things Costa Rican, including what was happening with the Guanacaste project.

Janzen had driven himself hard during those weeks, fighting fires, leading the field course, hosting VIPs from the IUCN meeting, and pushing the fundraising campaign. He was deeply fatigued but refused to rest. Randall García and other Guanacaste project staff members showed Hubendick around for a day. Janzen and Hallwachs joined her in the afternoon of the next day, and as they drove near the guard shack at the park entrance, Hubendick said, "Where's a book where I can read about all this?"

Janzen apologized and said there wasn't one. Then he added a half-serious but quite wild idea, whose wildness he later attributed to his near-burnout condition: "We need three-and-a-half-million dollars to raise more. You find me the three-and-a-half-million dollars and I'll use the time I would have spent on it to write the book for you." The conversation ended. Janzen thought nothing more of it.

Unbeknownst to him, when Hubendick returned to San José she went to see Umaña. The government of Sweden had promised Costa Rica $2.5 million for conservation in recognition of the Nobel. Hubendick told him that she wanted the money to go to the Guanacaste project, and she thought she could come up with another million. The next day, a Tuesday, she called Janzen and said, "How soon could you have a proposal that says what you would do with this $3.5 million?"

Janzen and Hallwachs immediately took a Macintosh and drove down to the Hotel Galilea in San José. They wrote the proposal on Wednesday and gave it to Hubendick on Thursday. She took it back to Sweden, talked Parliament into giving the other million, and a few months later called Janzen in Costa Rica to say, "You've got your $3.5 million."

Those words were all the "Green Magician" needed to hear to pack his "native dress" and head to Wall Street, sensing the ultimate victory in the Guanacaste project's fund-raising campaign.

Winnie Hallwachs photographing a flower, Volcán Cacao, July 1987.

Volcán Orosí (left) and Volcán Cacao behind Daniel Janzen as he stands on the Monument to the Heroes at Santa Rosa National Park, July 1987.

Volcán Orosí rises in the background as two members of the Guanacaste Conservation Area fire program, Alex Rodriguez and Didi Guadamuz, watch the landscape below for signs of smoke during the early 1996 dry season.

Fuel waiting for a spark: a field of jaragua at Santa Rosa, June 1986. George Godfrey takes a photo in the foreground.

The front of the Janzen-Hallwachs house at Santa Rosa, July 1987. Janzen and Frank Joyce record data in a study of trogons and the caterpillars they feed their nestlings.

The Casona—the historic ranch house of Santa Rosa, June 1986.

A remnant of Santa Rosa's legacy of patriotism: an armored vehicle left after the battle of Santa Rosa in 1955, in which a Tico militia fought off an invasion from Nicaragua led by Costa Ricans defeated in the 1948 War of National Liberation.

The once-secret airstrip at Santa Elena, March 1988. Note dead trees placed to block the strip.

A nursery at Cerro El Hacha early in the Guanacaste project, July 1987.

Playa Naranjo (Naranjo beach) on the Pacific Ocean, June 1986.

Ian Gauld of the Natural History Museum, in London, sets up a black light and sheet for a night of insect collecting on the Atlantic rain forest slope of Guanacaste Province, June 1986.

A seasonal river in Santa Rosa National Park. Canadian researchers Lauren (left) and Colin Campbell study how fish distribute themselves in the pools, June 1986.

Dormitory at the new Cacao Biological Station, July 1987. (Janzen climbing stairs.)

A view of Playa Nancite, site of arribadas of olive ridley sea turtles, June 1986. The white, ball-shape objects are remnants of turtle eggs.

The cliffs of Santa Rosa between Playa Naranjo and Playa Nancite, June 1986. *(Photo by George L. Godfrey.)*

Volcán Orosí (left) and Volcán Cacao, their tops partly covered by clouds, June 1986. Note pasture areas interrupting forest landscape on volcanoes and lowlands.

*All photos by William Allen except for the photo at the top of this page.*

# ・ *14* ・

# Touch of the Money Spider

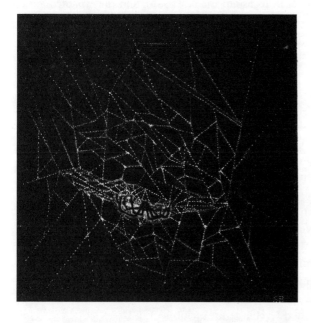

Money spiders, the common name for a group of web-spinning spiders known in many parts of the world, are members of the family *Linyphiidae*. Some species of *Linyphiidae* spin elaborate webs in the forest, hanging there until an insect flies in and becomes entangled among the sticky threads. The spider slowly pulls its captive through the webbing, ties it down with more silky lines, and goes back to repair any damage. Other species of *Linyphiidae* prefer dark places, scurrying beneath leaf litter on the forest floor, under the leaves of living trees, or in animal burrows. They play an important role as predators, and in turn serve as a food source for parasitic wasps and other enemies.

Spiders in general, and money spiders in particular, have long been associated with good fortune. In the folklore of Italy, France, Germany, and the United

States, sighting any kind of spider and letting it live meant good luck. As a school rhyme put it, "if you wish to live and thrive, let the spider run alive." In England, a money spider walking across your hand or thrown over your shoulder was said to bode well for future financial matters. In many parts of the world, touching a money spider without injuring it was believed to bring wealth.

As scientists, the Guanacaste project leaders cared little for such folklore, nor did they particularly focus on the fact that by restoring the forest they were saving countless populations of spiders. Yet when the Swedes pledged $3.5 million, what happened next might have stunned even the most ardent believers in the superstition of spiders and money. On getting that call from Sweden, Janzen began to push Alvaro Umaña and others to use the donation in a debt-for-nature exchange with commercial banks. The result was the largest single exchange in the history of conservation.

Debt-for-nature exchanges, or debt swaps, were a creative finance mechanism that conservation groups began to use in the late 1980s, and they marked a significant evolution in the way these groups used their skills and resources. Conservationists were now pursuing a more direct and sophisticated financial approach to protecting land, in addition to the standard publicity, fund-raising, and lobbying.

Applying a debt swap to conservation was the brainchild of the Smithsonian Institution's Thomas Lovejoy, who in 1984 first noted the link between the heavy external debt burdens of developing countries and the pressure to liquidate natural resources to pay the debt. "Stimulating conservation while ameliorating debt would encourage progress on both fronts," Lovejoy wrote. He proposed that conservation projects be financed by a debt-relief mechanism. The first debt-for-nature exchange was initiated on July 13, 1987, between Conservation International (CI) and the government of Bolivia. CI purchased $650,000 of Bolivia's debt in exchange for the government's pledge to set aside 3.7 million acres of forest and grassland in the Amazon basin. Dozens of debt swaps followed, including in Ecuador, Costa Rica, the Philippines, and Madagascar. The first one in Costa Rica involved a $5.4 million face-value debt purchase by the National Parks Foundation with funds provided by several conservation groups and foundations. It benefited several parks, including GNP.

Debt swaps were meant to relieve a small amount of debt, generate publicity, and create support for conservation within the home country. But the main motivation was to leverage conservation funds into a larger amount of local currency. In essence, such exchanges converted a country's debt, usually that held by a commercial bank in the developed world, into money for use in a specific national conservation project. The process often involved four parties: a debt-purchasing conservation group or government in the developed world, the commercial bank, the debtor country's central bank, and a nongovernmental conservation organization in the debtor country. For the commercial bank, the debt swap helped retrieve *some* return on a risky loan, and in some cases the bank got a tax write-off. For the central bank, it helped lessen, if ever so slightly, the poor nation's external debt burden.

There were many kinds of debt-for-nature exchanges. The following hypothetical example roughly illustrates a typical one for Costa Rica. Say the Central Bank owed a U.S. commercial bank $300,000. Because of declining export crop prices and consequent low foreign-exchange reserves, Costa Rica was having difficulty paying off the principal and interest. So the market value of the loan had dropped to 33 cents on the dollar, or $100,000. A U.S. conservation group bought the $300,000 debt notes from the bank for $100,000, a sum raised by various donations. The $300,000 in debt notes were returned to Costa Rica by the conservation group.

At a conversion rate of 75 percent, agreed to in advance by the U.S. conservation group and the Central Bank, the Central Bank issued Costa Rican Stabilization Bonds in colones, the local currency, to the National Parks Foundation, which worked in concert with the U.S. conservation group. That meant the bonds totaled 75 percent of the purchased debt's face value of $300,000, or roughly $225,000. Also in advance, the Central Bank and conservationists agreed on the specific conservation project to which the proceeds would go. The bonds would generate far more than $225,000 for the project as the government used tax-generated funds to pay interest and, at maturity, the principal.

During this process, assumptions were made about projected interest rates and other market variables, dollars and colones were converted at different times, and conservationists and Central Bank officials negotiated to balance concerns ranging from reducing the national debt to minimizing the threat of inflation. The bottom line was that the U.S. conservation group and Costa Rican conservationists got a better than two-to-one match of the original $100,000 in donations. Meanwhile, they could immediately borrow money from private banks to purchase land, using the bonds as collateral.

Some conservationists said that because a debt swap reduced foreign debt, it helped temper the drive for a poor country to liquidate forests. Some argued that it helped slow global warming because it left more trees to absorb carbon dioxide.

On the down side, printing local currency to buy the debt could feed inflation. The key, some conservationists countered, was to arrange the swap so that the poor country paid slightly less to the conservation project than it was already paying in interest and principal on the external debt. Some countries felt debt swaps were an intrusion on national sovereignty, since it appeared that politicians and conservation groups from rich nations were telling local officials how to manage threatened ecosystems. As Brazilian President José Sarney put it in 1989, "persuasion is giving way to attempts at intimidation." Proponents of debt swaps noted that a swap could not occur without the consent of local authorities.

Debt-for-nature exchanges were not a panacea, for either the debt crisis or biodiversity crisis. At best, they were a "marginal solution," as one conservationist put it, if for no other reason than the amount of debt relieved was minuscule compared with the $1.3 trillion debt owed by the developing world. Yet they were an important development in tropical conservation history, and the success of early exchanges engendered hope for broader use. Even if they weren't a dramatically effective response to global debt and environmental destruction, debt swaps could address big problems on a local scale. For the Guanacaste project they were crucial, helping raise capital to buy land and establish an endowment for the long-term operation of the conservation area.

Thanks to Umaña's pushing, Arias and other key advisers endorsed debt swaps for Costa Rica. In mid–1987, Costa Rica launched the debt-for-nature exchange as an institutionalized process for all its conservation and natural resource management programs.

As the Swedish debt swap appeared on the horizon, a nonadversarial environment prevailed in Costa Rica among the Central Bank, foundations, donors, and Ministry of Natural Resources. Umaña recalled that although he had second thoughts about the swap, partly because it would create within the national park system a "rich kid on the block," he nonetheless supported it because "on the whole it was a very great thing." An agreement that about 15 percent of the proceeds would go to other parks helped assuage that concern.

Umaña gave Janzen permission to pursue a swap, and Janzen and Rodrigo Gámez briefed the Swedes on the idea while Umaña, Gámez, and Pedro León began preparing the way in the halls of Costa Rican government. "We had to run, make phone calls, convince people, do all the things you need to do when you're working against a deadline," León told me. Randall Curtis, the Costa Rica program director for the Nature Conservancy during these years, assisted in both Costa Rica and the United States.

The motivation behind the Swedish donation was to conserve biodiversity for the socioeconomic benefits it could provide. "In their minds they were funding not the establishment of a huge national park, but a social development program for the benefit of people," Gámez recalled. This theme, which Gámez, Janzen, and others had sounded throughout the Guanacaste project, was a key selling point in San José.

The Central Bank, headed by Eduardo Lizano, was the main decision-making institution in the swap. The conservationists discussed with Lizano the Swedish money and resolved issues such as overhead, rates, and execution of the process. The Swedes picked Salomon Brothers, Inc., the New York investment house, to find Costa Rican debt in commercial banks at optimal discounts. Salomon would buy the discounted debt and deliver it to the National Parks Foundation through the Foundation's agent, the Costa Rican bank Bancoop, or

Banco Cooperativo Costarricencse. The Central Bank would then give approval to exchange the debt notes for colon bonds.

In the fall of 1988, Janzen gave a talk on the Guanacaste project at the Salomon Brothers office in the World Trade Center. His audience was a team of half a dozen or so young traders, lawyers, and others who would be involved in the debt swap. They were all in the prime of their running careers, so to speak. It was lunchtime, and as he spoke, other Salomon Brothers employees shouted orders to each other as they scurried among phones, computers, and desks. Janzen, the fund-raiser, had gone there to give his pep talk, and no doubt Janzen, the biologist, was watching this ecosystem in action.

"I talked to them about the idea, said we needed to do this debt swap, and asked if they would help us," Janzen recalled. "I told them I wanted them to do it for free, which they did. They fell in love with the idea." Although they had never even been to Guanacaste, and although this was the decade of unbridled greed on Wall Street, the idea inspired them. To use their skill and initiative for a cause like restoring a tropical forest seemed like a novel, strangely stimulating task. Something about the Guanacaste project, and the way the evangelistic Penn professor described it, made these masters of the universe want to go for it. As TNC's Curtis put it, "he appealed to their better side."

A few weeks later, they got on their phones and computers and, with unexpected energy and success, scoured the world for bargains on Costa Rican debt in commercial banks. As each piece of debt was located, Gámez worked in San José to obtain approval of its legitimacy from the Central Bank, which tracked the debt's location, registry number, and other financial information. When the bank gave approval, word went back to New York, and the team bought it. "Salomon would buy the piece of debt, and then go look for another piece, and then they would trade it amongst themselves to get better deals," Janzen said. "They were working all over the world. They were working on Sunday nights. It was incredible." The traders kept pouring their time and energy into the effort, pushing into early 1989.

Janzen shepherded it all. He and Hallwachs rose in Philadelphia at 4:30 each morning and went to their office at Penn. Janzen would spend much of the morning on the phone to Sweden, much of midday on the phone to New York, and much of the late afternoon talking with officials in Costa Rica. Eventually, the traders secured $24.5 million worth of debt, all on Janzen's word that he had the $3.5 million from Sweden to pay for it.

Near the end of the Salomon Brothers campaign, Umaña called Central Bank President Lizano and told him about the success. Lizano abruptly refused to go forward with the debt swap. He had anticipated handling only $5 or $6 million worth of debt. The $24.5 million bonanza for the conservationists turned out to be a terrible deal for the Central Bank, which had agreed to exchange the debt notes for currency bonds equal to 70 percent of the total debt purchased.

Put simply, the debt was worth something like 12 cents on the dollar, and the bank had agreed to pay 70 cents on the dollar.

Umaña went to work. One by one, he lobbied each member of the board of directors of the Central Bank, trying to convince them to outvote Lizano. They did, on two separate occasions.

Sweden sent the $3.5 million to Salomon Brothers, and the debt notes were transferred to the Central Bank. The bank issued the conservation bonds. Thanks to the Swedes and the young masters at Salomon Brothers, the Guanacaste project was the benefactor of the biggest single commercial debt-for-nature exchange ever. The project essentially squeezed more than $17 million from $3.5 million. It also now had the largest endowment in the history of tropical conservation.

Janzen called the debt exchange "a small ecological interaction" among all the stakeholders. The *Chronicle of Higher Education* dubbed the biology professor an "International Financier."

By this time, debt-for-nature exchanges had become an accepted mechanism to generate local currency for conservation in developing countries. They are still considered valuable, although the focus shifted through the 1990s from swaps involving commercial bank debt to swaps involving government-to-government debt. In such bilateral arrangements, debtor and creditor governments negotiated debt-for-nature swaps directly. Among actions approved by the Enterprise for the Americas Initiative, passed by Congress and signed by President George Bush, were debt-for-nature swaps with U.S. bilateral debt. Congress appropriated more than $90 million to pay for forgiving some $900 million of U.S. debt, creating more than $170 million in local currency for conservation trust funds in eight Latin American and Caribbean countries. As the decade closed, Congress passed and President Bill Clinton signed the Tropical Forest Conservation Act of 1998, which, among other things, established an office within the Treasury Department to administer debt reduction in developing countries with tropical forests.

The fund-raising part of the Guanacaste project was largely completed with the Swedish-Salomon bonanza, except for Santa Elena. Other grants and prizes trickled in during subsequent years, but nothing that approached the amount raised in this debt trade. By the mid–1990s, the efforts of Janzen and others had brought more than $30 million to the project. Most of that was invested in endowments that provided hundreds of thousands of dollars in interest a year for the Guanacaste Conservation Area operating budget. Some of it helped all but finish the purchase of key pieces of land.

There was a bizarre footnote to this debt-swap episode. Janzen came up with the idea to name several newly discovered species after the Salomon Brothers traders, a standard gesture of thanks in science. Ian Gauld of the Natural History Museum in London happened to be describing several tiny wasp species that he had discovered in Guanacaste and elsewhere in Costa Rica. "So I got Ian to name each of those wasps after one of these people on the debt-trading team," Janzen explained. That included John H. Gutfreund, then chairman of Salomon Brothers. The scientists named *Eruga gutfreundi* for him; *Polysphincta gutfreundi* for Gutfreund's son, Peter; and eleven other wasps for different Salomon executives.

This act of gratitude ended with an ironic twist or two. Janzen got a call from a reporter who was writing a story about the Guanacaste debt-swap success for a financial newsletter, and the scientist mentioned the wasps. The reporter asked what the wasps did. "I had no idea, aside from the fact that they were parasitic wasps," Janzen told me. "So I called Ian, and he said, 'Well, they're spider parasites.' And I asked, 'Well, what are the spiders called?' And he went and looked and discovered that the common name for the family of spiders that these things parasitize is the money spider."

To be precise, the wasp lays an egg on the back of the spider, and the larva feeds off its host's blood for several months before finally eating it. What bittersweet symbolism. The newly discovered wasps, now represented by the names of Salomon Brothers executives, live off a personification of wealth.

Gutfreund thought it ferociously funny, and he spread the word among acquaintances. *Fortune* magazine published a story on the debt swap on January 15, 1990, featuring photos of Gutfreund ("Gutfreund in his habitat, Salomon's trading floor") and *Eruga gutfreundi* ("*Gutfreundi* out of hers"), pinned, with purple light glinting off the wasp's wings. The headline was "Biggie Named in Wall Street Sting." How prophetic. Two years later, Gutfreund became the focus of a major Wall Street scandal. He resigned from Salomon Brothers after the firm admitted it cheated during an auction of Treasury Department bonds (a maneuver unrelated to the conservation debt swap).

From 1987 through 1989, the Guanacaste project achieved the lion's share of its land-acquisition goal. The peak year was 1988, when the project acquired about 44,920 acres, or 70 square miles. That was more than double the land gained in the next two most active years, 1987 (21,000 acres) and 1989 (20,400 acres).

Many of the land deals were not at all dramatic. Some negotiations were conducted amicably with the most reputable families of Guanacaste. Yet some were far from ordinary.

A new trend toward large sugar, cotton, and other farms had boosted land prices in the province, and many landowners knew that the conservationists wanted to buy quickly. This meant that some, especially the owners of large properties, could hold out for a higher price.

Early on, Janzen had run up the miles on the Land Cruiser as he made the necessary connections with key landowners, getting to know them and negotiating issues such as boundaries, prices, roads, and access. "Many days I spend just driving frantically back and forth in my jeep trying to find a landowner and talk to him and wheedle him and stand and argue with him for a half an hour," he told me in 1987. "Then I call San José and tell them the result of this, and then their land negotiator talks to their guy's lawyer in San José, and we sort of go ring-around-the-rosy in that way."

Before the major negotiations got under way, a commission was established to acquire land for the project and bring it into the orbit of the Park Service. The commission included the directors of the Park Service, Forestry Department, and National Parks Foundation. Janzen was the commission's scientific adviser, and Gámez its coordinator. In practice, Janzen quarterbacked the negotiations with the U.S. landowners, while Gámez did so with many of the Costa Ricans. Later, directors of the Guanacaste Conservation Area took over.

The actual bargaining was done by a National Parks Foundation team, headed by chief negotiator Leon Gonzalez and lawyer José Antonio Bustos. Janzen usually hovered in the background, playing the role, as he once described it, of "the naive, innocent and disinterested crazy scientist whose primary job is to step on outrageous claims about the agriculture or timber or water values of the land." He also could read a map or real estate diagram better than the landowners, and in most cases he was more familiar with the terrain and vegetation on their properties than they were.

Among the most important acquisitions in 1988 was a large series of parcels on Hacienda Orosí donated by Cecil Hylton, the U.S. millionaire who had donated Horizontes to the project in 1987. Hylton's Orosí parcels were transferred to the project anonymously using the Nature Conservancy as a conduit. Hylton died in August 1989.

Jimmy Swaggart Ministries sold essentially all of Hacienda El Hacha to the National Parks Foundation in 1988. These were the seventeen parcels of land— totaling about 20,000 acres—that Hylton donated in the mid–1980s to the ministry, based in Baton Rouge and known in Guanacaste as "Los Evangelistas." The property fronted on the Pan-Am Highway, and the entrance road was not far from the front gate of Santa Rosa. At the time it was sold, the site was occupied only by a manager and his family.

The Swaggart organization relied largely on money gained from appeals by Swaggart in his television shows, which reached millions of people and helped bring in some $140 million a year. Swaggart's appeals, and the Ministries' income, suffered greatly when in February 1988, the minister made a tearful televised confession that he had committed a sin against God and his family. Although he didn't specify the sin, another minister later reported that he had photos showing Swaggart with a prostitute. In April, the Assemblies of God, the country's largest Pentecostal denomination, defrocked Swaggart, but he continued his broadcasts. Although many still watched and contributed, income fell dramatically. Because of Swaggart's broad reach and incredible fund-raising success, some observers considered his fall one of the most historic events in American Christianity.

On the heels of the scandal, Janzen and a Swaggart representative conducted, as Gámez put it, a "very peculiar negotiation." Jimmy Swaggart Ministries sold the land in August "at a very reasonable price, a very low price as a matter of fact," Gámez told me. The Evangelistas kept a few acres, the swimming pool, and the house, which burned down in a fire that swept through the area in 1996.

Janzen deflected questions about the negotiations. Others involved with the effort said the American biologist, in order to get a better price for the conservation project, took advantage of the scandal. The way they saw it, the evangelical TV star's organization apparently did not want publicity about his ministry owning a large chunk of northwestern Costa Rica, including a house and swimming pool. Janzen said the scandal was "coincidental" and had "nothing to do" with the negotiation. As Swaggart might have said, only Janzen, the Ministries' representative in Costa Rica, and the Lord knew the details.

Also of note, Gámez led a lengthy series of meetings with Baltodano family members. The family had obtained Hacienda El Pelón de la Altura in the 1930s from descendants of Minor Cooper Keith, a U.S. contractor who had built a major railroad in Costa Rica in the late nineteenth century and in return had been given land grants covering almost 7 percent of the country. Keith also bought several Guanacaste properties. The Baltodano land was known as Hacienda Tempisquito, owned by Jorgé Baltodano, and Hacienda San Josecito, owned by his brother, Aristides. Tempisquito fronted on the Pan-Am Highway, and part of San Josecito covered the western slope of Volcán Orosí and jutted into the Orosí Forest Reserve.

Both pieces had large expanses of old-growth forest on the interface of dry forest and more evergreen forest on the slope. San Josecito was still 70 percent forested, and it contained, as Janzen put it, "the most valuable single large block of forest in the entire area to be purchased." Three ever-flowing rivers made it "extremely valuable biologically." Tempisquito's value included three such rivers, too, as well as "the best soils."

Jorgé Baltodano was both the owner of the major timber mill in Guanacaste Province, just outside Liberia, and president of the province's cattle ranchers association—"the typical enemy in conservation," Gámez recalled with a smile. "He said, 'O.K., I'm ready to sell. But I'm a businessman, you know.'"

Baltodano produced a detailed inventory of his property, including several pages showing the location, age, and size of every tree that had value as timber. He told the project's negotiators, "Look, my family has been saving this forest for years because we knew that this timber was going to be worth gold, or more than gold. So you want the forest? That's what you have to pay."

Gámez could not remember the specific figures, but the asking price was about ten times what the conservationists were willing to pay. "To our great surprise, he was the first guy who had very well-documented information about the price of his pristine forest," Gámez told me. During meetings over the course of more than a year, Gámez tried several arguments, bearing in mind that the land fell under the Zona Protectora declaration. First, he held up a coffee cup and told Baltodano, "Something has a price, of course, if you are going to use it for a purpose. This cup is worth a lot if I'm going to use it as a cup, but if I'm not going to use it, the price is different. That commercial price of timber is not a meaningful price for us because we are not purchasing your land to use the timber. We are purchasing your land for entirely different purposes. Consequently, the values we attribute to it are going to be different."

After several other meetings, Gámez came up with his most persuasive argument. He told Baltodano that he and his family would be making a contribution to the restoration of the old Guanacaste region, and that, "what you have done, unconsciously, was to preserve the seeds of what Guanacaste was. Your children and grandchildren will remember you for that, not for all the money you would make for selling the timber."

Gámez chuckled as he recalled the moment. "We ended up paying perhaps double what we expected, but not ten times." San Josecito and Tempisquito were bought from the Baltodanos in 1989.

Shortly thereafter, Baltodano, whom Janzen once called "perhaps the most honest local businessman I have met," became the first president of the local governing committee for the Guanacaste Conservation Area.

As pieces of the Guanacaste project landscape were assembled, teachers in the biological education program pursued their mission of biocultural restoration. By regularly bringing students to Santa Rosa for day-long immersions in field biology, the educators hoped to inspire biology-related activities at schools, as well as deeper interactions with tropical nature by teachers, students, and parents. Eventually, biologists were hired full-time to run programs on-site for several local schools.

The education program officially began shortly after two Costa Rican biologists joined the Guanacaste project staff in late 1986 to run field courses. The subject matter was the natural history of animals and plants, and how they interacted. Findings from the park's ongoing research programs were incorporated immediately into lessons.

The experience of one of these teachers, Giovanni Bassey, showed how the program aided the restoration effort beyond biological education. Bassey taught biology at Cuajiniquil, but because he was a marine biologist, he also became an unofficial scientific consultant to the small fishing village. First, fishermen came to him to talk about their problems at sea. Next, they sought his advice on where to set nets. Then, villagers asked him where he thought it best to build new houses. Within a month of beginning his teaching job, Bassey was invited to chair the Cuajiniquil civic committee. He had become a civic leader, park public relations man, and contact point between the park and the fishermen.

A typical day for one of the teachers based at Santa Rosa began when a class of about fifteen to thirty grade-school students arrived early in the morning. Together, they would walk through a patch of forest, stopping frequently to look closely at organisms and patterns. The teacher would tell stories about natural history. The rest of the day comprised a series of lectures, discussions, more hikes, and field experiments.

In one such experiment, the students would gather as many kinds of fruits and seeds as they could find in half an hour. They would return to a central point, separate the items into species, and discuss what a species is and why the fruits and seeds varied in shape, size, color, and other traits. "When we go into the field, I try to ask them questions to make them think, instead of just telling them about what they're seeing," said Liz Brenes, one of the first teachers. "I want to help them understand the importance of a national park and why they have to take care of natural resources."

Brenes was struck by how the visits appeared to be major events in the lives of children. Many of them got little sleep the night before, and they went through the day with looks of wonder on their faces. The local schoolteachers, who accompanied their classes, learned about the behavior of their students as well as about biology. "Often a child will teach the teacher about nature," Brenes told me.

I tagged along with a group of twenty-seven young visitors one day to see for myself. Within a few minutes of getting off the bus, the students had strapped on leather leggings to protect against snakebite, and they followed teacher Rosibel Elizondo into the forest. Before long, they encountered a group of spider monkeys feeding high in the trees. Elizondo questioned the students about monkeys, the fruits they ate, and the fate of seeds after passing through the animal. Many of the children raised their hands and gave good answers when called on.

"The monkeys are very important in the life of the forest," she told them. "Along with other animals and the wind, they disperse the seeds that become the trees."

Later, Elizondo stopped at an acacia tree, which, although only a bit more than six feet tall, towered over many of the students.

"What's happening here?" she asked them.

"Ants are walking over the branches," a child answered.

"Why are the ants doing this?" After a few moments with no answer, she said: "Defense."

She tied a string on the end of one of the branches and pulled, and the ants swarmed toward the string as if attacking it. She asked if anyone wanted to hold one of the ants. The children backed away, giggling.

"No," one of them said. "They sting."

Elizondo explained that the tree's thorny chambers provided a home to the ants, while the ants' stinging discouraged would-be leaf eaters. As she talked, a few boys picked ants from another acacia tree and flicked them at each other.

"This is a case of 'mutualism' among plants and insects," she said to those still listening. As the kids hiked, they learned about birds, wasps, deer, centipedes, guapinol trees, bats, caterpillars—and the ways they depended on each other to live. In most cases, the lessons came when the organisms were spotted by the students.

Near lunchtime, Elizondo led them out of the forest to the Casona, where they toured the historical museum. She guided them up the hill behind the house. "The goal," she told me, "is that they will become links between the park and the community, and they will teach their sons and daughters the importance of protecting nature."

The students were not allowed to climb the Monument to the Heroes, but from the hilltop, they could still see part of the Guanacaste project panorama of forest and pasture. Soon, most of them began to race around and down the hill, playing a game of tag. But a few lingered quietly at the top and stared into the distance, as if trying to perceive some mysterious spirit rising off the landscape.

PART III

# The Rising Phoenix

# · 15 ·

# Area de Conservación Guanacaste

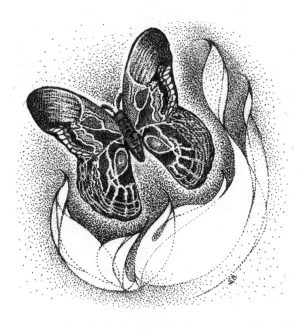

In August 1996, nearly ten years after my first visit to Santa Rosa, I stepped off the bus and onto the shoulder of the Pan-Am Highway a few hundred yards north of the entrance road. I'd been standing in the bus aisle, watching for the road and hoping the driver would remember to stop there. We drove northward along the highway at something like 45 miles per hour, and it seemed as if the surface over the years had been pushed inward by the trees on either side into a narrower version of itself, and that the bus was traveling twice as fast. There were still those classic Guanacaste Province pasture scenes, with vast fields stretching to east and west, dotted with a few cattle and isolated shade trees. However, along many stretches, like the last few miles approaching Santa Rosa, the trees seemed to almost form a tunnel. It was disorienting.

At about the same place where the park entrance ought to have been stood a solid wall of trees. It was a good thing I was watching. The driver sped right past it, the trees suddenly opening to reveal the road and guard house, then almost instantly closing.

"Santa Rosa, por favor!" I shouted several times before he heard me. About 500 yards past the entrance he stopped, grudgingly, I thought, perhaps not wanting to admit that he had made a mistake and forgotten. He even rolled his eyes when, as I dismounted, I reminded him about my backpack. He reluctantly rolled out of his seat, climbed down to the side of the bus, threw open the cargo door, and, as I swung the pack over my shoulder, jumped back aboard and pulled away, leaving me in a small cloud of diesel exhaust.

I smiled and waved to the passengers staring down from their windows, having long since learned that such near misses and seemingly inconsiderate treatment were just part of travel in Latin America. It was nothing personal, and it wasn't worth worrying about. Besides, as the rear of the bus cleared my line of sight, I was thrilled by what I saw: Where a decade ago I would have looked well into the park across a jaragua field, now at the edge of the road was a solid wall of trees, fully 12 to 15 feet high. I closed my eyes, imagined the 1986 scene, opened them again, and felt knocked backward by the sudden advance of foliage. It was unreal.

Many of these were, I later learned, *Lonchocarpus rugosus* and other *Lonchocarpus* species, a genus of wind-dispersed trees. I walked down the highway and in along the entrance road, and I could see that it wasn't just a wall of trees, but whole fields of them to the left and right.

Later that day, looking out from atop the Monument to the Heroes, I saw that Santa Rosa had become almost entirely trees. The rainy-season vista had changed dramatically in ten years, a change that seemed little short of a miracle. The tops of trees blotted out any sign of pasture as far as the eye could see. It wasn't until I looked well east of the Pan-Am Highway, to the flatlands at the foot of the volcanoes, that I began to see strips of pasture amid the mosaic of trees. To the west, a few small strips of pasture appeared to have been all but crowded out by young trees. These open spaces were fire breaks. Aerial photos shown to me a few days later confirmed these naked-eye impressions.

The casual observer, one who had never visited Santa Rosa or had visited only recently, might not have understood the spectacular shift that had occurred here over the past decade. But to someone who had been watching the Guanacaste project from its beginning, the difference was as dramatic as night and day. Here was the ultimate proof that the project was working. The forests of Guanacaste were, indeed, coming back, and along with them the watersheds for the eleven major rivers that served the region's towns and irrigation systems. This was not just ecological restoration, it was ecological revolution.

The advance of the forest had been matched with equal success in the Guanacaste project's other endeavors—financial, political, cultural, administrative, and land acquisition. The region was now called the Area de Conservación Guanacaste (Guanacaste Conservation Area), or, more commonly, ACG. It was the most recent, and perhaps final, name for what was once known as the Guanacaste National Park Project.

Whatever one chose to call it, the effort had led to major administrative restructuring. Now there were expanded programs in education, research, tourism, reforestation, fire control, and police protection. The research program included biodiversity prospecting. The ACG staff numbered about 100, all of them Costa Rican and four out of every five from the region in and around the conservation area. The annual operating budget was $1.6 million. This was quite a jump from Santa Rosa's handful of park rangers and $120,000 annual budget in the mid-1980s. The ACG income consisted of payment for services and interest from its endowment. Comparing it with Santa Rosa at the launch of the Guanacaste project, Janzen wrote that ACG was "ten times as large, costs ten times as much to operate, and generates a diverse and large basket of cash and barter for the region."

Biological research stations had been built or improved in several spots. The ACG itself was integrated with eleven other conservation areas in the newly reorganized government structure known as Sistema Nacional de Areas de Conservación (National System of Conservation Areas) or SINAC. The system covered about 25 percent of the national territory with various degrees of conservation status.

The ACG's education program had been teaching basic biology to the region's fourth, fifth, and sixth graders since 1987 and had recently added high school students. With several new teachers, the program used about 22 percent of the conservation area's annual operating budget. It reached 2000 students a year in forty-two schools, and by 1999 that had expanded to 2500 students. "It is widely rumored that the ACG has it easy because it is embedded in a 'tame populace,'" Janzen wrote. "I wonder why." The answer, of course, was in part the success of the education program.

As for the regrowth of the forest, when the fires of Santa Rosa were stopped, the natural successional processes of the forest had begun. It was as simple as that, yet not simple. There had been a massive influx of wind-dispersed trees. At the Santa Rosa entrance, eighteen of the twenty-three species of wind-dispersed trees big enough to cut firewood from had been the first to move in. Among them were the *Lonchocarpus* species, as well as trees of the genera *Cochlospermum*, *Rehdera*, *Gliricidia*, *Ateleia*, and *Machaerium*. The first few dozen species of animals—including monkeys and jaguars, which had rarely crossed the open land before—had begun to creep back. In their wake came a small contingent of animal-dispersed seedlings, perhaps only 2 percent of the whole stand. After a slow start, these would gradually overtake the wind-dispersed species, which don't do quite as well in the shade, and shade was now becoming a factor in the life of the new forest. The animal-dispersed trees, of which the region had more than 200 species, would manage here and there in the shade, gradually becoming adult trees, then slowly take advantage of their ability to disperse among treefalls. So after the massive entry of wind-dispersed trees into the jaragua field, and the slow succession of animal-dispersed species, the forest would become whole again. "It takes a long time," Janzen said, but across all of the dry forest region of ACG, "we see that over and over and over again." He estimated it was happening across at least 190 square miles.

The experiment to learn whether trees could regenerate in some of the area's harshest environments had demonstrated the power of nature. In these habitats, researchers had planted a few species of trees available to them in the mid–1980s, and these had by and large survived, although they were stunted. Meanwhile, natural regeneration surpassed everything as the forest threw hundreds of species onto the sites, and the ones best suited to the harsh environments took hold and filled in. "The whole issue of how you do research in effect vaporized because natural succession took over faster than anything we could have done, once we stopped the fires," Janzen explained. The natural succession observations in Guanacaste provided enough information to write "book after book," but, of course, he didn't have the time.

As to what exactly the forest would return to when succession was complete, tropical ecologists disagreed. Janzen and others sometimes argued over the basic nature of the tropical dry forest, and whether the Santa Rosa dry forest was a natural end in itself or was just a lengthy phase on the way to a moist forest, which some believed may have dominated the Guanacaste landscape millennia ago.

Douglas Boucher and a few other professors had debated this with Janzen in 1989 as they walked a group of college students through Bosque Húmedo. Janzen had argued, basically, that Bosque Húmedo was evidence that the forests of the region were once mainly semi-evergreen forest, that the Bosque Húmedo of today was a remnant of forests that were cut down and converted into pasture, that what had grown back as secondary forest was deciduous, that it was not the original forest of the region, and that if the dry forest was maintained long enough, the evergreen trees of a moist forest would return and dominate. In fact, young evergreens, such as *Hymenaea*, *Manilkara*, and *Swartzia*, were growing everywhere in the deciduous forest. The other scientists had argued that there was little evidence for a return to moist forest, that colonial-era descriptions painted a picture of a widespread deciduous forest in Guanacaste, and that Bosque Húmedo and other moist forests were there mainly because of localized differences in soil type and drainage. A creek, for instance, made the soil moist in Bosque Húmedo, but that moisture was not typical of the region. Janzen had countered that this was "obviously nonsense" and resulted from his opponents' failure to realize "that 'colonial' was in fact secondary succession."

"We went on for twenty minutes arguing about what kind of forest this was," Boucher recalled. "The students told us how entertaining it was, to hear the faculty, supposedly experts who knew all about the tropics, so much in disagreement." Many tropical ecologists familiar with the region had agreed that in the scheme of things, this debate was a relatively minor point. There simply wasn't time to research and resolve it, especially in the face of the speedy advance of ecological destruction. The more important point was that the Guanacaste forest was coming back. This was, everyone had agreed, an unprecedented ecological change in the era since humans came to dominate the landscape. And, perhaps, the experiment would, decades hence, provide a definitive answer.

Although they were still trying to acquire Santa Elena and a couple more parcels of land to the south of Santa Rosa, on either side of the Pan-Am Highway, the territory planned for the Guanacaste project was in hand. In his original proposal, Janzen talked about growing a more than 190-square-mile dry forest complex enough "to maintain both the species and habitats into perpetuity." That would have been five times larger than Santa Rosa alone. By the time the proposal was printed in the salmon-colored book, that figure had reached about 270 square miles of dry forest and other habitats, including cloud forest. Eventually the plan expanded far beyond that already wildly acquisitive idea, to the point where ACG, by the mid-1990s, comprised about 460 square miles of land, which comprised roughly 2 percent of Costa Rica, and about 290 square miles of marine national park.

The land portion of ACG was roughly eleven times the size of Santa Rosa. That was about one-seventh the size of Yellowstone and three-fifths that of Great Smoky Mountains National Park—quite a large area for a tiny country. If the ecological restoration succeeded, it would double the area of remaining dry tropical forest in Mesoamerica. This land contained about 235,000 species in the dry forest, cloud forest, rain forest, and transitional areas between them. Janzen estimated that to be 2.4 percent of the world's terrestrial biodiversity.

Components included Santa Rosa National Park, Guanacaste National Park, Volcán Rincón de La Vieja National Park, Horizontes Forest Experiment Station, Junquillal Recreation Area and Wildlife Refuge, and Bolaños Wildlife Refuge. About ninety private lands had been acquired, ranging, as Janzen put it, "from small farms to large ranches, from squatters to absentee landlords to land speculators."

When the Guanacaste project began, Janzen had estimated it would take about $11 million to purchase and manage the land surrounding Santa Rosa. By the mid-1990s, the total had risen, along with the horizon of the project, to $31 million.

One evening, I stood on Playa Naranjo, its immaculate sand glowing pink as the sunset seemed to light a fiery sky out to sea. Next to me was Róger Blanco, director of the ACG Investigations Program. He was a short, serious, and gentlemanly tough man, and he had worked his way up through the ranks at Santa Rosa. He had driven me down to Naranjo, along the same bumpy road I'd traveled on earlier visits. The Costa Rican tourism institute had offered money to improve the road, but the ACG administration had turned it down, fearing that a better road would encourage a damaging level of beach use. As we sped across the upper plateau, where young trees had begun to crowd out the jaragua, we passed a wide firebreak. A few minutes later we saw a carload of young North Americans returning from a day of surfing, their boards lashed to the top of a rented four-wheel-drive. Down through the *Cañón del Tigre*, our jeep bounced more violently. I ducked to keep from banging my head on the ceiling, and told Blanco about the jaguar I'd seen ten years before there. He had only seen pumas, but others had seen jaguars, and of late they were sighted once a month on average. "When you have more forest, you have more animals," he said. The number of crocodiles and other inhabitants of the estuary had increased greatly, too.

Originally from the San José area, Blanco started at Santa Rosa in 1985 as one of the ten rangers. His duties were typical: Depending on where the director sent him each day, he would either guard at the front gate, assist tourists at the Casona, patrol for hunters, clean trails, haul garbage, or care for the fifty horses maintained at the park for Santa Rosa and other parks in the system. He worked in the Guanacaste project tourism program for about two years beginning in 1987, and two years later began collecting insects and plants across the project's landscape for a new biodiversity program. He joined the ACG research program in 1991 and soon became its director.

Blanco was among the early opponents of the Guanacaste project. "In the beginning, we were against the change, because we didn't understand it," he told me. "Little by little, we understood the change, and we realized it was a better way to work. We started to have more responsibility, equipment, and facilities." In 1996, he considered this part of Guanacaste his home. He felt that by "preserving biodiversity for the future," he was making an important contribution to the country. "I believe I will die here," he added with a smile.

At the beach, Blanco squatted and poked his finger into the sand, drawing a map of the coastline, from the Gulf of Papagayo in front of us, around the Santa Elena Peninsula to our north, and along Murciélago to Cuajiniquil Bay and Junquillal Bay in the northernmost part of the conservation area. This was roughly 60 miles of coastline, including a set of islands known as the Islas Murciélago, just off the Santa Elena Point, and it all comprised the marine component of the conservation area. Each November, as wind pushed warm water out to sea, cold water from below welled up around Santa Elena Point, carrying nutrients that fed a cornucopia of marine life. Protecting this cradle of life and documenting its diversity had become a key mission of ACG, as was studying the ecological interactions between the marine and terrestrial habitats. Said Blanco, "We know that the land and the sea work together."

A marine biological station had recently been established on the Islas Mur-ciélago, and President José María Figueres had traveled there in November of 1995 for a dedication ceremony. Residents of Cuajiniquil and other Guanacaste fishing villages had been invited. They expressed concern that the government was going to ban fishing in the region, but ACG biologists emphasized that they wanted to work with them to ensure sustainable use of the marine ecosystem. Among other threats to marine life, a burgeoning tourism development along the Gulf of Papagayo a few miles to the south of ACG was creating new pressures in the region from scuba diving, sport fishing, and other tourist activities. The biologists wanted to hear from local fishermen on what parts of the marine zone to regulate. "We have to work together," Blanco said. "We don't want to act like 'the authority.'"

Parallel to the ecological restoration of ACG, its biocultural restoration had advanced. In addition to running the education program, encouraging visits, and routinely attending local meetings to explain how the forest protected the region's watershed, ACG officials hired as staffers many of the former workers on the ranches and farms that had become part of the conservation area.

All this helped to create a positive perception of the conservation area. A 1994 internal report by a committee of outside investigators found that much of the population in the surrounding communities felt the development of ACG had brought money and jobs into the region, thanks largely to tourism, and had helped improve to some extent the region's roads and other forms of communication. Most of the money in the annual ACG budget stayed in the region, Blanco told me. He cited salaries, purchases by employees, and spending by tourists and scientists. "Directly and indirectly, ACG produces $3.5 million a year for this region," he said. "That's a lot of money for a rural area in Costa Rica. It's more than the best farm in Guanacaste," and the public benefited.

In the report, the local population also attributed to ACG such benefits as "oxygenation of the air, healthy climate, watershed protection, river protection, and conservation of forest animals." Residents expressed positive comments about how the education program helped "future generations know their resources." In some areas near Santa Rosa, residents said they appreciated the efforts of park personnel in handling emergencies, presumably medical emergencies and fires.

The report noted, however, a wide range of social problems in Guanacaste Province that fueled a few negative perceptions about ACG. Paradoxically, one of the complaints had to do with tourism. Some expressed resentment that tour guides, many of them outsiders, and in some cases even foreigners, were benefiting from ACG's natural resources without giving anything in return to the local populace. Some felt that affluent tourists were driving up the cost of living, apparently through their purchases of food, clothing, and other items. Another view, showing that old traditions die hard, held that tying up large areas of land in conservation kept many campesinos from making a living off the land. Some residents resented what they felt were unfair restrictions placed on their access to fishing in waters now under ACG jurisdiction. For instance, the fishing restrictions at Junquillal had made the spot "detached from the community." Policies prohibiting the killing of bats and snakes—even venomous ones—bumped up against deeply engrained Guanacastecan impulses to kill what were believed to be mortal enemies.

Another expression of antipathy toward the conservation area, although perhaps crocodile tears, came in 1994 when several local leaders mounted a campaign to remove Sigifredo Marín as director. Among the attacks in this coup attempt was an ad in the September 7 *La Nación* in the form of a letter to President Figueres. It was signed by politicians and businessmen in some Guanacaste towns, including the development directors of Dos Ríos and Quebrada Grande. The letter complained that the conservation area was too isolated from the surrounding communities and the attitude of its leaders too remote. It was a strikingly baseless complaint, considering the crescendo of ACG outreach over the preceding six years or so. As it turned out, this was just another little dance by a Costa Rican political clique. The campaign, orchestrated by a local member of the Legislative Assembly, apparently was aimed at getting his cousin the director's job. In an odd way, the patronage motive demonstrated how important the conservation area had become in the eyes of Guanacastecans.

The contradictory views of the general populace had been shaped by the socioeconomic diversity of the communities around the conservation area, the panel of outside investigators found. The panel noted that a growing concentration of land in large-scale operations requiring little labor—many of them now owned by outside investors, foreign and domestic—resulted in a depopulation of the countryside as workers and owners of small farms were forced to other parts of the country to look for work. The relatively healthy economies of market centers such as Liberia, Bagaces, and Guayabo contrasted with economically struggling communities such as Colonia Libertad, Colonia Blanca, San Jorge, Dos Rios, Buenos Aires, Curubandé, Potrerillos, Puerto Soley, Colonia Bolaños, and Cuajiniquil. In and around many of these smaller communities, the roads, public transportation, and telephone service were in poor condition, and some areas still did not have water, electrical, or sewage facilities. Only one, Cuajiniquil, had a health center. Subsistence farming and ranching had been the norm in many of these communities, and the generally poor level of educational resources meant children often did not finish primary school. This was quite a contrast with the official version of Costa Rican social progressiveness.

The northern Guanacaste population was generally aware of ACG and the importance of conserving and protecting natural resources, yet it was a nascent consciousness. The biological education program was helping, but part of the population was still "very belligerent" toward the protected areas, the panel said. There was random garbage dumping and "indiscriminate destruction of ecosystems." The "depressed" socioeconomic condition of communities around the conservation area was creating frustration for the campesinos, social instability in the struggling towns and villages, and tension between ACG and some residents. Meanwhile, large sections of the Guanacaste forests continued to fall, as some landowners cleared trees for palm heart plantations on the eastern side of the volcanoes, while others cleared them to "improve" their land in hopes of bringing a better price when they put it on the market. The panel knew that if forced to choose between basic necessities and conservation, people would choose the former. To prevent such a choice, the panel recommended steps that were all too familiar by now to the ACG leaders: involve local residents in the tourism industry and other forms of production compatible with conservation *and* the economic needs of the populace.

Still, biocultural restoration was working. Only time would tell whether it succeeded "in perpetuity," as Janzen had put it.

As I walked around the administration area in 1996, I was impressed by the physical changes over the decade, mainly more and better buildings. The ACG Headquarters, Ecotourism Program, and Biological Education Program were housed in separate buildings near the point where the entrance road ran into the administration area, and the education building was a brand new structure. Also new were a storage shed for vehicles and a building that housed supplies and a financial office. (The garage had been moved to Poco Sol, now the headquarters for a large sector of ACG on the east side of the Pan-Am Highway.) Some of the older dormitories for park personnel still stood, as did the *comedor* built in 1987. The *comedor* fare had become much more varied than the plain meals of rice and beans I'd encountered ten years earlier. The kitchen had large ovens, a microwave, and an industrial-size refrigerator. You could go to the counter any time of day and buy a cold bottle of soda.

The old circular road originally marked by Alvaro Ugalde was still there, but another road had been built to the north, connecting the old administration area with a new complex of buildings called the Tropical Dry Forest Investigation Center, also known as the Santa Rosa Biological Station. Two large dormitories stood in almost majestic contrast to the crumbly houses that visitors had stayed in ten years earlier. A concrete walkway with a large overhang and fluorescent lights ran all around the dorms. Each had four bunk rooms and a bath facility with four toilet stalls and four individual shower stalls. Each bunk room had bunkbeds for eight people and four sliding windows with good, tight screens. The plastered walls were painted white, the smooth concrete floor gray. The rooms had storage racks, acoustic ceiling tiles, and electrical sockets every few feet along the wall. Some had desks and chairs. Instead of a dusty gray mattress or no mattress at all, each bunk had a new red mattress, a pillow, and a plastic bag containing clean sheets, a pillow case, towel, maps, and other information about ACG, including registration procedures and hours of operation for the *comedor* and a nearby token-operated laundry. Never mind that the material was all in Spanish, and that there might be a scorpion or two, usually dead, beneath the mattress— this was the Ritz by tropical biology standards.

Also along this road were a research laboratory used by resident scientists and students, a dorm for resident researchers, and another large building that housed more of the ACG administrative offices, computer rooms, a library, and an expansive conference room, complete with overhead and slide projectors, big-screen television, and VCR. Trenches had been dug between the office and lab buildings for an optical fiber network. In a few months the ACG would have its own Internet site (www.acguanacaste.ac.cr). Blanco and others routinely did much of their business—correspondence, information gathering, publishing, and so on—on high-end Macintoshes and other work stations. "Ten years ago in Santa Rosa the most important tool we used was the machete, to clear trails, and maybe the *carbina* [rifle]," he told me. "Today it is the computer."

Next to the forest near one of the new dorms, workers were installing an antenna for a global positioning system. There was talk of erecting a canopy crane—a huge construction crane similar to those used by biologists to study forest canopies in Panama and the Pacific Northwest. Despite all these high-tech changes, there was still only one telephone line. Plans were in the works for more.

The ACG staffers based at Santa Rosa, Poco Sol, and in more than half a dozen remote biological stations and other sites now kept in touch via two-way radios. Nearly half of them were women. Seemingly everywhere on the roads and paths in the administration area, you could hear radio chatter coming from different buildings, like calls of neotropical birds, as people in different locations hailed each other.

"Santa Rosa-Santa Rosa-Santa Rosa."

"Poco Sol-Poco Sol."

"Maritza-Maritza-Maritza."

Like the birds, the chatter was lively through the day and fell off at night.

A house for some of the ACG staff had been built where the road to Playa Naranjo once ran. Now, the road passed not through the administration area and in front of the Janzen house, but near the Santa Rosa campground and well to the south of the administration area.

What changes this place had undergone. And through it all, the little house where Janzen and Hallwachs lived more than half the year with their pet—a porcupine named Espinita that had replaced the dearly departed paca—remained about the same. If anything, the clutter inside and outside seemed more intense—more plastic buckets on the floor, more wooden specimen boxes piled on the picnic table out front, more plastic bags stuffed with larvae and food plants hanging from lines strung everywhere. As always, quite a humble-looking think tank.

Two signs hanging outside by the door told a lot. One said: "*En Mal Estado/Favor No Entrar,*" meaning, "In a bad state, don't come in." The other, above the first one, said: "*Sección de Metichismo.*" It looked like an official sign of the ACG, with yellow letters carved into brown wood. It had been given to Janzen and Hallwachs by Marín as a kind of administrative tease, and it roughly meant "busybody section," a good-natured reference to Janzen's continued strong presence as an ACG technical adviser.

When I walked up to the house the day I arrived at Santa Rosa, Janzen seemed as hurried as ever. He didn't have time to talk, and he was brusque about it. I forced myself to realize once again what everyone has to realize about Dan Janzen: He's on a mission, and if the schedule is tight, as it often is, forget about trying to see him. (On plenty of occasions before and since he gave freely of his time.) He'd spent the previous day in meetings, and this day he was rushing to catch up before leaving the next day for a week of meetings in San José.

I hung around for a few minutes, assuaged by the ever-diplomatic Hallwachs, and I saw him give the cold shoulder to several other people. Later, though, I heard that an undergraduate student had stopped by to see him, and, after getting the same treatment, the student pointed out that he'd been trying unsuccessfully to meet with Janzen for several straight days, that he was leaving Santa Rosa soon himself, and that he wanted to discuss a career in tropical ecology and how he might fit in. Janzen dropped what he was doing, sat down, and gave the student the better part of an hour.

In addition to its physical infrastructure, the ACG's scope and administration had expanded, too. In the early 1990s, the Guanacaste project conservationists had set up an innovative administrative structure that authorized local managers to make decisions on park policy that previously had been made in the capital. That, as then-director Johnny Rosales told me in 1991, "gives our people the opportunity to be more creative and respond better to local concerns and needs." Marín, the director in 1996, emphasized the same point when I asked whether ACG was the first conservation area in the world to restore its forest. "We are the first not only to restore," he answered. "We are the first to obtain our independence." That freedom, he was quick to point out, meant accountability. "If we make a mistake, we take responsibility. We take responsibility for the future."

Now, less than 5 percent of the budget came from the Sistema Nacional de Areas de Conservación (SINAC), and most of the rest came from interest from the endowment. The ACG was run by a director, a Regional Committee, and a Technical Committee. The Regional Committee served as the board of directors of a General Assembly of leaders and other representatives from local institutions and communities. In 1996, Regional Committee members made up a short list of important Guanacaste landowners and politicians. Among the committee's main responsibilities was to examine plans, programs, and proposals put forward by the ACG administration and to help link ACG programs and activities with communities and institutions in the region. These ranged from annual special events such as the National Parks Day celebration in August to proposals for constructing water wells at the research stations.

The Technical Committee consisted of the ACG director, subdirectors, and the heads of all its programs. This group's mission was to develop and make technical decisions about managing the conservation area. There were other committees, divisions, departments, programs, and sections. Over the years, the administrative structure had evolved, growing in some areas, shrinking in others, adapting to new conditions and concerns. By the mid–1990s, the conservation area had three main divisions: Administration, Ecodevelopment, and Restoration and Silviculture. These each had subsets. Ecodevelopment, for example, was responsible for tourist services, biological education, special events, and scientific investigation. The Administration division oversaw operations, fire prevention and control, land purchase and use, "security and vigilance," and other programs.

The conservation area had begun to publish its own magazine, called *Rothschildia*, after a genus of saturniid moth that plays a key role in Guanacaste forests as herbivore, food for birds, and host for parasitic wasps. The glossy, twenty-eight-page magazine has been published twice a year in Spanish since 1994 and was eventually put on the Internet at www.acguanacaste.ac.cr/rothschildia/index.html. It contains maps, color photos, scientific notes, and news and feature articles about the conservation area.

With the land expansion, ACG had acquired a new recreation site to manage—Junquillal, which included a popular beach and wildlife refuge, to the north of Cuajiniquil. It was donated by Cecil Hylton. The conservation area had upgraded several biological stations, like the one at Maritza, which now had a house for the site manager, another for a permanent biologist, a limnology lab, a conference room, a *comedor*, dorms for thirty-two people, electricity, potable water, and two-way radio. Entirely new sites also had been established, like Pitilla Biological Station in the rain forest on the northeastern slope of Volcán Orosí.

All these changes derived from the stated objectives of ACG: to serve as a model for bringing conservation management and society together, to generate the finances needed to guarantee future management, and, above all, as noted in a 1996 internal report, "protecting and regenerating an entire tropical dry forest ecosystem and its complementary cloud forest and rainforest habitats." Out of these objectives had emerged a kind of spiritual revival in the people who work in the conservation area. In 1986, the big responsibility for the Park Service personnel at Santa Rosa was making sure the guard was changed at the front gate every day. Not wanting to ride a horse or bicycle there, they often argued over the use of the park's only vehicle, usually Janzen's. They insisted on having two guards at the entrance shack through the night because, as one ACG official told me, "People feared the night, feared *el tigre*." A decade later, park personnel seemed excited, undaunted, filled with a mission. They eagerly explored the park. There were more vehicles now, of course. But there were also more bicycle riders.

As I traveled through the conservation area in 1996, its people talked passionately of their dedication to restoring nature, their pride in making it happen in this part of their country.

"We saw a problem, and we had a dream to solve it," said María Marta Chavarría, who at the time was science coordinator of an ACG-related project known as the All Taxa Biodiversity Inventory. We were driving down a road near Hacienda Los Inocentes, along the northern edge of ACG. "We accomplished that dream, and we are very proud of it."

Standing on the sand at Naranjo, Róger Blanco had said, "We don't know what will happen in the future, but we know it's our responsibility to work for the intelligent use of this area."

Beneath the roof of the Pitilla Biological Station, as rain fell on the thickly forested slope of Volcán Orosí, Juan Acosta Acevedo, the Pitilla Sector manager, spoke of how his role in the Guanacaste restoration fulfilled his love for nature. "It is important to restore the forest," Acosta told me. "It is important that we protect the animals and plants of the forest, the headwaters of these rivers. Water is life."

Dan Janzen's leadership in inspiring and nurturing this commitment was widely evident. Blanco called Janzen "a big, big encyclopedia for us, an important resource." Chavarría called him "our *guia espiritual*," which she translated as "soul guide." As she put it, "Sometimes, when you have trouble in your soul, you go and talk to Dan, and he gives you peace."

This was no false passion, no self-indulgence in a task accomplished. These people genuinely believed in what they were doing, and they believed they could pull it off. To spend a few moments with them was to be touched by an aura of hope, like standing in the shade of a guanacaste tree on a hot day and watching seedlings sprouting in the pasture around it.

As the ACG forests began to return to what they had once been, there were stark reminders in the landscape of the omnipresent threat. The most striking to me was the beautifully shaped *Nectandra* tree, the one I'd seen on my first trip to the Cacao Biological Station, with Janzen and Hallwachs in 1987. Coming down the mountain from the station in the dry season of 1996, I bounced around in the back of an ACG jeep and tried to keep from scratching a rash of chigger bites I'd gotten on my legs, hips, and waist during five days on the volcano. I looked out the window, and through the cloud of dust kicked up by the vehicle I saw the *Nectandra* along the road. This magnificent tree, which had so aptly symbolized the living dead, was still standing, but the crown had been killed by dry weather over the past several years. Once a thriving member of a forest of giants whose canopy covered this whole slope, the tree was destined, as it had been when the others were cut down years earlier, to drop in the barren pasture. The Guanacaste restoration had come too late for the *Nectandra*. Weathered, leafless, lifeless, it no longer conjured the image of the phantom forest. A living dead had become a truly dead, a stark reminder of how close the forests of Guanacaste had come to obliteration, and, perhaps, how close they remained.

# · *16* ·

# A Time of War

There is nothing more difficult to take in hand, more perilous to conduct or more uncertain in its success, than to take the lead in the introduction of a new order of things, because the innovator has for enemies all those who have done well under the old conditions, and lukewarm defenders in those who may do well under the new.

—Machiavelli, as quoted by Janzen and Hallwachs (1993)

Calixto Moraga stood like a cat on the shady rain forest trail, watching a leaf that a sunbeam had illuminated, like a searchlight in the night. The leaf was wet from a brief shower that had just swept through this mountain forest, and it glistened against the backdrop of the dark leaves behind it. All around us the forest dripped. Moraga knew that the sun fleck on this leaf would make it particularly attractive to a passing insect, so he watched, holding his long, white collecting net steady underneath. A tachinid fly alighted. Moraga snapped the net up deftly and flipped the cloth back on itself to prevent escape, all in one skillful motion. Tucking the handle under his arm, he worked the net with his hands until the fly dropped into a small glass vial he pulled from his pocket.

Moraga, a short, lean man in his twenties, was a parataxonomist. He was one of dozens of specially selected Costa Ricans with a basic education who had received unique training, mainly at ACG, to collect and prepare plant, insect, and other specimens from the nation's wildlands.

The term parataxonomist, created by Janzen and Hallwachs, was a spin-off of "paramedic." It signified the parataxonomist's ability to work independently in the forest and, even though he or she was not a university-graduated "specialist," to understand the basic technical aspects and philosophy of taxonomy and specimen collecting. Janzen and Hallwachs had conceived parataxonomy as a practical solution to several problems. For one, Costa Rica's biodiversity needed to be inventoried, and the nation had neither the time nor the money to train the Ph.D. scientists to do it. Nor could it rely on foreign scientists. To meet the urgent need, why not turn to an abundant, underutilized resource: the rural population, especially the one around the conservation areas? Job by job, this also would help the wildland's neighbors view it as a resource, not just a walled-in protected area.

The concept challenged a long-held view in science that only university-trained experts knew how and what to collect. This, in the view of the two biologists, was an overestimation. In their years at Santa Rosa, they had employed local rural residents as field assistants and found them to be tougher, more enthusiastic, and more helpful than university students. They were more "logistically competent," meaning the instinct and experience they had gained by growing up rural made them at home in the field. They had found that two enthusiastic students from the University of Costa Rica, Isidro Chacon and María Marta Chavarría, performed brilliantly during a year of mainly unsupervised field work for the Moths of Costa Rica project. Besides, as Rodrigo Gámez once told me, comparing the biodiversity crisis to armed conflict, "When you are at war and you need to get an airfield ready, you use whatever resources you have."

So, with cooperation from others, Janzen and Hallwachs developed an intensive, six-month course for parataxonomists combining classroom and laboratory work with field trips. In dawn-to-dusk-and-beyond sessions that in some ways resembled an OTS field course, the students—a blend of former park guards, farmers, hunters, housewives, and students, from teenagers to middle-agers—learned the basics of plants and animals and how to collect and process them. They were taught, as Janzen, Hallwachs, and others later wrote, how to "drive a car, operate a chain saw, care for and use horses as pack animals, use a computer and a topographic map, use a field guide in a foreign language, manage a budget and petty cash fund, and fathom and tolerate foreigners." Janzen and Hallwachs played the roles of teacher, parent, and drill sergeant.

Parataxonomists were trained at ACG every two years beginning in 1989, with funding by a wide range of sources, from the U.S. AID to the Liz Claiborne and Art Ortenberg Foundation. By the mid–1990s, forty-two parataxonomists plied the nation's conservation areas, working out of more than two dozen field offices and funneling dozens of select plant specimens and thousands of mounted insects a month to a central collection near San José, along with carefully recorded information about their natural history. Many in the earlier classes, like Róger Blanco, went on to other conservation jobs. The annual salary of a parataxonomist ranged from about $3000 to $4100, which was considered good though not great for a rural resident without a university education.

Moraga's story is typical. He grew up on the family farm in the Santa Cecilia area of northern Guanacaste Province, believing in the traditional view of the forest: clear it and farm. He went to high school for three years and worked in a shoe factory and on the farm. He served as an evangelical preacher on the Costa Rica-Nicaragua border. A friend told him about parataxonomy. Moraga took the first course and returned to his home region to work as this new breed of frontier scout. Now he was "bioliterate," and rather than viewing nature as something to be conquered, he was one of the forest's most fervent defenders in Santa Cecilia. He even wrote a guide to his favorite topic: beetles and palms. He enjoyed the fact that each day on the job he learned something new. "Now I understand that this is very important for the country and the world," he told me. "With time, we can say to the world what we are trying to protect."

We had walked a few hundred yards along sloshy brown mud paths leading from Pitilla Biological Station, the most rain-forested part of ACG. The station was built on a hill that jutted out of the northeastern slope of Orosí, about four miles south of Santa Cecilia. With its small lab, *comedor*, conference room, and dorm for twenty people, it served as a base for scientists, parataxonomists, and an ACG teacher who used the region as a classroom for students from all around the conservation area.

One way parataxonomists collect, Moraga said as he hiked, was to simply walk and look until they see something of interest. At night, another way was to shuttle among several black lights and pick off specimens from the mass that came to the sheet. There were other kinds of traps for insects. When gathered, all of these specimens needed to be pinned, dried in a stove at Pitilla, and placed in boxes for shipment to the central collection. Moraga proudly told of finding several species never before collected in this part of Costa Rica, including a beetle that was completely new to science. He noted that he had met his wife, Petrona, while she was training as a member of the same parataxonomist course. Just the year before, their infant son, José Mario, had been born, in a way a product of the Guanacaste restoration.

Moraga stood at the edge of the path, watching leaves illuminated by solar pulses between the cloudbursts, trying to catch more tachinid flies. "They are not well known, and they are very difficult to collect," he told me. "But I can stay a long time watching." It was easy to imagine the Moraga family as establishing a kind of beachhead in the long-term biocultural restoration of Guanacaste.

The Guanacaste project was a crucible for parataxonomy and other revolutionary and evolutionary ideas in conservation biology. They came about during anguishing financial and political struggles from 1988 to the mid–1990s, as the administrative structure and philosophy inherent early in the project gradually overcame resistance from the Park Service, triumphing in the form of SINAC. The ACG, SINAC, parataxonomy, the National Biodiversity Institute, and a project known as the All Taxa Biodiversity Inventory were all major experiments in institution-building that emerged along with, and partly out of, the push to restore the forest. Some innovations were built in parallel, some on the one before. It was a gradual process with an uncertain ebb and flow.

A detailed description of the remarkable transformation of Costa Rican conservation institutions from 1986 to 1996 is beyond the scope of this book. However, discussion of a few concepts illustrates the central role of the Guanacaste project in these changes.

The SINAC structure for the conservation areas grew out of meetings in the late 1980s organized by the National Parks Foundation. Progressive conservationists continued to push for a change from the fortress mentality for protecting nature to one exemplified by ACG. The philosophies they developed, and which shaped the conservation areas system, included several familiar to ACG, among them hiring local residents and educational outreach to help neighbors see parks as a resource. Out of these discussions also came the idea to expand the management role of conservation areas beyond their wildland borders. By more active involvement in the "agro-ecoscape" around the areas, conservationists could work to lessen destructive outside influences.

Through various presidential decrees and legislative resolutions during the period, SINAC materialized, but not without continuing tension between the Park Service central office and the new areas. SINAC oversaw eleven major conservation areas, each gradually moving toward the decentralized administrative example set by ACG. All eyes were on ACG, the pilot project, watching how it handled its money, political decisions, ecological strategies, and relations with local residents.

As SINAC matured and Costa Rican conservation became further decentralized, the environment ministry gave ACG new responsibilities, including conservation land management and environmental law enforcement in a so-called geographic-administrative area that covered about 1700 square miles in far northwestern Guanacaste. This agro-ecoscape comprised ranches, small farms, plantations, cities, timber mills, and forestry activities in more than a dozen districts from Liberia to La Cruz. However, ACG officials felt it wrong—morally, politically, and socioeconomically—to hold such sway over the private agricultural landscape.

As this political structure gelled and the money rolled in for ACG, recognition and honors mounted for Janzen. He became, in the words of Robert Goodland, author of *Race to Save the Tropics* (1990), "arguably the world's foremost practitioner of applied tropical ecology." It was a widely held view.

Among students of tropical biology, he approached mythic stature. In 1996, as I rode into the Santa Rosa administration area with a busload of Latin American university students taking an OTS course, we passed Janzen, who was walking near the *comedor*, a patch of sunlight making his gray hair appear snow-white. "*El Viejo!*" several of the students shouted to each other, chuckling, but clearly with respect for "The Old Man." Most had never seen him before, but, as the saying goes, his reputation preceded him.

In July of 1989, Janzen was among twenty-nine people named a MacArthur Fellow by the John D. and Catherine T. MacArthur Foundation. Popularly dubbed the "genius" awards, the intent was to give individuals of "exceptional creativity and independent thinking," as the foundation put it, the freedom to "pursue their own creative, intellectual, and professional inclinations." Among the short biographies for the artists, writers, composers, scholars, labor organizers, and others selected that year was one for Janzen, then fifty. It called him "a pioneer in restoration ecology of tropical forests" and noted his "struggle against deforestation" and his work with the Costa Rican government and people. Janzen's MacArthur was worth $350,000. The MacArthur Foundation, of course, had given the Guanacaste project its first major grant, for half a million dollars, in July of 1986.

Janzen also won the Joseph Leidy Medal of the Philadelphia Academy of Natural Sciences in 1989. A year later he was elected a fellow of the American Academy of Arts and Sciences and in 1992 became a member of the National Academy of Sciences. The academy memberships are two of the highest honors short of a Nobel attainable by a U.S. scientist. In 1993, Janzen was decreed an honorary member of the National Park Service of Costa Rica and given, with Hallwachs, who had just finished her Ph.D. at Cornell, the University of Costa Rica's Award for Improvement of Costa Rican Quality of Life. By then, Janzen's official role was, as it always had been, unpaid technical adviser. In 1997, he was awarded Japan's Kyoto Prize in the Basic Sciences, a $430,000 prize honoring his broad contribution to humanity and science. The Inamori Foundation, which issued the award, called him the "world's foremost pioneer in the field of tropical biology." As he had with every other moneyed prize (in total close to a million dollars now), all the award money—and all his salary—went into the service of ACG, except for what he and Hallwachs needed for food, clothing, and a few other necessities. "It's the way we operate," he said. To avoid U.S. income tax on the Kyoto Prize money as it went to the conservation project, he arranged with the prize's sponsor, the Inamori Foundation, to deposit it straight into a tax-exempt endowment fund, registered in California as the Guanacaste Dry Forest Conservation Fund. "Anyone can contribute," he pointed out when we talked about it shortly after his return from Japan.

Yet even as Janzen's prestige grew in international scientific and conservation circles and his financial contributions to ACG accumulated, the bullish role he had played in the restoration campaign continued to anger some Costa Ricans. Although several Park Service employees had failed in 1987 to have him declared persona non grata, they kept fighting him at every turn. He continued to offend people, creating "unnecessary attentions," as Umaña put it. Melania Ortiz Volio, director of the National Museum, summed up the anti-Janzen sentiment when she told me: "He lives in Costa Rica, and he has absorbed from Costa Rica what he likes, but he has not absorbed Costa Rica, and Costa Rica has not absorbed him."

Many of the same critics, some of them ignorant of what was actually happening in Guanacaste, sniped at "the rich kid on the block," as Umaña had put it. They accused ACG officials of abusing government property and skimming money out of the operating budget. One rumor that circulated into the mid–1990s even claimed that Janzen was a CIA agent bent on buying up Guanacaste Province and engineering its secession from Costa Rica. ACG leaders admitted later that all was not perfect—some personnel hired for certain jobs didn't work out, some of the land purchases could have been handled better. But conspire to treason? Even with Guanacaste Province's historic links to Nicaragua, it would have been quite a feat for a tropical ecologist to pull off.

Behind much of this wrathful rumor-mongering, of course, was the resistance of the traditionalists in the Park Service not just to change, but, in their minds, to sleeping with the enemy. They bitterly criticized the hiring of former campesinos and making them the new custodians of the park. As Gámez once told me, "these were, in their view, the enemies, and we were bringing the enemies into the system." The fact that they lived with their families on the land, and kept chickens, hogs, and other animals, was anathema to old-style Park Service cadre. So, too, was buying land, changing fire-control practices, setting tourism policy, and taking other steps without consulting the central office.

Prominent Costa Rican conservationists such as Gámez, León, and Umaña were behind these changes as much as Janzen, but because of his personality, and because he was a gringo, Janzen bore the brunt of the criticism. A case in point was naming the new protected area. Janzen originally proposed calling the expanded park Guanacaste National Park, and under that name he included all the newly acquired land as well as Santa Rosa and Murciélago. "Guanacaste National Park" was the title of the little salmon-colored book, and there it was, at meetings, in newspaper and magazine articles, in letters and grant proposals, in scientific publications, in his conversations and speeches. Even in the mid–1990s several scientists and conservationists in and out of Costa Rica used the term "GNP project." Many Costa Ricans thought the Guanacaste name was a wonderful way of emphasizing the independent nature of the province.

What Janzen did not understand at the time was the depth of reaction many other Costa Ricans would have to the proposed name change. Specifically, they hated the idea of bringing Santa Rosa under the new name, mainly because Santa Rosa had such a special historical meaning to Ticos. It had been Santa Rosa for centuries and was the site of the patriotic battles. Even many Guanacastecans didn't like the idea of the new name. To them and other Costa Ricans, the place and the name were sacred. To change it would be, well, an offense. "Dan has been here many years," León said, "but I think he'll admit that despite the fact that he's fluent in Spanish and so forth, there are some things about the culture that are sort of baffling, and he also baffles people here in some ways."

Janzen knew he was treading on sensitive toes. Some people, he knew, didn't want to call it Guanacaste National Park simply because *he* proposed the name. Others opposed it because it appeared to give Guanacaste too much autonomy, and there was a certain resentful sentiment nationally that did not like to see the northwest's regional identity strengthened any more. Early on, Janzen anticipated the "patriotic flag-waving" about Santa Rosa's history, as he put it, but he felt that the opposition to the name change was overestimated. In a 1987 interview, he was still confident it would come to pass. "It's very delicate, a lot of infighting," he admitted. But the press would keep calling it GNP, and that, despite the protests of the Park Service, would eventually make it a fact, in that peculiar way of making facts in Latin America.

In short, what happened was this: Janzen pushed for the name Guanacaste National Park and others pushed back, some of them quite angrily. Something so tied up in Costa Rican patrimony should be a Costa Rican decision, they argued, and Janzen was getting into "dangerous waters," as one of them told me. Late in the Arias administration, as the idea for the conservation areas system rose, a handy compromise emerged tangentially: keep the names of the national parks, but bring them under the unifying name of the conservation area that contained them. As a result, four national parks—Santa Rosa, Murciélago, Rincón de La Vieja, and the newly established Guanacaste National Park on the east side of the Pan-Am Highway—became units of the Guanacaste Conservation Area.

That fight among the conservationists ended, but it had created many hard feelings. The victors, including Ugalde, later downplayed the conflict. It was, he once told me, "one of those unnecessary disagreements that bring noise to something very successful."

These and other debates had made Janzen enough enemies that on July 25, 1989, when Arias, Umaña, and other officials returned to the Casona for a ceremony formally creating the new Guanacaste National Park on the east side of the highway, Janzen was left out of the proceedings. He told me he was "now smart enough to opt out." It was an odd contrast with the 1987 ceremony at which Arias had called him up in front of the crowd and thanked him for his contributions. In the 1989 ceremony, the Costa Ricans celebrated the newly enlarged protected area, the restoration effort, the entire concept of taking back the forests of Guanacaste that Janzen had been largely responsible for, but they snubbed him. This bothered a few Costa Ricans, but Janzen claimed to feel nothing. Years later, he attributed the circumstances to Costa Rican politicians playing a chess game with each other. "If in their internal politics they don't want me on the stage, cool," he told me. "I couldn't care less. I'm not in this game to be honored. I'm here to make this forest work. What matters to me is the budget, the land, the staff."

In a separate interview, however, Janzen revealed something of the frustration he felt during this struggle, frustration at least with the time wasted on it. He told me that when he began the Guanacaste project he had no idea how big the changes he was setting in motion were. In fact, the changes were so big that the reaction to them at one point made him feel as if he were about to be "burned at the stake." That did not mean he feared the pain of dying, but rather feared losing the forests of ACG.

In 1990, the administration of President Rafael Angel Calderón replaced that of Arias, and among the significant changes was the return to government service of the two great heroes of Costa Rican conservation, Mario Boza and Alvaro Ugalde. Boza was appointed Vice Minister of Natural Resources, and he brought in Ugalde in 1991 for another term as director of the Park Service. Ugalde served for two and a half years, a period marked partly by what some say was his attempt to revert the conservation areas system back into the park system. Ugalde denied that, saying he worked hard with the Legislative Assembly to try to get a law approved establishing the system and took the job only on the condition that Boza support the concept of regional conservation units, later to be labeled conservation areas.

But much had changed, and Ugalde and other Park Service leaders during and after his administration bristled at some of the changes. In turn, they encountered tremendous resistance. Advocates of both systems thrust and parried. "It was," one Costa Rican observer told me, "a time of war for the Guanacaste Conservation Area."

The local advisory committee of politicians, businessmen, university leaders, and less well-to-do residents had been established as the de facto board of directors for ACG, and it behaved as if the Guanacaste forests were *its* responsibility, which they were. The central office of the environment ministry, its control undermined, questioned this "nontraditional" view and argued—persuasively to many who already held a grudge—that it was fundamentally unfair for ACG, with its more successful fund-raising, to grow stronger while other conservation areas struggled. Some accused ACG of being a private entity, unaccountable to the government in the way it handled budget, hiring, fire control, and other programs formerly under central jurisdiction. In their view, ACG and other conservation areas were becoming Balkanized, with independence that went well beyond decentralization of a national system.

One episode dramatically symbolized the rift. It centered on logos that ACG and a few other conservation areas had designed, displaying them widely on signs, vehicles, and brochures. The ACG logo was an image of the broad-winged, colorfully patterned *Rothschildia* moth. Ugalde prohibited use of the logos, and ordered that the Park Service emblem be displayed instead. One by one, the *Rothschildia* logos were removed from signs and scraped off the doors of trucks, replaced by the words Servicio de Parques Nacional. Today, the logos are back and Ugalde says he is happy with them, but in the eyes of many, the episode demonstrated his failure to realize that the old war of Costa Rican conservation was over and that a new, more peaceful approach to the populace had begun. "Alvaro Ugalde was a great general in the war," one observer told me, "but he was not the right minister in peacetime."

Throughout the early 1990s, the old Park Service system came close at times to regaining control of ACG and restructuring it. The attempt to integrate conservation and decentralize it faced a continuing threat of reversal. "You shouldn't underestimate the resistance to change from the system," Gámez told me in 1996. "It comes back and comes back and comes back through time." There were, for example, attempts to bring ACG under the umbrella of a conservation unit comprising all of Guanacaste Province, to put a traditional San José bureaucrat in charge, and to make the ACG endowment available to the whole conservation areas system. To advocates of this move, it would remedy the uneven level of development in the conservation areas. To opponents, it was an attempt to loot the endowment. The effort failed, but the fight continued over the endowment, personnel, jurisdiction, land management, and administrative structure.

Gradually the Park Service adjusted, and by the mid–1990s, the momentum of the conservation areas system seemed unstoppable. The leadership changed again with the new administration of José María Figueres, elected in 1994, and the evolution continued. While about 260 people worked in the service's headquarters in San José in early 1995, only eighty remained the following year, about twenty fewer than worked in ACG alone. Some resigned, some retired, and some moved out into the conservation areas.

Despite this bureaucratic triumph, Gámez, Janzen, and other progressives realized that it would not be enough in the long run. They could not now sit back and say that ACG was "safe." The main threat still existed, the threat to biodiversity in ACG and throughout Costa Rica. It would not be effectively countered, really, until biodiversity was linked to the quality of life of the people. Somehow, the conservationists needed to demonstrate the value of wildland areas, to create a "consciousness-raising" about biodiversity that had not yet been widely achieved.

Out of this intellectual and political landscape, in parallel with the evolution to a decentralized conservation areas system, came INBio (pronounced "IN-bee-oh"), the common term for Costa Rica's Instituto Nacional de Biodiversidad, or National Biodiversity Institute. The key figure in that development was Gámez, who, thanks to a MacArthur Foundation grant, worked with Umaña as a "bio-diversity adviser" to the Arias administration. Gámez developed what was, in 1987, termed a "biodiversity program," and it evolved in 1989 into INBio, of which he was founding director.

Funded by a wide range of sources, INBio was headquartered in the San José suburb of Santo Domingo de Heredia and quickly grew into a small army of parataxonomists and scientists who collaborated on collecting projects all over the country. It was a nonprofit, nongovernment organization, and its mission, put simply, was to determine the flora and fauna of Costa Rica's conserved areas (an estimated 500,000 species, 80% of which had yet to be described and named), catalogue them, and make this information available in ways that allowed it to be used sustainably by society. (Its home page on the World Wide Web is www.inbio.ac.cr.)

The intent was to find economically profitable uses for this biological inven-tory, but also to raise the "bioliteracy" of the general public to a level of profound understanding about "why humans need to maintain a proper balance with nature," as Gámez put it. This is the biocultural restoration and "use-it-or-lose-it" lesson that Janzen and the others brought out of the Corcovado affair, in which unemployed workers and ordinary citizens became miners and created an ecological crisis in the park: Organisms must be used as sources of economic and intellectual wealth, so the people who might otherwise destroy them realize it is better to save Costa Rica's rich storehouse of biological wealth than to convert it to boards, cattle ranches, and banana and coffee plantations. "We don't have gold," Róger Blanco once told me, harkening back to the history of Costa Rica's original misnomer as the "rich coast." "We don't have petroleum. But we do have biodiversity—green gold."

Beneath this concept lay the notion that tropical wildland biodiversity could be conserved over the long term only through its nondestructive use by a wide array of "social sectors," the term the conservationists used to describe agriculture, ecotourism, education, and other dimensions of modern society. True conservation of a wildland came about only when humans decided to pro-tect it, find out what biodiversity it contained, and put that to work for the other sectors. "No one of these three actions is independent or sufficient," Janzen once wrote. "All three should operate simultaneously, and all three rein-force each other."

INBio's specific goals included producing field guides to all Costa Rican organisms and creating a computer database of all the work and a complete national research and reference collection, which could serve as one of the bases for taxonomic organization of all neotropical organisms. It was a pioneering effort that other countries, including Indonesia, Kenya, and Mexico, have since tried to replicate.

The biodiversity institute idea had simmered in conversations among Janzen, Gámez, and other biologists since the Guanacaste project began. In 1988, Gámez attended a Global Biodiversity Strategy meeting in Bogotá, Colombia, where leading conservation biologists outlined their ideas for blending biodiversity and sustainable development. When he returned, he arranged a meeting in San José of more than fifty biologists and officials with Costa Rican government agencies and institutions that worked with biodiversity. The participants agreed that their fragmented efforts should be replaced by a unified program. "Coincidentally," Gámez and several colleagues wrote in *Biodiversity Prospecting*, Janzen had just returned from a Guanacaste fund-raising trip "with the news that the international community was ready to consider financing a national commitment to understand, manage, and sustainably use biodiversity. In the enthusiastic discussions and planning that followed, the idea of an independent biodiversity institute emerged."

INBio was governed by a fifteen-member Assembly and six-member Board of Directors. Although its leaders hoped it would one day be self-supporting, it relied on financial backing from more than two dozen foundations, government agencies, scientific institutions, corporations, and individuals. INBio was created as a *private* institution for strategic reasons, meaning it would collaborate with Costa Rican government institutions, but its mission, and especially its financial resources, would not be exposed to bureaucratic and political manipulation.

One of the early political landmarks in the life of INBio was the decision by Hernán Bravo, the Minister of Natural Resources during the Calderón administration, to permit the institute to conduct the national inventory as well as bioprospecting activities within the conservation areas. That legal standing to pursue those two activities was crucial, especially as the institute's success engendered jealousy among institutions that increasingly saw themselves as competitors, such as the National Museum and the University of Costa Rica.

Walk through INBio on any given day in the 1990s, and you would see dozens of technicians and scientists sitting at desks, some pinning strikingly beautiful insects into storage trays, some mounting plants on high-quality 11-by-17-inch backing paper, and others typing information about each specimen into a computer file for the Internet—its identification number, species, date and location collected, and much more. The institute had pioneered using computer technology to organize biological information, even attaching bar codes to its specimen mounts, an idea Hallwachs came up with in a Philadelphia supermarket. In offices and labs, taxonomists worked with microscopes, "keying out" specimens, that is, identifying them in a step-by-step process that involved comparing their characteristics against those in a taxonomic guide written by a specialist. There were classes, planning meetings, and workers preparing supplies for parataxonomists in the field and packages to send to taxonomic specialists worldwide.

During one of my visits, in 1995, I witnessed a good example of what INBio produced, and the international cooperation that made it possible. Sitting at their desks, Barry Hammel, Silvia Troyo, and Cecilia Herrera worked on one of the institute's major publication projects, the "Manual to the Plants of Costa Rica." Troyo was illustrator of the book series, and Herrera was page editor. Both were Costa Ricans. Hammel, an associate curator with the Missouri Botanical Garden and a U.S. citizen, spent most of his time in Costa Rica as an adviser and volunteer botanical curator to INBio. He was among those who helped train the parataxonomists, and he led the manual project in Costa Rica with his colleague Michael Grayum, based at the botanical garden in St. Louis.

The manual, in four volumes, was to be an illustrated compilation of all the nation's plants, designed for use by scientist and nonscientist alike, in both English and Spanish. The project, supported by the U.S. National Science Foundation, was a collaboration of INBio, the National Museum, and the Missouri garden. Contributors, not including the parataxonomists, included Costa Rican botanists Nelson Zamora, Quirico Jimenez, Francisco Morales, Jorge Gomez-Laurito, and Jose Gonzales, as well as a host of researchers from international institutions.

While Herrera edited text on her computer, Troyo drew intricately detailed illustrations of plant specimens with a black ink pen. These would be copied into the computer file of that specimen and would become one of hundreds of illustrations in the manual. Hammel turned to the computer to check whether a plant he had collected recently on a field trip was new to science. He knew it was a species of *Anthurium*, a member of the philodendron family, but he needed confirmation by a specialist, who in this case was Grayum, 2000 miles to the north in St. Louis. He put a cutting from the plant on a computer scanner, punched a button, and a second later a sharp color image of the cutting appeared on the computer screen. It showed the stem, the light green underside of a leaf, and the waxy, dark green top of another. The leaf ribs and edges came through even more clearly as Hammel clicked a few buttons to adjust the frame and shading. "This is quicker than sending a fax and thousands of times better quality," he said as he pushed another button to shoot the image immediately to a computer in St. Louis via the Internet.

Grayum soon weighed in with his analysis: The specimen was a species known as *Anthurium tonduzii*—not new to science but seldom collected and never before seen from the region where Hammel found it. Even if not new, all of these collections were valuable to Costa Rican teachers and researchers, who used them as a comparative library to help identify specimens collected in different areas. It was part of the process of learning about the distribution of species throughout the country.

Such knowledge is scientifically interesting, but it also can be profitable, as INBio learned when it became involved in a watershed in the history of "biodiversity prospecting." Also known as bioprospecting or chemical prospecting, biodiversity prospecting meant to search among various organisms for compounds that would lead to new medicines and other products. The watershed came in September 1991 when officials with INBio and Merck & Co., the world's largest pharmaceutical company, announced that Merck would pay more than $1 million over two years to the institute. In turn, INBio would provide the company with a limited number of samples from plants, insects, microorganisms, and soil in the conservation areas. To do this, biodiversity prospectors froze the samples, and others drove them in a former ice cream truck from the field to the institute's headquarters. There, technicians ground them up and extracted compounds which, when dried into a powder, were flown to Merck's offices in New Jersey. If commercial products resulted, the institute and the government would receive a share of the royalties. The government got 10 percent of the up-front payment, too. The government money went into the conservation areas' budget. Although no commercial product had resulted by the mid–1990s, Merck and INBio renewed the agreement twice, and the company provided equipment and training to Costa Ricans to help them establish their own drug-screening program.

Thomas Eisner, a chemical ecologist at Cornell and rain-forest mafia member who coined the term "chemical prospecting," helped engineer the Merck agreement with Costa Rica. Eisner put together a small conference in Ithaca in October 1990 that examined, among other questions, how to get the pharmaceutical industry to pay for conservation up front, long before any possible royalties might accrue. It was at this conference, sponsored by the MacArthur Foundation, that Gámez met Paul Anderson, a chemist who had been a postdoctoral fellow in another lab at Cornell and had become Merck's vice president for medicinal chemistry. Eisner recalled, "The meeting was only one day long, and we knew by noon that there were sufficient grounds between INBio and Merck to launch some kind of collaboration. By dinner time, at a small Italian restaurant, the broad strokes of what was to become the eventual agreement were sketched out in conversation."

The agreement was based on the widespread belief that, despite the steady advance of modern science, Mother Nature was still the best chemist. Put simply, some of the compounds designed naturally, say, by a plant, hold promise for defeating human diseases because the structure or activity of the disease organism might resemble that of the plant's adversary. Tropical plants are a major focus of today's bioprospecting because they have produced large and varied chemical arsenals, a consequence of the army of fungi, insects, and other plant enemies whose species diversity is greatest in the tropics. The diversity of chemical compounds in nature is beyond what scientists can even imagine, much less synthesize. So the pharmaceutical industry puts extracts from plants and other organisms through a battery of screening tests that indicate possible "bioactivity" against disease. If testers find a "hit," the compound goes through a long process of analysis, molecular redesign, and other study. Out of this process, perhaps in one in 10,000 tries, comes a marketable product years later. Even with such small chances, the potential for developing a new medicine was great.

In essence, Merck was paying Costa Rica to do something that drug companies had previously done for free. However, critics in Costa Rica lambasted INBio, saying that as a private institution it had no right to sell the nation's patrimony—its biodiversity resource—and that even if INBio did, the price was too cheap. Some of this, legitimately, came from the biologists on a shoestring budget at the University of Costa Rica, who were struggling just to get into the field, while the institute was "rockin' and rollin'," as one researcher put it. Even so, many concerned about the biodiversity crisis considered the INBio-Merck agreement a great step forward for conservation. It also was a test case in whether tropical countries in the future could rely on forests for their wealth of chemical and genetic information, or would continue to log them and substitute plantation agriculture.

INBio steamed along, concluding by 1998 half a dozen other collaborative research agreements with companies in agriculture, biotechnology, and the cosmetics, flavor, and fragrance industries. For example, the institute agreed to work with the British Technology Group and the Royal Botanic Gardens in Kew, England, to test a chemical extracted from the seed of a particular plant. The biodegradable chemical was found to kill nematodes, a worm that plagues bananas and other tropical crops. In another contract, the Swiss firm Givaudan-Roure, the world's largest fragrances and flavors manufacturer, planned to use fragrances from flowers found by biodiversity prospectors who sniffed the air and visited promising flowers at different times of day and night to determine when a scent was most powerful. The fragrances would be used in perfumes, shampoo, and soap. The institute's Alvaro Fernández told the *Tico Times* that in a mangrove at Santa Rosa, he had discovered a flower that "smells clean" and might be a candidate for use in a washing powder.

In 1996, officials from Yellowstone National Park visited INBio and ACG. They wanted to study and apply to the U.S. national parks the Costa Rican method of negotiating agreements with bioprospectors. The fact that North Americans had come to *them* to learn was a source of great pride for the Tico biologists.

The All Taxa Biodiversity Inventory (ATBI), a "biodiversity moon shot," as Janzen and Hallwachs termed it, was conceived in the early 1990s but collapsed in 1996. The ATBI was aimed at putting scores of specialists from around the world into the field to conduct the world's first complete survey of all types of organisms in a specific area—the Guanacaste Conservation Area. With an initial price tag of $90 million over seven years, the inventory would require an unprecedented amount of money for a field biology project. It would command the attention and collaboration of the entire "Taxasphere," a Janzen-Hallwachs term for the people and institutions involved in taxonomy. It would do for field biology what the Human Genome Project had done for molecular biology and the Apollo missions had done for NASA. It was a huge, marvelous attempt that would yield an immense amount of biological information.

In many ways, the ATBI plan looked like a scheme to leverage resources into the flagging science of taxonomy, a goal Janzen had espoused at least since his agitation in the pages of the *Bulletin of the Entomological Society of America* in 1984. The same need that drove his plea for more taxonomy then, the same one that drove Linnaeus, was driving the ATBI proposal. Some mistook the inventory as simply an intensified effort at bioprospecting, or at pumping more money into ACG.

But it was much more than these things. Janzen had long since concluded that, as he and Hallwachs put it in a 1994 paper on the ATBI, "The important questions become not how does one get the maximum amount of dollars into ecology, into the Taxasphere, into biodiversity prospecting, into conservation land purchase, etc., but rather, how does one most wisely and widely share the biodiversity resources among all sectors of the society, in the most non-destructive manner[?]"

In essence, the ATBI was a bold attempt at consciousness-raising about biology, about the complex interactions among organisms, and about the value of a wildland area. Although focused in Guanacaste, this message was not just for Costa Ricans. The inventory, Janzen wrote in a proposal to the World Bank, aimed "to inspire other tropical countries to: (1) consider conserving large wild-land areas as an explicit form of productive land use and (2) actually carry out bio-diversity development where conservation initiatives have already been enacted." One of the ATBI documents written by Janzen and Hallwachs was intended to be "of use to any nation that considers doing one." In other words, the hope was that an ATBI would trigger a worldwide acceptance of a new view of nature's value, an acceptance that would transcend the inventory, transcend Costa Rica, transcend even science. They wanted it to cause a global, public paradigm shift.

The push for an ATBI was born out of discussions among biologists in Costa Rica, but the effort expanded in April 1993 at a three-day workshop at the University of Pennsylvania run by Janzen and Hallwachs and sponsored by the NSF.

The fifty-some leading taxonomists, ecologists, and conservation biologists who represented universities, museums, and conservation organizations all over the world discussed a wide range of problems of an ATBI, no matter where it might be conducted. For one, the goal to inventory *every* type of organism in a region seemed unattainable. It would have been a tall and costly order even if there hadn't been a dearth of trained taxonomists to handle the work.

Another problem involved how to start such a massive inventory. In some taxa, as the major divisions among groups of organisms are called, that might be possible. Bird species, for instance, are fairly well studied and relatively few. However, in less-well-studied taxa, like bacteria, only a small percentage of the potentially millions of species had been found and described, much less their ecological relationships sorted out. The attendees also realized the ATBI challenge included finding ways to collect not just a monkey, but the microbes and parasites in its gut, and the parasites of the parasites. Even if they could solve these daunting technical problems, some wondered where they would get the millions of dollars needed to organize, equip, and execute such a project?

Despite the challenges, the idea energized the participants. No such survey had ever been done. Although questions about its reality abounded and the project's development was rife with politics, they were attracted to it because of its huge scale. "The origination of ATBI is classically Janzen," said one of them, a veteran tropical biologist previously unconnected with ACG. "I can't think of a single person who would even think of such a thing other than Dan. A lot of people can sit back and see the big picture, but Dan is the only person I know who has the commitment and energy to do this—to be crazy enough to do this."

Janzen, Hallwachs, and Gámez moved to launch the first ATBI in Guanacaste, and Costa Rican officials agreed in July 1994. Officially, ACG was chosen by INBio largely because of the conservation area's level of organization and its diversity of ecosystems, which housed an estimated 65 percent of all the species in Costa Rica. Yet who could discount the fact that an inventory was essentially the idea of Dan Janzen and Winnie Hallwachs, and that ACG was their scientific backyard? Besides, what better way to ensure the safety of the forest they had fought for than to focus an international, $90 million project within its boundaries? It would shine a brighter spotlight on ACG as well as provide dozens, perhaps hundreds, more jobs for local residents.

Many of the workshop participants signed on as leaders of the twenty Taxonomic Working Groups (known as TWGs) that would divide and conquer the task according to taxa. Among the TWIGs were those for vertebrates (with an estimated 940 species in ACG), fungi (50,000), nematodes (7000), Hymenoptera (13,000), and Coleoptera (20,000). The magnitude was put into perspective best perhaps by Robert Langreth, who wrote in *Popular Science* of his impressions while standing on a path in a patch of ACG forest teeming with life.

"There are thousands, probably tens of thousands, of species within 20 feet of where I'm standing, and Janzen wants to catalog everything within *20 miles!*"

The basic goal of the ATBI was to do the following for each species:

- determine what it is and how to tell it from others;
- find out where at least a portion of its population lives and how to get a specimen when desired;
- acquire a rudimentary understanding of its natural history;
- put this information on the Internet in a large, dynamic, searchable, easily accessible format in the public domain.

Even that relatively complicated list is just the beginning of the intricate scientific and technical dimensions of the inventory. Consider, for example, that even determining the species—the first item on the list—required attention to the administrative differences between a species and an individual specimen. Each specimen in the ATBI would require a unique identifier, a "license plate," so that members of the inventory, and future consumers, could collect, sort, store, recover, and analyze information about it. This individuality was necessary because each member of a population had different variations that were crucial to understanding its natural history; such variations might include basic traits like sex, size, age, developmental stage, health, and behavior.

Three specimens of the moth *Rothschildia lebeau*, for instance, might have the same genetic sequence, but one might be an egg, another might be a newly emerged caterpillar feeding on a certain kind of plant, while the third might be an older, wounded adult moth resting on yet another kind of plant. Such details could be crucial to a teacher, farmer, or drug company. As Janzen and Hallwachs explained in a memo, "each of these different individuals [moths] will have different visibility, different chemistry, different impact, etc., and biodiversity development based on this species must be sensitive to these inter-individual differences." Róger Blanco emphasized this point when he spoke of an Internet home page for each species containing its picture, natural history, and other information. Even 235,000 home pages was just a start. "An organism is a book, a box of information," he told me. "We need to organize this in ways that make it useful."

ATBI was the crowning extension of the INBio theme—save, know, and use—except that it was applied more intensively and to a smaller place than the entire country. The leaders hoped it would become a pilot project for biodiversity development throughout the tropics, and perhaps everywhere around the world. It would create a synergy, both across many scientific disciplines and among people from the social sectors involved. It would be an enormous tool box, library, a living museum, and even a "mine canary," providing a way to alert society to environmental problems, from degraded local water quality to global impacts of climate change. "We are trying to prove a new concept," said Róger Blanco, "that biodiversity should not be preserved just to be preserved, but that there is value in each specimen. We have to show that a park can be changed into a new industry."

The ATBI was to be run by INBio and ACG, and was planned as a two-step process: (1) to establish the infrastructure and training programs and organize the TWIGs and (2) to do the inventory. ATBI leaders held a series of workshops, some of which helped the working groups choose their sites, select members, and figure out methodologies. In other workshops, representatives of Costa Rican agriculture, ecotourism, and education learned what the inventory might do for each social sector. For instance, it could provide natural history information for tour guides and genes for improving crops. Backed by President Figueres and other national officials, the ATBI concept brought in more than $20 million in actual and pledged planning and action grants from Norway, the Netherlands, the World Bank, and the NSF. If that kept up, the inventory was to have begun in 1997.

Five of the working groups visited INBio and each conducted a two-week workshop in ACG planning its taxon-specific actions. To meet the demands of the project, the inventory leaders planned a six-month course for seventy parataxonomists to begin in January 1997. Two new dorms and four more labs would be built at Santa Rosa. The field stations would be renovated, and computer connections would link them with the new network at Santa Rosa. Roads to the stations would be improved, and automated weather stations would cover the entire conservation area. Aerial photos would be taken to develop geographic information system (GIS) maps of soils, habitats, and other data "layers."

The cadre at ACG worked hard to convince skeptics at INBio. There were skeptics, too, at the National Museum, the University of Costa Rica, other conservation areas in Costa Rica, in other Latin American countries, and in the United States. Some scientists believed the intense approach of the ATBI was wrong, that instead a broader investigation of a select few of the world's species or higher taxa was better. Some believed an inventory should be done on a more reasonable scale, say, of one or a few sites only a few hundred yards on a side. That would allow ATBIs not just in Guanacaste, or even just in Costa Rica, but in many parts of the world. Some within INBio worried that the tremendous flow of money for the Guanacaste inventory might somehow drain the institute, and others saw a problem with ACG becoming an even richer kid on the block.

Some viewed Janzen as having become unreasonably obsessed with the ATBI. "He only sees that you have to do the blanket approach, and it has to be in Guanacaste and nowhere else," said one Costa Rican critic. Meanwhile, Tico scientists bickered about where the specimens would be kept. Although the idea of holding all the collections in a central location had been fairly widely accepted at INBio's birth, the great gap in wealth between INBio and other local science institutions had worn that feeling down. Some TWIG members felt the inventory offered an opportunity to bridge the gap, to allow the collections to be held in different—and worthy—locations, thanks to an influx of money. Officials at the biodiversity institute remained adamant about controlling the specimens.

While five TWIGs readied for launch, Janzen and Hallwachs shuttled frequently between Santa Rosa and San José for meetings about ATBI organization, funding, informatics, and other issues. But it was not enough. The critics prevailed, and in November 1996, the inventory plan fell apart. More precisely, Janzen pulled the plug, mainly because he and INBio officials concluded that the inventory didn't match INBio's agenda.

Janzen, Hallwachs, Gámez, and several other Costa Ricans signed an electronic mail sent to TWIG members announcing the demise. Amid the thanks and diplomatic language was this admission: "We have come to the conclusion that the necessary political, economic and institutional conditions to actually conduct an ATBI of the ACG are not present, either nationally or internationally. It is not the right combination of the time, place and circumstances in the history of tropical biodiversity development to carry out this initiative. An ATBI may well appear somewhere in the future but it would be non-productive to continue to try to force it to appear here now."

INBio issued an official statement saying that it and ACG had encountered "important institutional and financial difficulties that justify a change in strategy" and that "the project's original focus must be reoriented." The time and effort had not been wasted, and the two were renegotiating a new direction that would "reorient the project towards strengthening the national inventory and developing sustainable uses in a parallel effort." Behind the officialese was the institute's feeling that the inventory focused too much on science and not enough on social and economic aspects such as bioprospecting and education. Some TWIG members said part of the problem was INBio's insistence on using ATBI money in the general budget of the institute. INBio later said it planned to use the money that had already arrived to conduct less intense biodiversity inventories in five conservation areas, focusing on organisms that could provide more immediate, practical benefits.

A few days after the collapse of the ATBI, while we sat in his office at Penn, I asked Janzen what had happened. "My short version is that the ATBI died," he said, tersely. "The longer version goes on for"—and then he paused to chuckle—"four thousand pages, you know." He laughed. It was not a cynical laugh. Rather, it was the laugh of a man in the early stages of the grieving process, somewhere between shock and sadness. "It's only been dead for a short time and the funeral procession is still marching down the street, so it's hard to know what to say."

Still, he tried. It was like having a bunch of dancers—the ACG, INBio, the Taxasphere, individuals—each having a different way to operate and a different agenda. Trying to get them to dance to the same music, on the same dance floor, using the same steps, with the same level of sophistication, resulted in "chaos." Switching metaphors, Janzen said, "I wasn't about to load up hundreds of taxonomists on an airplane, and hundreds of Costa Ricans on an airplane, and take off, unless I thought that airplane really would fly. It got to the point where one had to take off or kill it. One couldn't keep standing around and planning and thinking anymore. Something had to happen." He made a few more cryptic comments about the need for the money for such a project to be in the bank, and for the people conducting the work to be the ones who controlled the money. He hoped that the ACG would find a way to conduct a similar program on its own.

Some scientists other than Janzen took a realistic view, saying the word "failure" didn't really describe what happened to the ATBI. "They were trying to get to the moon on the first shot," said one researcher, implying that there was no shame in falling short of such an audacious goal. Besides, the ATBI concept would eventually succeed. Others pointed out that the process itself had triggered concrete advances, including new collaborations among members of some of the TWIGs. Plus, everyone seemed to be talking about ATBIs, and not just in Costa Rica.

A year later, when I spoke with Janzen at a National Academy of Sciences forum on biodiversity, in Washington, his spirits seemed buoyed by the possibility that the U.S. National Park Service might conduct an ATBI in the 800-square-mile Great Smoky Mountains National Park. Janzen had been asked to serve as an adviser, and it clearly gave him a sense of satisfaction. His father had surveyed the boundaries for that park, he told me. In 1999, scientists started the Great Smoky inventory, and one writer called it "the biggest ecological scientific undertaking in American history."

# · 17 ·

# The Vigilant

*Se Vende Este Finca* (Farm for Sale)
> —A common sign along highways in northwestern
> Guanacaste Province, 1986

*Evitemos Los Incendios* (Let's Avoid Fires)
> —A common sign along highways in northwestern
> Guanacaste Province, 1996

Alex Rodriguez and Didi Guadamuz stood calm and firm in the buffeting wind, their binoculars drawn up to their eyes, as they scanned the length and breadth of the Guanacaste Conservation Area. The men, both in their early twenties, seemed eerily connected to the ground, as if long roots extended from the bottom of their feet deeply into the soil. They were posted atop a grassy knoll known as Cerro Pedregal, a hill that jutted like a boil out of the western slope of Volcán Cacao about 3000 feet above sea level. I stood with them. It was February 1996, and the dry-season wind raced across the hill so powerfully that it flattened the brown grass atop the hill. I felt as if at any moment the wind would pick me up a few inches, and, if a gust should come along, send me on a flight out into the gray dust-haze that partially obscured the dry landscape of brown and green literally at our feet. Perhaps I would make it a bit farther to the west, and into the Pacific.

Rodriguez and Guadamuz were the early warning system of the ACG fire-fighting program. They were part of the area's eleven-member dedicated fire team, and they and other members took turns from January through April watching for fires on the land below from their lookout "tower" atop Cerro Pedregal. They stood for hours, or lay on cots raised up to face the windows of their small green hut. The hut was built to keep from being blown away in the fierce wind. It was set in a two-foot-deep hole in the ground, and extra-long nails held it together. The roof only had a short overhang, with "*Bienvenidos A Pedregal*" (Welcome to Pedregal) painted in thick white letters on the front facia. Like mast stays on a storm-tossed sailboat, guy wires anchored the ten-foot radio antenna fore and aft, starboard and port. Solar power cells were fastened tightly to the roof.

Their role was that of forward observers in an artillery unit, watching for the enemy and, when spotted, calling in the big guns—specially trained ground crews. Both men had grown up in the region, and they knew it well—the roads, farms, ranches, homes, and land forms. They knew it so well, in fact, that even without consulting their maps they could usually tell the crews below whose farm or house was nearest the fire. When they saw smoke or flame, they radioed the main fire office, at Poco Sol, below by the Pan-Am Highway, and relayed the fire's location, the direction it was burning, and what roads and paths to take to get there. If it was after dark, fire fighters on the ground often shined a light toward Pedregal if they couldn't find the fire. The lookouts could see the light in relation to the fire. "We tell them to go right, go left," Guadamuz told me.

On this day, they already had seen one fire, but they quickly realized it was nothing to worry about. The smoke had come from a small dump where someone was burning garbage. The most recent threat had come four days earlier, when a fire started along the Pan-Am Highway only about 500 yards from the Santa Rosa entrance. "When we see a fire, we're anxious to go to it and put it out," Rodriguez admitted. However, he and his partner knew their job was to stay put, observe the fire, and communicate crucial information to the men and women below.

As I leaned hard into the side of the hut for balance and talked with the two lookouts, it occurred to me that they were prime examples of the ACG policy of hiring local people to protect the forest. If not for that philosophy, they might otherwise have pursued traditional Guanacaste ways of life—farming, ranching, poaching, logging, burning. Instead, they had become part of a team dedicated to protecting the land where they had grown up from its greatest enemy. I asked how they felt about their mission.

"It's very important," said Rodriguez, with subtle pride.

"Clearly," said Guadamuz.

Among the keys to the rebirth of the forests of Guanacaste had been improved fire fighting. The main weapons were early detection and tenacious counterattack. Since lookouts had been posted at Pedregal several years earlier, the fires of Guanacaste had been dramatically curtailed. The ground crews, too, attacked their foe with pride and passion, and, put simply, their training and experience had made them very, very good at putting them out. Part of this success derived from the game-like atmosphere that now surrounded fire fighting in Guanacaste. Rodriguez and Guadamuz, for instance, kept notes on the time it took a crew to get to a fire and extinguish it.

They are no longer teasing Julio Díaz for his stern expression on the fire lines. Díaz, whom I had seen set the *contrafuego* in 1988, was the boss in 1996 of what Róger Blanco called "the most important fire prevention program in all of Latin America." As director of the ACG Program for Prevention and Control of Forest Fires, Díaz, then in his mid-thirties, was still long-legged and deerlike, but his frame had filled out a bit, and he wore glasses. He took a break from his duties one day during the dry season to describe how the Guanacaste fire-fighting effort had evolved.

It all began in 1988, when a core group of fire-fighters decided to take things more seriously. They lingered after extinguishing a fire to study it. They talked about what they could do to better fight the blazes destroying Santa Rosa. The group coalesced, and soon they launched the official ACG fire program. Mistakes were made early on, such as the time the crew forgot about their new fire truck and a fire destroyed it behind their backs, but by and large a new commitment and esprit d'corps defined the team, which was now made up of men and women. The program developed three general objectives: to reduce the initiation and propagation of forest fires in ACG, to get instruction for the staff on how to better combat forest and structural fires, and to raise awareness of the fire problem among ACG personnel and the local population, pushing toward a more intelligent use of fire.

Team members traveled to fire-fighting workshops in Spain, Honduras, and Costa Rica. They constructed a database with information about the fires of Guanacaste. This fire history included the origin and extent of each fire, the weather conditions at the time it occurred, and details about how it was extinguished. The data were used to investigate fires and to try to predict where they might ignite again. The team identified areas with the largest amount of potentially combustible biomass in and around ACG and made maps of access routes to these areas. The maps had red, orange, and yellow zones showing high, medium, and low fire danger, respectively, and they indicated sources of water that could be used in fighting a fire. The team learned how to read maps and interpret the influence of weather and topographic conditions. The main map was used not only when fires struck, but also when deciding which areas to cover with routine vigilance patrols.

The team launched a public information campaign that expressed concern about fires and elicited help in preventing and controlling them. A phone number to report fires was disseminated through radio, television, and newspaper ads. Team members initiated a volunteer program and distributed hats, T-shirts, and other items to get the message out about the dangers of fire to the Guanacaste forests. They kept in contact with local fire fighters, the Rural and Civil Guard, and other organizations, in case extra help was ever needed.

The crew organized its year from October to October. Members started the year tending to 86 miles of key roads, trails, and firebreaks in the conservation area. They mowed firebreaks and paid a local bulldozer owner to repair roads that gave them access to remote areas. In December, they conducted controlled burns on pasture that had become dangerously susceptible to fire, mainly in Sector Santa Rosa, Sector Poco Sol, and Horizontes. They cut grass along the Pan-Am Highway and repaired equipment. In January, with the advent of the critical part of the fire season, they settled into their base at Poco Sol—and the Cacao lookout—and prepared for battle.

By January 15, the seven core team members were required to be on duty twenty-four hours a day, seven days a week. Four additional professional fire fighters from nearby towns and farms joined the team under a January-to-July contract. Some crew members patrolled on certain days and in certain areas that had been identified as particularly prone to fire. For example, Saturday was an active day for fires around beaches and rivers, largely because of people out on weekend getaways. At these times, bonfires were restricted. Hunters tended to set fires on Tuesdays and Wednesdays. Meanwhile, Rodriguez and Guadamuz—and periodic replacements—maintained their high-altitude surveillance in shifts around the clock. Others constantly assessed weather conditions so that, if a fire was detected, fire fighters had an edge on predicting its behavior. Still others watched for fires from atop the monument behind the Casona.

The physical tools of Guanacaste fire fighting remained much the same: brooms, shovels, chainsaws, and plastic backpacks filled with water. A truck carried around a larger cistern of water than before, and the fire fighters wore yellow protective jackets, but the biggest change was radio communication. All team members were linked by two-way radio and carried maps. If anyone reported a fire, the team swung into action. "Some will go by car to look," Díaz said, keeping an ear close to his walkie-talkie. "They may call on the radio: 'To extinguish this fire, we need the water tank.'" If reinforcements were needed, all ACG workers were expected to volunteer if possible, and eleven other local volunteers trained during the rainy season were called in. "They all understand that they *must* control every fire in the area."

The most exciting moments came during battles with large, dangerous fires, although these had become much fewer in number. At times like this, Díaz saw most how the fire-fighting attitude had changed over the decade. "When the crew knows it *must* extinguish a big fire, everyone works together," he said. "They sing together." He remembered a particularly challenging few days in April of 1993 when the crew battled a set of blazes in the mountains of Murciélago, then immediately after extinguishing these rushed to a large fire in the fields of Altamira on the road to Maritza. There, they had another close call with the truck carrying the cistern. The truck became surrounded by flames. "The only way out was to drive through the fire as fast as possible," Díaz said.

When the rainy season began, team members stayed in place for a little over three weeks to make sure nothing flared up. They went to a doctor for a physical, took three weeks off, then returned in July for meetings to analyze the fires of the previous season. "We study them all," Díaz said. "If we made mistakes, we ask why." It reminded me of a professional football or basketball team, studying videotapes of previous games to prepare for the next. Some of them traveled to get more fire training, while others gave educational presentations about fire to local community groups and schools.

Although they had been doing this for several years, fire crew members still had much to learn about fire, its causes, effects, and the kinds of areas most vulnerable to it. They felt they needed to work more with people in surrounding communities to educate them about the devastation caused by fires and show them how to prevent and fight fires. "One of the most important things we can do is education," Díaz said.

The Guanacaste Fire Committee, made up of community leaders and ACG staffers, erected signs along roads throughout the northern part of the province that read, "*Evitemos Los Incendios*" (Let's Avoid Fires). Now, anyone who wanted to burn a field first had to get a permit from the committee. Nonetheless, many unauthorized fires were started. Throughout the province, the main source of fire remained farms and ranches. It was still cheaper to burn a field than to pay two or three people to cut it, and the new grass that rose out of the ashes was better for cattle. In ACG, however, hunters or vandals started most of the fires, and arsonists who simply disliked ACG would start fires inside or just outside the park.

Only a month later, Díaz and his charges would battle one of the largest fires since ACG had been established. Set by arsonists, a series of at least three fires swept across the old grass pastures of Sector Poco Sol on the east side of the Pan-Am Highway and, despite the efforts of more than 200 fire fighters, Rural Guards, and volunteers, jumped the highway and burned into the Santa Elena forest. In all, an estimated 7400 acres of ACG burned, including regenerating forest. The fire affected mainly grass and shrubs, since many of the new trees were old enough to resist damage. A notable casualty was the large house still owned by Jimmy Swaggart's Evangelistas.

About 3700 acres of ACG burned in a seven-day series of fires in April 1997. Areas of regenerating forest also were among the casualties of these fires, fought by some 260 people. One official told the *Tico Times* that arson was suspected. "Nineteen different fires were started in a line one mile long, and at regular distances from each other," the official said. Within a few days, ACG guards detained three hunters suspected in the fires. They were found within the conservation area carrying dead animals, guns, hunting dogs, and kerosene.

Despite such setbacks, Díaz was unrelenting, like a warrior on a patriotic mission. Stopping the fires of Guanacaste had now become his singular purpose in life. He was in charge, and he was succeeding. Since the program began, fires had dropped by more than 90 percent annually. "The main enemy of this forest is fire," he told me. "If we allow fire to destroy the forest, it will contaminate the rivers, affect our resources, affect the communities."

While one group of ACG staffers worked to halt the main destructive force in Guanacaste, another group worked to better understand and enhance the restorative force. After early trials at Santa Rosa and Cerro El Hacha, the Horizontes Forest Experiment Station had become one of two centers of the ACG effort to study and disseminate knowledge about restoring the forest. I rode there one day in August 1996 with Alejandro Masis, then coordinator of parataxonomists for ACG and the Tempisque Conservation Area to the south. Masis, in his mid-twenties, was short and wiry, and he had an easy-going manner and playful sense of humor. He had graduated two years earlier from Kansas State University with a degree in biology. We drove south from Santa Rosa to Liberia, then west several miles before turning north through the Tempisque River valley, where the landscape varied from scrubby ranchland dotted with white Cebu cattle to muddy black fields being prepared for the upcoming rice crop. As we neared the station, an increasing number of saplings and medium-size trees rose amid the jaragua.

On a map, Horizontes, the 28-square-mile former rice, cattle, and sugar cane farm donated by Cecil Hylton, looked somewhat like an anvil. The northern part touched Santa Rosa, Hacienda Rosa María, and the Pan-Am Highway. The southern section was surrounded by farms and ranches. The Río Tempisque, the largest river in Guanacaste, ran through the eastern part of both sections. The station itself consisted of an office, dormitories, meeting room, and lab, some converted from the former hacienda. Closely cut grass and gravel roads ran between the buildings, and one of them still had a ramp which had been used to fill up cropdusters with fertilizer.

We were shown around Horizontes by the administrator, Fermín Mendez Miranda, who headed the ACG Reforestation and Silvicultural Program.

Mendez, in his early thirties and lean like Masis but a bit taller, was matter-of-fact but friendly. He told us that when the ranch was donated in 1987, Janzen immediately decided to make a forest experiment station of the southern part, closest to the agroscape and farthest from the established ACG, and regenerate the forest in the northern half. A restoration program there began in 1989, aiming to replace the dominant jaragua pastures with forest through natural processes and by planting trees, both supported by stopping fires. A nursery that had been started at Poco Sol was moved to Horizontes in 1990. Program members gathered seeds from the forest floor in Santa Rosa and around Poco Sol, then reared them in the nursery at Horizontes.

The seeds were planted in row upon row of dirt-filled plastic bags the size of coffee cans. Some, like the typically animal-dispersed species, needed special treatment to ensure they sprouted. The guanacaste tree seed, for example, was boiled in water, then soaked for two days before being planted about half an inch deep. Guapinol seeds had to be scratched with a file, then soaked. Wind-dispersed trees, like the *Tabebuia* species, were simply placed atop the soil. All the seeds were doused regularly with fertilizer, fungicide, and water. If planted in February, most were ready for transplantation into the field in June or July. Each year the nursery produced 30,000 seedlings or more of native species, most of which were transplanted in reforestation sites in Poco Sol, northern Horizontes, and Junquillal, or given to neighbors.

Horizontes's mission was that of a forestry experiment station. In this sense, although it was part of ACG and, thus, a protected area, it was not a traditional protected area. Trees were being encouraged to return, but they were planted, charted, and graphed under a variety of conditions that might shed light on which species propagated best and under what circumstances. The goal was to provide information and examples for Guanacaste Province, an agricultural and ranching landscape now desperate for alternative uses of pasture land.

This experimentation included the application of insecticides, for example, which normally would not be viewed by conservationists as compatible with pure conservation. The leaders of the Guanacaste restoration took this approach, though, because they wanted a complete silvicultural investigation on the most important dry-forest tree species for both their ecological and economic potential. If private land around the conservation area could be used for sustainable commercial forestry, that would mark an improvement over farming and pastures in the province's land management, from the perspective of the conservation area. Eventually, it was hoped, Horizontes would serve as a seed bank of native tree species that land owners could use as a resource if they wanted to start commercial forests in any of the wide range of the province's dry forest habitats. "It's like a farm," Rodrigo Gámez once told me, "only you extract elements of biodiversity."

About a dozen people then worked in the Reforestation and Silvicultural Program, including a few in the strip of land between Cacao and the mountain to the south, Volcán Rincón de La Vieja. In 1989, Mendez and the others started the work at Horizontes by planting seedlings of ten species in thirty-six plots. Each plot had 140 trees of the same species. This hopeful beginning was cut short during the ensuing dry season when a fire suspected to have been started by hunters swept through the plots on its westward rush to the ocean. The seedlings got torched. "After that first experience, we saw that we needed to get rid of the jaragua," Mendez said, as we drove around the fields where the tragedy occurred. "So we rented parcels to ranchers to get the jaragua down. The cattle didn't eat the seedlings, except a little in the dry season. After that, the regeneration process boomed."

The year after the fire, they planted seedlings in even more plots, totaling 250 acres. Each plot contained more than 1000 trees. Some of the plots were, again, homogeneous, but some were mixtures of as many as fifteen species. "The idea," Mendez explained, "was to study the behavior of the species competing with each other." In another experiment, several exotic species were brought in to see how they fared separately and in competition with the native trees. "Our goal is to gather as much information as possible and make it available to anyone interested—the government, private interests, schools," he said. One product of this effort was published in 1996, a small, illustrated book describing fourteen major dry forest tree species in Guanacaste, how to get seeds, and how to grow them.

The trees of Horizontes were returning, and Mendez and his co-workers kept a watchful, hopeful eye on their progress. "My hope is to see this area all in forest again," he said. "I want to walk all the way from the gate here to Santa Rosa and see all forest, and not worry about the fire anymore."

A few days later, I stood on the veranda of the Casona an hour after sundown. Strong thunderstorms had rolled through the region that afternoon, and now countless frogs and insects croaked and chirped and peeped a raucous cheer in the humid air of the nighttime tropical forest. In the black foreground before me, in the corrals beneath the old guanacaste tree, thousands of fireflies traced a chaotic aerial ballet with minuscule dots and dashes of blue-white light. Above the dark outline of trees to the south, west and north, a panorama of storm clouds floated far out over the Pacific and Nicaragua, illuminated only by a hint of the moonlight they let through. Heavy with moisture, they seemed to sit in thick, claylike piles, as if someone's pottery wheel had gone wildly spinning, throwing its contents about the sky. Lightning and thunder still played games along the murky horizon, and the distant flashes called to mind the fury of battles past around the Casona. At the same time the flashes gave form to the peaceful silhouettes of the trees.

These thoughts were unexpectedly interrupted when something fluttered past my head. Ah, a bat.

I had experienced tropical bats enough to enjoy what seemed to be a near collision in the darkness, but which was actually just an agile bat steering around me. Rather than trying to harm me, this and other bats were intent on finding insects, fruit, and other foods, and as they pursued this, they played a key role in forest ecosystems. As it flew by, it softly touched the edge of my ear. Or was that just a puff of air rolling off the wing?

I walked down the veranda to where María Teresa Fernández Monillo untangled a bat that had flown into one of her mist nets. The black nylon fiber of these nets is so fine that bats sometimes fail to detect it until too late. Fernández and her partner, Marco Hidalgo Cheverri, both students at the Universidad Nacional, in Heredia, had set up two mist nets on the veranda, each about eighteen feet across and seven feet high. Wearing headlamps, they tended the nets and worked with bats already captured and hanging in white cotton bags pinned to a nearby clothesline.

As Fernández carefully removed the nylon strands caught around its head, feet, and wings, the bat emitted a noise that was part squeak, part growl. "Sometimes this requires patience," she said, not smiling, but not frowning either. She was calm, determined, in control. The bat quieted down, then resumed growling when Fernández freed one of its cellophane-thin wings. She clicked her tongue at the creature and made a "cht-cht-cht" sound. A distant flash of lightning illuminated the strikingly beautiful face and lean body behind her headlamp.

Using mist nets and other tools, biologists over the past few decades had gained new insight into the surprisingly complex world of tropical bats, including their diversity, behavior, and ecological roles. Among other findings, researchers had discovered when and where bats feed, the size of their home ranges, their patterns of flight, the sound signals they use to locate objects, and their roles in forest regeneration. Based largely on studies done on Barro Colorado Island, a research site in Panama run by the Smithsonian Tropical Research Institute, their work had begun to paint a picture of the incredible diversity of bats in the New World tropics: bats that fly high, bats that fly low, bats of the deep forest, bats of the forest edge, bats that snatch fish from the water, bats that grab lizards from high branches, bats that pluck fruit, bats that eat frantically dodging insects, and the vampire bats, which use their teeth to make a painless incision in cattle and other animals and, with the help of an anticoagulant in the bat's saliva, lap up the blood.

Fernández, like other bat researchers I've talked with, was quite upset about the popular notoriety of bats, which she blamed on misinterpreting vampire bat behavior. In Guanacaste Province, where cattle had been so much a part of local culture, hatred of all bats was common. In a way, who could blame the average rural Tico, considering that they sometimes saw the dark red evidence left by a nighttime blood meal on the side of the family cow?

The myth of the evil vampire bat had been reinforced in the mid–1990s by the cult of the *chupacabras*, a fictional beast that was said to fatally suck blood from goats, chickens, and other animals in the middle of the night. *Chupacabras* attacks were reported in Costa Rica, Colombia, Mexico, Puerto Rico, California, Florida, and Texas, although none was ever captured. "Witnesses" described the *chupacabras* as a cross between a monkey and a space alien, with spines atop its head. To biologists, such nonsense would have been just a funny cultural quirk, in a class with UFO sightings and the Loch Ness monster, if it weren't for the fact that *chupacabras* reports had fueled widespread fear and killing of bats.

Even before *chupacabras*, rural Guanacastecans had routinely killed individual bats they came across. Sometimes they burned caves full of bats. "We know of more than thirty species of bats here, but only two feed on blood, and all the innocent species are dying for it," Fernández told me, a trace of anger in her voice. "It's very important to give people an ecological education, to tell them that bats have a very important function in the ecosystem, and that we have to preserve them."

The field work done on tropical bats has revealed the nighttime forest as a kind of mirror image of the daytime forest. During the day, the forest was a green tapestry dotted with throngs of colorful birds flitting about, eating insects and fruit and filling the air with a symphony of chirps. When the sun went down, the birds largely disappeared, and the bats came out, carrying on much the same activity. The only difference was, we couldn't see or hear them.

Bats are the only true flying mammals, and the world is home to nearly 900 species, most of them in the tropics. They come in many colors, shapes, and sizes, ranging in wingspan from only a few inches to a few feet. Perhaps what distinguishes them most is the sensory tool they have developed to navigate the world of the night. Known as echolocation, it is a process in which bats produce high-frequency sound, then listen with finely tuned ears for the echo. This enables them to locate nearby objects and get around the forest as easily and efficiently as a bird in broad daylight.

Despite what was known about bats, biologists admitted they had much to learn, especially about the influence of bats on tropical ecosystems. They did know that, as a whole, bats feed on a wide range of foods. They are major seed dispersers and pollinators, and they apparently reduce the impact of herbivory by preying on katydids, cicadas, and other insects. However, the biomass of bats, their relative abundance, and finer details about their interactions with each other and with other organisms were still poorly understood.

The most closely studied group was the fruit-eating bats, many of which specialize in eating figs. The fig tree's fruit is usually ripe only for a few nights, so the bats waste no time, swarming on fruiting trees like piranhas on a chunk of raw meat. Some researchers, using night-vision goggles, have watched virtual clouds of bats rustle continuously over a tree through the night. A fruit-eating bat depends on a well-developed sense of smell and follows its nose to a fig, using echolocation signals to "see" and eventually recognize the fruit when it gets close. The bat picks the fig and, in a behavior evolved to lessen the chance of attack by a predator lurking in the tree, carries it back to a dining roost. There, it chews the fruit, swallows some seeds and juice, then spits out the pulp. Later, the bat defecates the seeds in mid-flight. This dispersal mechanism spreads a particular tree's seeds far and wide through the forest in return for letting bats strip the tree of fruit.

Among the bats Fernández and Hidalgo were investigating was the long-tongued bat, *Glossophaga soricina*. It is one of many species believed to play an important role in pollinating and dispersing seeds of interest in the ACG restoration process, and other scientists had expressed concern about dwindling populations of *G. soricina* in Costa Rica. These scientists already had found that this bat, in addition to eating insects, seeks the fruit and flower nectar of several tree species, and when it drinks nectar from the flowers, pollen sticks to its body. As it visits flower after flower and tree after tree, it spreads the pollen and aids reproduction of the trees. Later, it spreads the seeds after eating the fruit.

One potential food source for the bats—and a tree of particular interest in the study by Fernández and Hidalgo—is *Bombacopsis quinatum*, also known as pochote. Researchers in the ACG Reforestation and Silvicultural Program were having a problem getting enough pochote seeds for the nursery at Horizontes, and they suspected that the low productivity might be the result of a scarcity of pollinating bats. Fernández and Hidalgo aimed to help solve this problem. If they could learn enough about the basic biology of *G. soricina* and other bats, they might be able to suggest how to attract greater numbers of them to Horizontes and other areas where they were needed to help catalyze reestablishment of pochote.

The researchers knew they needed to better understand the bats' distribution, foraging patterns, social structure, interaction with plants at different times of year, and other aspects of behavior. Since pochote flowers in February and March, they wanted to learn, for instance, exactly how *G. soricina* survives at other times. Now, during August, they worked at the Casona because it served as a daytime refuge for many kinds of bats that flew out into the surrounding forest and fed on what was available.

First, Fernández and Hidalgo planned to trap bats and assess the existing population, weighing, measuring, tagging, and releasing them. Next, they would attempt to find out what these bats were up to on their sorties. A few nights hence, they would begin attaching radio transmitters to the bats and spend dusk to dawn tracking their foraging behavior in the forest around the Casona. Their hope was to put enough pieces together in the bat puzzle that they and the ACG leaders could apply this knowledge to restoring the forest.

I watched for an hour or so as they netted and measured bats, then, while Fernández worked to free another one that had gotten itself deeply entangled, I climbed down the front steps of the Casona, walked past the old guanacaste tree, and down the pitch-dark road through the forest to the administration area, shining my flashlight on the ground ahead for snakes. Along the way, I smelled the sour, musky scent of an animal upwind. Peccaries, perhaps. A few hundred yards farther down the road, a distinctly pleasant patch of air hung like a cloud, permeated by the rich, sweet smell of a tree in bloom. Could it be that at that very moment a legion of long-tongued bats were quenching their thirst deep in the wells of nectar amid those blossoms, just upwind? I smiled at the beautiful odor, and at the thought that within a few weeks of nighttime surveillance, Fernández and Hidalgo might have the answer to that question.

The big exception to the steady restitching of the natural landscape of northwestern Guanacaste Province was the shady case of Santa Elena. The conservation leaders knew that for ACG to be a complete dry forest ecosystem bridging the Pacific side and the Atlantic lowland rain forest, Santa Elena would have to be part of it. While nearly every other property owner in the conservation area willingly negotiated deals with ACG, lawyers representing the powerful U.S. owners of Santa Elena frustrated the conservationists' attempts to reach a settlement. According to the Costa Rican constitution, the conservationists could not make Santa Elena part of the park until they paid for it, and the payment dispute was tied up in litigation more than a decade after Arias signed the decree and nearly two decades since the original decree in 1978. Despite great care and attention by the conservationists—here, after all, was the big prize, the one major element needed to complete the dream—the case was marked by delay, obfuscation, and unusually bitter confrontation.

An arbitration tribunal was picked in June 1997 to settle the case under the auspices of the International Centre for the Settlement of Investment Disputes, a body overseen by the World Bank. The three-man panel was expected to issue a ruling in late 1999 or early 2000. Janzen remained a technical adviser to the government on the case, but he had long since stopped talking about Santa Elena, and others said it was because the Santa Elena lawyers had taken his earlier statements, twisted them, and made them part of a strategy of delay.

The main issue before the tribunal was the value of Santa Elena. By the beginning of arbitration, the Costa Ricans wanted to pay $2 million to $3.5 million for the roughly 39,000 acres of expropriated land, $2 million being the fair market appraisal by the government in 1978. The owners wanted more, as much as $41 million. They claimed this high value on the idea that the land could be developed into a tourist resort. That was anathema to the government, which wanted to protect and restore the land all the way to the ocean, especially the highly sensitive and endangered mangrove habitat near the beach.

Documents filed in the case showed that the partners owning Santa Elena, in addition to Joseph Hamilton, included Swiss businessman George Livanos and Edward Carey, a U.S. oilman and brother of former New York Governor Hugh Carey. Hamilton apparently let a purchase option for the land to Udall Resources, a purchase never realized following exposure of the secret airstrip. Whether the partners were involved was unclear. Ugalde remembered a meeting during his fund-raising campaign in the early 1980s when he and Spencer Beebe of the Nature Conservancy met in Hamilton's office in Manhattan. Ugalde asked Hamilton to donate Santa Elena to the park system. "It was a very nice meeting," Ugalde recalled. "He's a nice fellow, very humble looking." Ugalde chuckled, tinged with regret, as he recalled how Hamilton declined the request. "He knows how to fight, for sure."

A nasty twist to the expropriation dispute, according to the conservationists, arose when Hamilton enlisted the aid of a friend from his home state, Jesse Helms, the ultraconservative Republican senator from North Carolina and long-time supporter of right-wing governments in Central America. Helms launched a campaign to pressure San José to resolve several cases in which Costa Rica and other countries had "confiscated" land owned by U.S. citizens, as his staff put it. He made it clear that foremost among his concerns in Costa Rica was Santa Elena. Helms, minority leader and later chairman of the Senate Foreign Relations Committee, got the State and Treasury departments to withhold $10 million in AID funds to Costa Rica and threaten to block a $175 million loan from the Inter-American Development Bank.

A 1994 Republican staff report to the Foreign Relations Committee made the case that Central American governments were "abusing" the rights of Hamilton and other U.S. citizens who owned property there. Since 1978, Hamilton had been barred from developing a "profit-making, eco-tourism project" on the hacienda, which the report noted had 19 miles of "undisturbed coastline," a mangrove swamp that was home to "many endangered species," and beaches where olive ridley turtles nested. Hamilton lived in a house on the land, but the owners were losing "a great deal of money."

Before the 1978 government appraisal valued the property at just under $2 million, an appraisal by the owners, anticipating the expropriation, set it at $6.3 million. Because of this discrepancy and the "legal uncertainty of the expropriation," the 1978 decree was not implemented, the report said. It cited "many good-faith efforts" by the owners to negotiate. For example, they offered to turn over three-fourths of the hacienda to the government if they could keep the rest. Of the released portion, half would be donated and half compensated at a price determined by arbitration.

In 1992, Helms wrote President Rafael Angel Calderón, suggesting he drop the plan to make Santa Elena part of the park. "Until this and other American properties are returned to their rightful owners, I will be in the unfortunate position of being obliged to oppose the release of aid or debt relief for Costa Rica," the senator wrote. The report claimed that Calderón then pledged to lift the expropriation. It also claimed that the Costa Rican president invited Hamilton to his office only to tell him that he had changed his mind and was launching a "national crusade" to acquire the land. "For the next several months," the report said, "the Costa Rican government directed a negative propaganda campaign that depicted the owners of Santa Elena as being the unreasonable party and the government as being reasonable—the exact opposite of the truth. This smear campaign caused Joseph Hamilton to be harassed in Costa Rica on numerous occasions."

Calderón called the U.S. position an attack on Costa Rican sovereignty. Charges and denials flew back and forth. The dispute was submitted to international arbitration.

An undercurrent of resentment flowed strongly among many Tico conservationists. Some critics insisted that the U.S. pressure was a kind of reward to Hamilton and the others for helping in the contra effort. One rumor claimed Washington was working behind the scenes to retain control of Santa Elena as part of a plan to build a deep-water naval base on the peninsula. In the early 1990s, some Guanacastecans could be seen wearing a T-shirt that said, "Santa Elena Is Our National Heritage; We Must Not Lose It." Conservationists, trade unions, and the Costa Rican Human Rights Commission ran full-page newspaper ads calling on San José to defend national sovereignty. The tourist development might be built, but it would, as Ugalde put it, have a hatchet hanging above it. "They have the Costa Rican presidents saying since 1978 that it should be part of the park," he told me, "and they've got the Costa Rican conservation community saying that, and they've got the Guanacaste people ready to go to the streets for that."

Pedro León, as reasonable and decent a man as I have ever met, recalled the negotiations as "very sad, very depressing." León was head of the National Parks Foundation when it began to bargain for the land. "These guys have used this to create enmity between our two countries. It's been a source of major, major conflicts. We know of at least two [Costa Rican] ambassadors whose mission was to straighten out the situation of the properties."

When I asked him in 1996 if, looking back, he would have done anything differently in the Guanacaste project, Janzen demurred. He felt muzzled over the Santa Elena affair, and the fact that he couldn't talk about it obviously frustrated him. However, he did admit that until the case was resolved, the first phase of the restoration could not be claimed complete and successful. "Maybe that's just in my head, but—" He paused for several painful seconds, then added, "I don't feel like I can legitimately say, 'Hey, we did it,' until Santa Elena is out of there."

During my last trip atop Cacao, in 1996, I watched the shadows shift across the Santa Elena mountains as the sun sank toward the horizon. I was tired from a hike to the top of the volcano that afternoon with the group of Latin American students in the OTS field course, and now many of us stood or sat on the porch of the main dormitory at the Cacao Biological Station, freshly showered, awaiting our evening meal with the satisfaction of having done a hard day's work. Below us, we could see the patchwork of dry-season browns and greens on the plains of ACG stretching out to our west, then the dark water of the Pacific. The sun's course would drop it down behind the mountains, which blocked part of our view of the ocean. The chit-chat gradually ceased, and soon everyone focused their complete attention on the stunning spectacle. This was, indeed, a magical place.

Finally, the moment came. Just as the last sliver of the fireball disappeared behind the mountains, one of the students, a young woman standing next to me on the porch, whispered, almost imperceptibly, "*Fué.*" It meant "gone," and the sound sailed across her lips so serenely that its beauty rivaled the sunset itself.

Pink light flashed above the mountains—this was the final reflection off the water as the sun descended below the sea horizon. In the rapidly fading light, I looked just below us at the little green hut on Cerro Pedregal, where Alex Rodriguez and Didi Guadamuz were settling in for their nighttime vigil. Across the landscape, I knew the bats of Santa Rosa were preparing to venture out once again into their dark world, and on the plains of the Tempisque valley, I suspected that Fermín Mendez, after spending hours monitoring the progress of trees in the Horizontes nursery, was sitting at his supper table, dreaming of the day when he would walk to Santa Rosa and see nothing but forest.

# · *18* ·

# Life Jackets

On the slope of Volcán Rincón de La Vieja, I learned another way to measure rainfall. One rainy day, I hiked with María Marta Chavarría along the soggy paths in the thick cloud forest on the western slope of the volcano. Chavarría, a veteran field biologist, was a strong and spirited woman in her thirties. A braid of rich black hair swung freely on her neck as she set a formidable pace. We waddled through mud, stepped over medleys of tree roots that threaded the trail, and splashed through puddles of different depths. The forest was dark, and as I swung my leg over a particularly big root, something strangely large and round and white darted away from the dingy spot where my foot was about to splash down.

"Jesus," I hissed. In a split second it scurried three feet away, pressing back against a tree trunk, and I jumped an equal distance to the opposite side of the path. My visual pattern-recognition system had been programmed to distinguish between roots and snakes, but this was something completely different. I brought it into focus: stalk-like legs, beady eyes, and claws extending like fists from a six-inch-wide shell.

A crab.

Here, some 5800 feet above the Pacific estuaries where you might *expect* a crab to jump out from behind a root, there was enough water sitting in pockets on and along the trails to provide homes for them. Well, of course. How silly of me, I teased myself. High in the cool tropical mountains, a creature of the sea.

As we walked on, past more puddles, more roots, and a huge, deafening waterfall, I watched for crabs, and pondered them. Somehow, over ages of evolution, they had worked their way up to these heights from the ocean, all because there was plenty of water here year-round. Much of the water was delivered directly by rain, and much of it by several magnificent waterfalls that carved out pools and fed a network of clear, rushing streams. This is the way upland crabs measure rain, not by inches of liquid falling from the sky, but by the rise and fall of streams and the expansion and contraction of pools and puddles.

Chavarría had earned a bachelor's degree in biology from the University of Costa Rica and taught introductory biology classes there. She joined Janzen's Moths of Costa Rica project in the early 1980s, working at the La Selva Biological Station, the renowned OTS lowland rain-forest site on the Atlantic side of the country. Already a prototype parataxonomist, she took the first parataxonomy class in 1989, then became coordinator of the new graduates, darting all over Costa Rica as liaison between them and INBio. She had long worked in the field, had a deep knowledge of moths and beetles, and helped teach new waves of parataxonomists. Simultaneously she worked as a plant curator at INBio, and when the ATBI proposal came along, Janzen asked her to help. At INBio, she had been active as liaison with ACG, and now her main job was to do the same for ATBI and the international corps of taxonomists.

Graciously honoring my request for a trip to Rincón, Chavarría had brought me to a part of it known as Sector Pailas, the Spanish word for a large, shallow pan that is used to boil sugar cane juice down to crystalline sugar. In this case, the word referred to several bubbling-hot pools, craters, and other openings on this part of the volcano. Before we began our hike, we talked about the trails with Javier Sihézar Araya, the tall, stout caretaker for Sector Pailas. His favorite was a tough, five-mile trek up to the windy lip of the volcano's main crater. The trail was only 20 inches wide in some spots, and after it passed through four miles of forest it opened onto a moonscape of rock and ash that led for one more mile to the summit. The lip of the volcano was constantly shifting, owing to volcanic activity, and it was often shrouded in sulfurous fog, the soil at the edge unstable. Once, nearly blinded by these conditions, he had crawled to the edge and looked down. The fog lifted for a moment, and he saw deep into the crater, but couldn't detect the bottom. "If you fall," he said, "you fall forever."

It had been an eventful year at Rincón. The volcano had erupted only nine months earlier, at three in the morning on November 6, 1995, followed by several days of repeated eruptions. A few weeks before the eruption, Sihézar and another caretaker, Freddy Villavicencio, had climbed to the summit several times and noted that the level of the boiling water in the crater had risen from 600 feet below the top to only 90 feet. When it finally happened, the eruption was one of the largest known for Rincón. It shook the ground like an earthquake for miles around, showered the mountain with rock and ash, and broke trees in half. I asked Sihézar if he was concerned about more such violence. He smiled, said that a vulcanologist had told him there were "more or less" five years between eruptions, and admitted that staffers in other sectors of ACG had asked him, "How can you live there with a volcano on your back?" The sector caretaker for Islas Murciélago had once said he would never go to Rincón. Responded Sihézar, "I will never go to Islas Murciélago because I am afraid of ocean storms."

Volcán Rincón de La Vieja National Park had been established in 1973, about twenty-five miles north of Liberia. Its two sectors, Pailas and Santa Maria, comprised just over fifty-five square miles, mostly of high-elevation rain forest and cloud forest. Geologists had identified nine craters on the mountain, only one of them still active. Its watershed included thirty-two rivers and sixteen seasonal streams.

Shortly after the park was created, squatters invaded. The Park Service didn't have the money to hire rangers, so President Daniel Oduber sent the Rural Guard to order the squatters out and to patrol the park. When the conservation areas system was created out of the archipelago of Costa Rican parks, Rincón was absorbed as part of the new ACG.

Sihézar reiterated a theme I heard everywhere I went in the conservation area, from the beach to the east side of the volcanoes: that scientific research needed to be encouraged. "If we know more about what we are saving, we will have more arguments to save it," he said. He liked how the area had been divided into sectors, and that many of the ACG employees, like him, were assigned to work in the sector where they had grown up. "Before, with park rangers, everyone was in charge of everything. It's better the way we do it today. Everyone has their own responsibility. They live in the place, they know it, and they have an obligation." His sister, Gloria Sihézar, had become an ACG paraecologist, the ecology version of a parataxonomist.

The other major event at Rincón that year had been the death in May of two Dutch hikers who failed to return to the sector headquarters station one rainy day. A couple of days later a search team found their bruised and battered bodies in the Río Blanco, a short distance to the north of the station. They apparently had tried to cross the river at night without flashlights just as heavy rain had made it rise unusually high. "It was a shock," Sihézar said, looking up from an official report, which he pulled from a file when I asked him for details. The event had caused a policy change. No longer would the ACG staffers simply warn hikers that precipitation had made conditions dangerous; they would ban them from hazardous areas on rainy days.

Sihézar lived in a small house near the sector headquarters station, and on the rough wooden front porch of this house I had a conversation with his father, Santos Sihézar Cid. In his early sixties, he looked nearly the same as his son, save for a few extra pounds and the gray of his mustache and hair. He was a former Guanacaste *sabanero*, and his skin was wrinkled and leathery from a lifetime in the sun. His eyes glistened as he described his hunting and cowboy skills. He had enjoyed hunting in this region all his life, but he had noticed a steady decline in rainfall and the number of forest animals over the years. "This is very good, to regrow the forest, so young people will have a chance to enjoy it," he told me.

He spoke of his days as a *sabanero*, working on Santa Rosa, Orosí, and other haciendas as a horseman and helping clear forest for pasture. He helped drive herds as large as a hundred head down to market in Liberia. He remembered seeing the first car in Guanacaste Province in the early 1950s. With the Pan-Am Highway, more and more cars and trucks came, and horses were used less and less. He spoke of the pleasure he got from riding bulls at fiestas, even after he broke his two front teeth when he was sixteen at a fiesta in Quebrada Grande and spent three weeks in the hospital. He explained how his father had taught him the skills of a horseman and an axeman, but that instead of learning those skills today it was more important for young people to go to the university to learn "what is necessary to make a living." There was a slight sadness in his voice, as if he hadn't quite finished mourning the end of a lifestyle, the extinction of a tradition. He finished by saying, "This is something that you should accept, because this is life."

A short time later, Chavarría and I left on our hike. I had requested something less arduous than the trail to the summit, so she led us on a path to the south, in the opposite direction chosen by the ill-fated Dutch hikers. In a mild but steady rain, we walked through a stretch of forest, then crossed the nearly knee-deep, swift-running Río Colorado, balancing on large rocks that lay just under the surface. After clearing the river, we slipped and slid along a mud trail over a ridge. As carefully as I tried to place my feet, a few times I found myself riding hard on my derriere down a muddy slope.

We came out of the forest into an open, rocky area covered with splotches of yellow and white. Trees were scarce, the plants small and stunted, and pockets of air that smelled like rotten eggs wafted across our path, taking our breath away and making us cough. Steam rose up from tiny holes and cracks in the soil. We had entered the region of the fumaroles. They were vents on the face of the vol-cano—bizarre, bubbling ponds of sulfur, water, and mud so hot that if touched, the skin quickly burned. I knew Chavarría wasn't kidding when she told me this, but I still needed to try it. I stuck a finger into the side of a hill from which small streams of whitish liquid gurgled, holding it only a split second—long enough to feel the heat and turn the finger red, but fast enough that it wasn't a serious burn. Still, for a moment, it felt as if I had connected with some strange, primeval form of energy and intelligence miles below.

The pond-like fumaroles had different colors—gray, yellow, cream, chocolate brown—depending on the mix of sulfur, mud, and other components. On the shores of the ponds, and sometimes on small islands in the middle, the bubbling action had gradually heaped mud into high piles which, over days and months and years, the rain had patiently carved into odd, chunky sculptures. It really was like being on another planet. As we stood near these sculptures, these open-air stalagmites, listening to the fumaroles hiss, burp, and "bloop" pockets of sulfury air, it was hard to escape the thought that we were listening to messages whispered from the very center of the Earth: "Be careful. Be careful. Be careful."

A day later I was in the passenger seat of an ACG four-wheel-drive guided by Roberto Espinoza. Also the son of a *sabanero*, Espinoza had grown up in Cuajiniquil, where he held jobs on fishing boats, as a cowboy, and as a field hand swinging a machete. He worked as Janzen's field assistant for a dozen years and was in the first class of parataxonomists. Now in his early thirties, he was resident botanist of the conservation area. The father of two children, Espinoza had a lanky body and an easy smile. We were heading east off the Pan-Am Highway on a rocky road that slowly climbed through ranches and dairy farms toward a gap between Volcán Cacao, on our left, and Volcán Rincón de La Vieja, on the right.

The dark-green forested slopes of the two towering volcanoes hung above bright green pastures. A storm roared overhead and unleashed a bolt of lightning and bone-rattling thunder. Farther up we crossed rushing streams on old wooden bridges barely wide enough for one vehicle. Other streams, most shallow enough not to be dangerous, we crossed by driving straight through. However, as we neared our destination atop the gap, one last watercourse remained to be crossed, the Río Cucaracho. We slipped and shimmied down a road of burnt-orange mud beneath an archway of trees, and Espinoza stopped at the edge where the road disappeared into the churning water.

He pondered the noisy torrent for half a minute. Rain had swelled the river, and it seemed to rush like a rapid. I couldn't see the bottom, but the water had to be at least a foot deep. I shook my head, thinking of people back home who make the mistake of driving through flooded creeks and get swept away by the surprising force of even slow-moving water. I also thought of the Dutch hikers.

"I don't know," I said in Spanish. "It looks dangerous."

"Let us hope," Espinoza responded, putting the jeep into gear.

I clasped my hands as if in prayer and made the sign of the cross, only half in jest, then grabbed my notebook and camera and considered how I would get out of the jeep as it tumbled downstream. He drove into the river slowly, and, when we were about a third of the way across, gunned it to reach the other side.

"You have good luck," I said with a grin.

"No," he said, laughing. "*We* have good luck."

We had arrived at the San Cristóbal station, the other center, with Horizontes, of the ACG Reforestation and Silvicultural Program. The station was at about 2300 feet altitude and roughly twenty miles east of Horizontes, and here researchers were experimenting with wet-forest trees in the strip of land that dipped between Cacao and Rincón. The gap was commonly called the Rincón-Cacao Corridor or Rincón-Cacao Bridge.

Much of the forest in the gap had been cleared over the past several decades, leaving behind a mosaic of original rain forest, disturbed forest, and small dairy farms with African star grass. The air was wet and chilly, and the long, silky ribbons of clouds that hung over the dark rain forest at the foot of the peaks, and the way the mist seeped out from the cloud bottoms and slowly sunk into the canopy, brought an aura of mystery and calm to the landscape. We were only a few hundred yards from the spot where Ian Gauld and I had spent a night at the black light ten years earlier.

Rincón was the nearest protected area in the *cordillera* to the south of Cacao. When Rincón was added to ACG, its boundary was separated from the Cacao sector by about four miles of these private farms and pastures. The ACG leaders immediately wondered: Why not buy that land and connect the two wildlands? Why not build a biological bridge across the gap?

Janzen, awakened by the potential impacts of global warming, outlined the rationale for linking Rincón and Cacao in a 1992 proposal. As the climate warmed, the bridge, and the forest it linked on Rincón, Cacao, and Orosí, would offer high-elevation cloud and rain forest for organisms "ecologically pushed" out of the increasingly hotter, drier habitat of western Guanacaste. It would be, he wrote, a "global-warming life jacket" for the ACG dry forest.

In 1992, the land in the gap was distributed among sixteen owners. "None of them are presently interested in making this area their permanent home and primary source of income," Janzen wrote, but he added that it would be "unwise" to start talking with the landowners about buying their properties until funding was assured. He estimated that a minimum of 2700 acres would be needed for a significant bridge, and that it would probably cost about $180 an acre. Experience had shown that land could be obtained for 40 percent less, "if one is extremely careful in the land purchase strategy."

After some hesitation, Janzen and the ACG staff steadily filled in the bridge with purchases over the next few years using money from several donors, including the Children's Rainforest groups of England, Germany, Sweden, and the United States. Another Children's Rainforest fund, this one in Japan, along with the Japanese government, contributed money for restoring the forest in these pastures.

The final link proved extremely difficult. A 343–acre farm owned by the Brizuela family would have closed the connection. The parties couldn't agree on what both felt was a reasonable price. However, another landowner, Hernán Rodríguez Arce, who owned the largest hardware store in Liberia, agreed to sell his farm for about $400 per acre, and in June 1997 Sigifredo Marín signed the agreement for ACG to buy the property, unofficially estimated to be 1317 acres. The link was complete.

The farm had been used for pasture and easily could have become a palm heart plantation, but it still had large patches of old-growth and regenerating forest. Even more important, it formed a broader bridge and, since it was to the east of the Brizuela property and at lower elevation, boosted the critically short supply of Atlantic-side lowland rain forest on ACG's eastern edge. Janzen estimated that it increased the number of species in ACG by another 3 to 5 percent.

In sum, from 1992 to 1997 the conservation area purchased a strip of land about two miles wide stretching between the two volcanoes. This brought the total area of ACG land in the bridge to about 7400 acres and made the entire eastern face of ACG a 20–mile band of forest atop the continental divide, from the northern slope of Orosí to the southern slope of Rincón, interrupted only by large patches of pasture in the gap.

(The gap eventually became a link in a plan called the Mesoamerican Biological Corridor, funded by the Global Environment Facility, a fund set up by the World Bank and United Nations. The aim of that project was to connect the most species-rich areas in the Mesoamerican land bridge between North and South America. Identifying and protecting forested corridors that connected parks was part of the plan.)

Restoring the forests of the bridge was an intriguing conservation concept, but unlike the essentially automatic regeneration of the dry forest in the *cordillera's* rain shadow, the rain forest pastures proved far more difficult to restore. In these wet areas, the problem was not recurring fires, but rather a relatively slow succession rate in the wet pastures. This difficulty was really no surprise. In 1988, Janzen told me that as one went up the volcanoes and into wetter forests, pastures didn't fill in like those in dry forest habitat. "I'm very curious to know whether that's because there are no dispersal agents up there or whether the seeds get out there but the seedlings don't survive," he said.

At San Cristóbal, restoration manager Felix Carmona, a thin but muscular man in his thirties, drove us around a sloshy two-track road through a dense, waist-high carpet of the emerald-green star grass. "We don't have much experience with restoration in the wet areas," Carmona said, pensively. "It is more difficult." He and his colleague, David Morales, found that more of a human hand was required. The thick grass was generally too formidable a foe for tree seeds to overcome, and everything suggested that very few were arriving.

Studies had shown that even those plants that sprouted generally died within a year in the wet pasture. The researchers had been experimenting for about two years, transplanting seedlings of moist-forest species after rearing them in a nursery, and placing them in rows and other patterns, trying to find the best combination to form an island of regeneration. To speed the process, they mowed and in some cases applied herbicides to the grass to give a fighting chance to transplanted seedlings, root stakes, and seeds dropped from trees already standing in a few small patches.

There was hope. They were adding 30,000 seedlings a year, and many of these, when mature enough to attract birds, bats, and other animals, would begin to expand naturally. In the gap, there would be no quick restoration. Still, islands were taking hold, and dozens of tree species were becoming established, some of them growing nine feet a year.

At one point along our drive, Carmona got out of the truck and waded into the grass that surrounded one such forest island. Behind the island rose the mighty cloud-draped cone of Rincón. He talked about the painstaking work of getting the islands started, and of grooming and nurturing them. As he spoke, I looked south to Rincón and north to Cacao, then back again. Their lush forests seemed so close, as if you could reach out, pull, and tie them together. I asked Carmona how long he thought it would take until the forests were joined, until animals could make use of a full, forested corridor.

"That's the big question," he said, letting out a part-chuckle, part-sigh. Birds already frequented the corridor, and recently coatis and coyotes had been spotted. More animals would come eventually. He paused, then added, "Perhaps in five years."

It seemed that here, in this former wet ecological desert, the Guanacaste restoration was being put to its toughest biological challenge. Patience would be required.

Camilo Camargo pulled a leaf off the orange tree and held it up to us. Camargo and Mónica Araya, both in their twenties, stood with me amid thousands of orange trees whose branches held small green balls that would grow quite a bit larger before they were harvested, still green. The leaf had been damaged by a tiny caterpillar that, still at work inside the leaf, between its upper and lower layers, had mined an odd, curved maze of tunnels, visible by the trail of white excrement that exited the end of the creature even as it pulled in new leaf material through its mouth.

This was *Phyllocnistis citrella*, a major pest of oranges and other citrus fruit worldwide, a leaf miner, a microlepidopteran, and the object of Camargo's study.

*P. citrella* laid its eggs in newly emerged orange leaves, and the tunnels that the caterpillar ate inside the leaf hampered the tree's ability to produce oranges. Growers often spray their groves to kill this and other pests.

Camargo, of Colombia, was one of Janzen's graduate students at the University of Pennsylvania. Tall, thin, and thoughtful, he had taken me and Araya, a Costa Rican graduate student, into the groves of Del Oro, S.A., an orange juice company. It was August 1996, and the groves extended in long rows across fields just to the north of Volcán Orosí. Camargo had spent much of the year trekking these groves as part of his master's research, which was aimed at learning whether the orange orchard, which ran along the northern edge of ACG, was protected by natural biological control agents coming from the conservation area.

Araya was researching her master's thesis in economics at the International Center for Economic Policy for Sustainable Development, at the Universidad Nacional, in Heredia. An assertive woman with a sharp mind and broad smile, she had been looking into issues surrounding the international orange market, the competitiveness of "green" products, the "green" strategy for orange juice exports, and the economic value of ACG to Del Oro. She had come along this day with Camargo, like me, to gain an understanding of his research, but she also provided a wealth of information about the economics of oranges.

Araya had been searching for a subject for her thesis a year earlier and had run into Norman Warren, the local chief operating officer of Del Oro. Among other things, Warren confirmed what ACG biologists had earlier told her: that although Del Oro had been growing oranges in Guanacaste for about a decade, they hadn't had to apply many pesticides. "That was unusual," Araya told me. It meant big savings to the company. "I was looking for something not pessimistic, something realistic but optimistic," she told me. She dug with a passion into Del Oro, orange production, and issues of international trade and the environment.

We stood on a ridge in the midst of the orange plantation, and to the north we could see Lake Nicaragua and the 5000–foot cone of Volcán Concepción (also known as Ometépec) in the middle, halfway to Managua. "The lake is too large to be called beautiful," Thomas Belt wrote, describing much the same scene from Nicaragua more than a century before, "and the mere glimpses of its limits and cloud-capped peaks appeal to the imagination rather than to the eye." To the south, clouds concealed Orosí's peak, typical of the rainy season. They gave way to dense cloud forest below, and, as pasture entered the scene near the foot of the slope, the forest separated and pushed down in five gradually narrowing fingers into the orange plantation, like rivers of dark green lava. The fingers were part of a rare habitat, a transitional lowland forest that stood between the Pacific dry forests and Atlantic rain forests. The boundary of ACG had once cut straight across this landscape, as if across the knuckles of the forest fingers. Del Oro still owned much of the forested land, but it had swapped about 990 acres of it with

ACG in 1992 for 86 acres of pasture in the conservation area better suited for orange groves.

Del Oro, S.A., was comprised of two entities: Del Oro Juice Marketing Inc., of Winter Haven, Florida, and the Commonwealth Development Corporation of England, a financial development institution which has helped set up hundreds of commercial agriculture operations in dozens of countries worldwide. In the 1980s, Del Oro had acquired low-grade cattle pasture in northern Guanacaste and grown orange trees. The acquisition was part of a trend, as orange growers, encouraged by the rise of juice orange production in Brazil after killer frosts in Florida, began to push into Costa Rica. In half a dozen years, Del Oro began harvesting commercial-quality fruit.

At first, it sold the oranges to a processing company, but in 1995 Del Oro built a plant to make frozen concentrated orange juice, for sale on the international market, mainly in Europe. By 1996, when the plant began large-scale operations, the company operated five farms in Guanacaste Province, three of them along or near the northern border of ACG. In the ACG region, it owned about 17,300 acres of land and had 7400 acres of it in oranges. The rest was in forest or field, and much of the field part comprised regenerating forest. The fruit was harvested from December through May, and since the production plant could handle far more fruit than it was getting, there was talk of expanding the harvest season using orange varieties that mature earlier and later, growing fruit like grapefruit and pineapple, and encouraging local farmers to grow oranges.

This pending expansion was not what made Del Oro interesting to Camargo, Janzen, and others associated with ACG. Rather, it was the way the company was willing to think about controlling its pests on the Guanacaste plantations. Like other orange growers in the tropics, Del Oro used an insecticide to attack one of its biggest problems: leaf-cutter ants. The company had experimented with other means of control, such as planting vegetation around ant nests to see if that, in a kind of sacrificial way, would satisfy them and keep them from slicing and carrying away orange leaves. Workers also applied insecticides in Del Oro greenhouses to control leaf miners and sprayed agricultural oil on its trees to protect against fungus. These were relatively light uses of agrochemicals, and yet orange production proceeded quite well. Del Oro was not being forced, as growers confronted with mass insect attacks usually were, to make the tough cost-benefit decision about increasing pesticide use to avoid crippling crop damage.

The question, then, was: Why were the orange tree's most voracious enemies, in some of the enemies' most hospitable tropical growing conditions, not assaulting the new orchard? Janzen saw an opportunity, and he began to suggest that the ACG forests were providing pest control—that a wide range of animals kept orange pests in check naturally. These included parasitic wasps and flies and predators such as birds, dragonflies, crickets, ants, beetles, spiders, and lizards.

Because of this extensive blanket of biological control, the argument went, Del Oro did not have to spend money for heavy applications of pesticides. There was some debate about whether biological control was actually working in the Del Oro orchards, with doubters saying the orange trees simply hadn't been around long enough for pests to establish a sizable population. However, Camargo gave proponents of the biological control idea their first solid piece of evidence in 1996 when his study revealed that parasitic wasps were significantly more active in killing *P. citrella* on orange trees near the conservation area.

Camargo had set out to discover what impact wasps and other parasitoids might have on populations of *P. citrella*. (A parasitoid is an insect that develops as a larva within the body of another insect, and in so doing kills the host, then lives free as an adult.) He selected three areas within the Del Oro plantation: the first, well up on the slope of Orosí and near the full body of ACG mainland forest; the second, farther down the slope but near a peninsula of forest; and the third, farther down and well isolated among the orange trees. He collected mined leaves that contained a caterpillar, put each leaf in a plastic bag, and watched them daily to see what emerged—a moth or parasitoid. If a moth emerged, he knew that it had not been attacked. If a wasp emerged, he knew it had killed the caterpillar. From May until August, he collected more than 1000 mined leaves.

The laboratory where he kept the plastic bags was an uninhabited house that nearby residents had dubbed Casa de Brujas (Witch's House). The house stood on Del Oro land, and local legend had it that a previous owner had made a pact with the devil to become rich, and when he left, evil spirits had stayed. It was an absurd fabrication, of course, yet the house was, indeed, a spooky place, dark and shadowy beneath the trees that surrounded it. In the screened porch off the back of the house hung more than a hundred plastic bags, fastened with clothespins to several rows of line. Camargo made his way, one by one, along the lines, peering into each bag for a few moments with the aid of a flashlight, looking for the moth or wasp. Finding one seemed to me an incredible feat, for both creatures were tiny, the moth white and smaller than a rice grain, the wasp black and smaller than a poppy seed. That day, during a half-hour search, he found four wasps and one moth. The next day he would return and check again.

Camargo's study found that roughly 40 percent of the leaf miners had been attacked by parasitoids in the two areas close to the forest, whereas roughly 10 percent had been attacked in the area isolated amid the orange trees. A second intriguing trend he discovered was that the diversity of parasitoids increased closer to the forest. Eight or more parasitoid species were found active at the sites near the forest, while only two or three were found active in the distant site. These results were preliminary and limited, and they left open the question of whether the leaf miner had been effectively checked by biocontrol agents from the forest, or whether the moth simply hadn't had a chance to establish itself as a power in the orchard. His results suggested the former, and they were good evidence, but much more research was needed before a definitive conclusion could be made. "Only time will tell," he said.

Camargo was about to finish his research project and return to Colombia. He hoped, however, that others would continue looking at the many dimensions of biological control in the orchards. Despite the small size of the individual organisms, this war raged across a broad and deep front well beyond the perimeter of the forest, as a variety of species engaged, legion against legion, in age-old, single-warrior combat. The net effect appeared to be pest control, and, like the constant supply of water that flowed year-round out of the cloud forests above, this was but one of the services provided by the biodiversity within ACG.

Janzen had long been developing ideas that would make economic use of ACG's position as an environmental pioneer. In the early 1990s, he proposed to the Ministry of Agriculture that it lead a program to produce beef low in fat from cattle raised in an environmentally friendly way on select pastures near the conservation area. This "green beef" would become a specialty product that could find a welcome market among conscientious consumers, and it could bring profits for local ranchers who touched their land lightly. Ministry officials, to put it nicely, found the plan amusing.

In 1995, Janzen approached Chris Wille, a U.S. environmental writer and conservationist based in Costa Rica, and broached the idea of "green oranges." What if, Janzen proposed, the Del Oro oranges could be certified under the ECO-O.K. program for tropical agriculture? Wille and his wife, Diane Jukofsky, also an environmental writer, ran the Tropical Conservation Newsbureau from their home office in Moravia, a suburb of San José. The bureau was a unit of the Rainforest Alliance, a New York-based environmental group, and Wille and Jukofsky helped direct the Alliance's ECO-O.K. program in Latin America. The program set environmental standards for bananas, coffee, and other tropical commodity crops, and it sent experts into the field to evaluate farms to see whether they met the standards for ECO-O.K. certification. The program required protection of forests, soils, wildlife, and watersheds. In the marketplace, the certified products could display a label that set them off as having helped conserve the tropical environment where they were produced. The program won the Peter F. Drucker Award for Nonprofit Innovation in 1996.

Wille recognized that with the decline in the beef market, cattle ranchers in Guanacaste Province were looking for alternative crops, and the deforested pastures were prime land for orange farms. He also recognized that orange farms would be an improvement, ecologically speaking, for biodiversity, soil, and water. Finally, he knew that orange farming provided good jobs, and that the crop could be grown by family farmers and cooperatives in a potentially sustainable way. Wille and Fundación Ambio, the Costa Rican ECO-O.K. partner, worked with Janzen, Del Oro, and other stakeholders to develop the guidelines for growing oranges and processing the juice. The guidelines were built on experience with standards already in use for banana and coffee farming. They included a prohibition against deforestation; strict control of pesticide use; management of solid wastes; worker health and safety precautions; and protections of rivers, watersheds, and wildlife. After several inspection visits, the Del Oro orange groves near ACG were certified in 1998, and the juice became available in Costa Rica and, through the distributor Johanna Foods, in the eastern United States.

This served as a model for the Rainforest Alliance's plan to expand the ECO-O.K. certification program to all orange growers in Costa Rica, and there was good cause for hope, since orange production was a relatively nascent activity, unlike coffee and bananas, whose plantations had been working for more than a century. Once tested in Costa Rica, the alliance planned to export the program to other Latin American countries. The ACG-Del Oro relationship had helped break the ice with the orange industry, and if the expansion of orange growing in the tropics turned out to be a sustainable one, historians might again look back and see the Guanacaste project as the root.

In 1997, Janzen and Warren, a fourth-generation orange farmer from Florida, proposed to the Del Oro board that the company enter into an agreement with ACG to buy "environmental services" for the next twenty years. These services included biological pest control, water, ecological consulting services, and what Janzen called "orange peel degradation." That last provision was an especially novel idea. It gave Del Oro the right to dump hundreds of tons of rinds from the processing plant onto land in the conservation area, where it would decay naturally. The contract called this a "Biodiversity Processing Ground."

Each year, the company would dump the peels in a fifty-acre plot of flat pasture marked by ACG officials between Cerro El Hacha and Volcán Orosí, somewhere on La Guitarra, the former ranch of Luis Roberto Gallegos. The peels would be spread into a thin layer. This would occur during the dry season, when processing peaked. The rotting would spring into high gear with the onset of the rainy season, when microorganisms would flourish. In turn, populations of three species of wild flies commonly found on fruits fallen on the forest floor would explode, fed by the boom in microbes. The plot would slowly transform into a thin, soupy lake, churned by the action of the flies plowing amid the decaying peels. Since no one lived nearby, the odor wouldn't matter. Within nine months, the mess would be gone, the orgy over, the flies having gone on to become food for birds, frogs, spiders, and other creatures up the food chain. The Biodiversity Processing Ground would rotate from year to year within that area.

According to the contract, all these biodiversity and ecosystem services amounted to a $24,000 annual value to Del Oro. Payment for twenty years of services would come in the form of 3000 acres of that biologically scarce forest in the five fingers on Orosí. At the rough market equivalent of $160 per acre, the land deal was worth $480,000 to the conservation area. In typical Janzen eco-think, he termed the proposal a "mutualism."

Del Oro, of course, didn't have to pay for a thing, since ostensibly it could just sit tight, harvest oranges, and collect freebie environmental services. After all, ACG was not in a position to withhold the biological pest control or water that came into the orange groves. However, Del Oro was a progressive company, and its competitive strategy was to market itself as a "green" orange juice producer. The question was, how does one judge the value of these environmental services and the public relations value of paying for them? It was new territory. Before the meeting, Wille told me: "If Janzen can pull this off, he will have set another important precedent." The Del Oro board did, indeed, approve the proposal, and so did the Costa Rican government. The contract was signed in August 1998 by officials from Del Oro, ACG, the National Parks Foundation, and Ministry of the Environment and Energy. The following month a clique of rival citrus growers and their legislative allies attacked the deal as hurting farmers and endangering health, but the charge appeared to be little more than jealous scare-mongering. A study released in 1999 by scientists at the National University backed the Del Oro-ACG agreement. Next, the Supreme Court ruled the contract illegal and ordered the orange rind removed. The microbes and fly larvae had beat them to it, and the controversy continued.

Janzen used the Del Oro case as an example of the value of biodiversity to agriculture. He claimed it reinforced "the contemporary Costa Rican attitude of taking virtually its entire agroscape into sustainable development." Well-managed plantations with a comparatively low environmental impact, such as Del Oro oranges or other ECO-O.K.-ed commodities like shaded coffee or cocoa, would serve as better wildland neighbors than the standard mix of marginal farms, or even the mixed-activity buffer zones that conservationists had promoted for more than a decade. In effect, Del Oro had replaced cattle ranching and subsistence farming with what appeared to be sustainable orange farming in land bordering ACG. "If we can figure out how to do this right, so it's effective for both the grower and the environment, it will be an important planning tool for the future," said Wille. It would be another kind of life jacket for the animals and plants in ACG and in other places where it was applied.

It also would fulfill a paradigm shift that Janzen, Gámez, and other conservationists argued was necessary for the survival of wildlands: the idea that conserved areas must be perceived as a social endeavor, something that benefits society economically and otherwise. This was the use-it-or-lose-it philosophy of tropical conservation, the latest spin in the intellectual evolution that began with the plan to restore the dry forest in the mid–1980s. "How to Grow a Tropical Forest" had been replaced by "How to Grow a Wildland," the title of a paper Janzen delivered at a National Academy of Sciences Forum on Nature and Society in 1997. "Sustainability of a wildland will only be achieved by bestowing garden status to it, with all the planning, care, investment, and harvest that implies," he told the audience at the Washington meeting. This "gardenification of nature," as he termed it, required what Janzen had proposed long ago: detailed understanding of the wildland garden's biodiversity and ecosystems, and "integration of its socioeconomics with the neighbors at all scales," local, national, and international. In short, biocultural restoration.

The gardenification of nature required a big garden—a big conservation area. If you were going to integrate with the surrounding economic and social milieu, you first had to accept that the entire wildland would not be completely pristine. You had to make sure it was large enough to absorb what Janzen simultaneously called the "footprints" and the "opportunities for presence and socioeconomic integration." A pasture fire, a poached deer, a soccer field, a picnic ground, a field of grass to feed the work horses, land on which to let orange peels rot, a plot for "carbon farming" where trees could be grown as part of international agreements to reduce atmospheric carbon dioxide—these and other footprints were rarely absorbable in a 42-square-mile park, like Santa Rosa in 1986. But they were in a 460-square-mile wildland, like ACG. A small percentage (Janzen put it at 5%) of the area's biodiversity and ecosystems could be sacrificed to guarantee the rest. It was an acceptable price. As Janzen said, "this is the ACG wildland peace treaty that is being negotiated with the agroscape and the urban landscape."

# Epilogue: Lessons from a Tropical Kitty Hawk

Set aside your random research and devote your life to activities that will bring the world to understand that tropical nature is an integral part of human life. If our generation does not do it, it won't be there for the next.

—Dan Janzen, "The Future of Tropical Ecology," 1986

Bette Loiselle led the students through the forest on Volcán Cacao, looking at the great book of nature for lessons to teach. Loiselle, tall and thin with an angular face, stopped before a small tree and pondered it. She pulled off a leaf, pinched it between her thumb and forefinger, and sniffed.

"Try it," she said in Spanish, passing the leaf back to the students, who were from different Latin American universities. The leaf had a gingery smell, like sarsaparilla. As Loiselle rattled off its scientific name, the students strained to commit the leaf to memory—its name, shape, scent.

"This tree is interesting because, as you see, no other trees are growing beneath it," she said. "Why not?" The students, intent on her every word, hesitated. Finally, they began to rattle off some reasons.

"Because of the shade it casts," said one. "The chemicals it produces inhibit growth," said another. Loiselle neither nodded yes nor shook her head no. She merely asked more questions, guiding more discussion. It was 1996, and she was playing her part as one of the teachers in the legendary OTS tropical ecology course. Step by step, the students were learning the specifics of plants and animals in the forest, as well as their connections.

Loiselle, in her late thirties, was a biologist with the International Center for Tropical Ecology at the University of Missouri at St. Louis and a popular teacher of the OTS courses. She had been chief coordinator of four of them and a visiting teacher on half a dozen more. On this one, a seven-week tropical ecology and conservation course for Latin American students, she joined for the final week, including a five-day stay at the Cacao research station.

"Peter-Peter-Peter"

"That's a green shrike-vireo," Loiselle told the students. They pierced the thick forest with eyes and ears in search of the bird, but it was hidden. So she pulled a copy of *A Guide to the Birds of Costa Rica* from her rucksack to show them: a small and bright green bird, with a dash of blue on the back of the neck. They listened to its repeated calls.

Farther down the path, Loiselle stopped the group at a sunny clearing. A decade earlier, Guanacaste project leaders had removed the cows from this pasture when the land was acquired. They had stopped the annual prescribed fires that killed tree seedlings, but the pasture grass now grew high and thick, blocking any chance for a tree to sprout.

"After 10 years, there's still no regeneration here," she said, setting up another Socratic discussion. "What should be done?" One student suggested bringing back the cows to eat the grass. Another countered that cows would carry ticks, which would infect the forest's endangered monkeys with disease.

On it went, over the next few hours and days, with that theme—"What should be done?"—resurfacing several times as they walked the forest, knelt in leaf litter, looked out on the valley from windy ridges, listening, smelling, learning.

ACG provided plenty of room for students to learn the lessons of tropical ecology, and it had plenty to teach about how to restore tropical ecosystems, too. Its success was not perfect, and there were legitimate questions about issues such as whether jobs for local residents would be enough to sustain lasting conservation, and whether decision-making power about ACG was truly in their hands or the hands of elites. But clearly ACG was a major success in biocultural restoration, tropical or otherwise.

What follows are some insights on the Guanacaste project's achievement, the lessons learned from it, and the relevance of the project to conservation throughout the world.

Why did the Guanacaste project succeed? I asked this question of many people in and out of the project and got an abundance of opinions. Rodrigo Gámez may have summed it up best when he told me that "transcendental people" had made a series of "transcendental decisions." He meant transcendental in its classic sense: transcendent, extraordinary, excelling. However, there also was a hint of its additional meaning, as Webster's dictionary defined it: "beyond the limits of possible experience."

Alvaro Umaña put it this way: "There's a saying here that only madmen and mystics change the world."

Among the transcendental people, Oscar Arias created a broad set of political circumstances conducive to the project. He did not do so because of the project per se. Rather, he did it as a consequence of his policy initiatives, which ranged from peace in Central America to expanding the parks. "I am very fond of Guanacaste," Arias told me in a 1997 interview. "In my campaign, I pledged that I would try to save as much land for future generations as possible. That's why we were so committed to restore this dry forest."

Many people close to the Guanacaste project felt that the conflict over the Santa Elena airstrip helped. It focused a significant part of Arias's attention on the idea of acquiring Santa Elena as a conservation measure. Then came the Nobel Prize, which, as Arias put it, created "a sense of euphoria among all of these marvelous governments who wanted to help me personally." This support boosted the Arias Peace Plan and translated into money for the Guanacaste project from countries such as Sweden and Holland. As a result of the Nobel, Sweden gave Costa Rica $2.5 million, which grew to $3.5 million, which became earmarked for the project and, through the magic of debt-swap, grew into $24 million. Arias created the circumstances for this.

Meanwhile, Arias backed the recommendations of Umaña, Gámez, and Janzen for debt-swapping, restructuring the Park Service, and various pro-Guanacaste project appointments in the national government. "What Arias respects the most is knowledge, and he knew Janzen was knowledgeable," Umaña told me.

Umaña, an extraordinary administrator of natural resources, and a clever politician in his own right, played several crucial roles. He persuaded Arias to create the Ministry of Natural Resources, Energy, and Mines, and he gave the Guanacaste project major support, maintaining a close relationship with Janzen. He absorbed the rougher aspects of the headstrong scientist when other politicians would have recoiled. The time Umaña spent at Stanford getting his doctorate in engineering exposed him to U.S. academics, including "crazy" ones. Janzen, he said, was like many others in that respect. "I knew how to talk to him, and he knew how to talk to me, too," Umaña said, adding that Janzen called him what seemed like two or three times a day. He viewed as an important part of his job "keeping Janzen under control."

Gámez, equipped with his political connections, scientific background, and dedication to conservation, threw a tremendous amount of personal energy into making the Guanacaste restoration a reality. He helped Janzen and Hallwachs become sufficiently aware politically to survive the struggle with the Costa Rican bureaucracy. He contributed to the project's central ideas. He, perhaps ultimately, held the conservation alliance together as it negotiated a series of problems, including some that were exacerbated by Janzen's personality and the news media's focus on the Great White Explorer. Gámez kept the strands that bound the group from snapping apart, shrewdly assuaging the egos of Alvaro Ugalde, Mario Boza, and other highly motivated conservationists who had brought the Park Service into being on little more than pure guts.

Clearly, Janzen, backed by his omnipresent intellectual shadow, Hallwachs, was the most transcendental figure of them all. He has shied away from the spotlight, and spread credit around. But ask anyone else about the project, from any perspective, and their comments either begin with Janzen or shift to him quickly.

Key were Janzen's conversion from scientist to activist, his steady accumulation of data that informed restoring the forest, his insistence on preserving not just species, but also interactions. Also critical were his skills as a persuader, in both English and Spanish. Several Costa Ricans had the sales and language skills needed to push conservation ideas to international organizations and foreign governments, but Janzen was widely viewed as taking it to another level, a level rare not just in Costa Rica, but anywhere. "Dan is a great communicator," Pedro León told me. "He is great at convincing people, because he is convinced of what he says." Umaña put it this way: "He's the best salesman that Costa Rica has. We gave Janzen the support he needed to do his very enthusiastic sales job."

When I asked Janzen about this, he shrugged it off and repeated something he's often said: "It isn't that I'm a good salesman. It's that I have a good product to sell." True enough. He wasn't representing a completely denuded landscape in an impoverished country torn by civil war, with all that negative ecological, social, and political baggage. Janzen could show his prospects a nice piece of forest, and a strong conservation movement supporting it. Yet he also brought his reputation, his credibility, his self-confidence, and drive. It was a partnership, as Arias put it.

That drive. There weren't many others around like Janzen who could force something like the Guanacaste project. Doug Boucher once told me, "It's one of these things that's so ambitious that before somebody like Dan came along and said, 'Why don't we do this,' they would have said, 'No, that's impossible.'" Early in the project, Janzen himself told me, "These dreams have been there. The difference is that I'm a pushy, Protestant gringo who says, 'Well, I believe it, so therefore I'll make it happen.'"

A dedicated crusader is a common element of successful conservation projects, and Janzen was nearly in a class by himself. Jonathan Berger, author of

*Restoring the Earth,* told me that Janzen had the classic virtues of a restorationist: "He is public-spirited, courageous, self-reliant, and tough, and he is willing to do what needs to be done no matter what the sacrifice." David Brower, arguably the most important North American conservation crusader of the mid- and late twentieth century, said that in terms of "helping nature heal," the Guanacaste project was "one of the most dramatic stories of all." Brower, eighty-three when I spoke with him in 1995, was known for his zealous leadership of the Sierra Club in the 1950s and '60s and for founding several conservation organizations. He had long been impressed by Janzen's description of the "scattered biotic debris and the living dead" in Costa Rica. "The living dead," Brower repeated. "Those are not inspiring tools to work with, but he did."

Perhaps Janzen's greatest addition was the idea of tropical restoration. Since the 1960s, leaders such as Rachel Carson, Donella Meadows, and Paul Ehrlich had highlighted problems of environmental destruction and related issues. Conservationists became focused on identifying and stopping harmful trends. The idea of restoration surged in the 1980s. Books such as Berger's *Restoring the Earth* illustrated the new trend in restoration ecology. The Guanacaste project leaders applied and demonstrated the concept in its most grandiose way, and in the tropics at that. It was such an attractive idea. León told me: "Not only are you talking about stopping the trend, implicitly you're actually saying we can turn the tide in our favor. In this endangered habitat, the dry tropical forest, we can turn things around to the point where we can restore the environment, promote its reappearance. That has a tremendous appeal." Others described the project as an historic bridge between old-style conservation projects and new.

Several observers noted Janzen's political acumen. He walked a political tightrope within the Costa Rican government, pushing the Park Service to restructure, garnering support for the Guanacaste project from national and local powerbrokers, and avoiding potentially debilitating personal involvement in the airstrip issue.

Janzen excelled at finding money. The basis for gathering money to support conservation had been laid by Ugalde, the National Parks Foundation, and others long before Janzen entered the scene. But he took it an order of magnitude higher, from the $5 million range of the Ugaldes and Bozas of the early 1980s to the $20 million-plus range. Despite this accomplishment, Janzen was the first to credit a host of people and events for the success. "Gudren found the money in Sweden, Salomon Brothers went out there and bought the debt, Umaña argued with the president of the Central Bank, Gámez did all the running around inside and made it happen. So you say, 'Dan Janzen raised $24 million.' No, a *team* of people did."

Yet Janzen played the catalytic role, and Hallwachs listened, watched, poked, and throttled him when he slid too far off base. He was on the front line of engineering the project in all its facets: ecological, political, financial. Thus, he adhered to a fundamental principle of fund-raising: the most successful were those who pitched a project they would execute personally. In other words, he wasn't just asking for money for a conservation project. He looked the potential donor in the eye and said, "Give me the money and *we* will do it." For foundations and philanthropists, Janzen told me, "that's 90 percent of the story." Those who gave money to the project knew that they were giving it, essentially, directly to Janzen and that *he* would buy the land, guide the restoration, and move the necessary long-term administrative and financial structures into place. Their basic decision was not whether to trust a bureaucracy, but whether they believed Janzen.

Behind it all was the commitment that Janzen and Hallwachs had to bringing nature back for humanity. This was a manifestation of their personalities, their training, and their focus on that single goal. It was easy to see and hear this commitment. Even a speech that might otherwise have been canned, delivered for the hundred-and-first time to just another crowd down the meeting circuit, was loaded with energy, feeling, dedication. One longtime observer described the lives of Janzen and Hallwachs as a daily process of defeat and self-renewal, driven by that goal and by each other. "Dan and Winnie have this sort of homeostasis where they get beaten down but then pull back up to the point where they're able to deal with the politics and the human aspect of the struggle. It's really impressive. They don't seem to be defeatable. Aside from the biology and the creativity and the determination of getting the funds and making it all happen, they're made of steel."

Over the course of the Guanacaste project and its evolution into ACG, several people—friends and enemies alike—posed the following questions to me. What would happen if Janzen suddenly were to die? Wouldn't that be the end of ACG? Wouldn't that put a stop to restoring the forests? Wouldn't that show how the whole thing had been shoved together and propped up by one man? These implied, of course, that ACG didn't have sufficiently strong support to function indefinitely.

Although they had seen major shakeups and a few backward steps, supporters of the project felt the conservation area and its philosophical underpinnings were on relatively firm ground. Janzen and Hallwachs, to be sure, had spent a great deal of time and energy teaching and inspiring the people who were by the mid–1990s running ACG.

There had been a time when Janzen's departure might have doomed the project. Potent enemies had long existed. However, he was needed less and less as the cadre committed to the principles of restoring Guanacaste grew, the concept of conservation through utilization of natural resources took hold, and the endowment and management plan were put in place. He still pushed megalomaniacal ideas like ATBI, but by the mid–1990s Janzen was effectively just a science adviser to ACG and INBio. As Gámez told me, "it's very much institutionalized now."

Róger Blanco, María Marta Chavarría, Sigifredo Marín, Julio Díaz, Alejandro Masis, Fermín Mendez, and Roberto Espinoza represented a new generation of commitment to ACG. "Dan produced many new ideas, but Dan alone can do nothing," Blanco once told me. "We have a new way of thinking, new information, a new structure. People say, 'When Dan Janzen dies, the ACG will die, too.' No. The ACG is a living institution in our country, and it is not supported by money, it is supported by people." Blanco likened the process to building and starting a car. The ACG car was built and started with Janzen's help, and he had also helped press down on the accelerator to get it moving. "Now all we have to do is decide where it will go," Blanco said.

I once asked Janzen whether he thought the project would have happened without him. He shrugged. "If I hadn't existed, some other individual would have filled roughly the same role," he said.

Hidden between the lines of some critics' complaints about ACG was a nationalistic resentment of Janzen. The project was viewed by many as to some extent a foreigner's project in Costa Rica. In many respects this was understandable. There was widespread resentment in Latin America of gringos, whether in the form of banana conglomerates, the U.S. Marines, or Great White Explorers, and rightly or wrongly, Janzen had long been identified as the sole inspiration and engineer of the project.

Much of this resentment was personality clash, and much of it was a feeling that Janzen had too much power. But some believed a gringo had rammed the Guanacaste project down Costa Rica's throat. "Basically, Dan Janzen is a North American, and he hasn't learned anything about Costa Rican idiosyncrasies," Melania Ortiz Volio, director of the National Museum, told me, reflecting that sentiment. Janzen, she said, had narrow scientific interests and didn't understand the broader needs of Costa Rican society. "He has built and followed a North American way of developing things without really taking consideration of the Costa Rican point of view."

Ortiz had been outlining for me the historical context of scientific and conservation institutions in Costa Rica. The museum, the Park Service, and other institutions had grown out of a traditional philosophy of state management. She and others were upset by what they saw as a radical shift away from the tradition of state-controlled institution building, with ACG and INBio leading the way in pushing major institutions into "private hands." "The state has given away a lot," she said, the irritation growing in her voice. Costa Rican "cultural patrimony" was now being managed "from the outside." "You have to applaud someone who has done so many things," she said. "The only thing is, we do not totally agree with the method."

Those who agreed with the method, like Gámez, were quick to point out the Costa Rican tradition of being open to ideas from outside. Tico culture—including science, politics, philosophy, public education, health care, and religion—had been influenced by many people and institutions from France, Germany, Belgium, Spain, England, Italy, Switzerland, the United States, and several Latin American countries, he noted. Dozens of U.S. institutions and scores of U.S. scientists had influenced Costa Rican science in the twentieth century alone.

Kenton Miller once told me that the Costa Rican government and people were "unusual in that their country, like ours, had a constant stream of immigrants, including scientists. So they don't reject foreign thought. They don't just buy it; they digest it. The point is, Janzen's advice was listened to by presidents one after the other. And that was a major contribution to establishing the value, importance, and seriousness of that project."

Janzen interpreted this Costa Rican tradition very well, Gámez said. "Because of his scientific, ecological, and cultural perceptions, he came up with an idea that was totally viable. It was accepted and carried out by local people. It never would have worked if it had been a foreign imposition."

I once asked Janzen his thoughts on the criticism that a gringo had come to Costa Rica to set up a conservation project that Costa Ricans should have set up. It might have been the term "gringo" that set him off, but he got angry.

"What business did the Spaniards have to invade this country?" he shot back. "I'm no more an immigrant than they are." His point was that Costa Rica itself consisted almost entirely of European immigrants, drawn from the same continent as his ancestors.

He admitted that he got tired of hearing that criticism, and it usually upset him.

"I get this question from Canadians, for example, who say, 'Well, we can't respond to your request for money because you're not a Latin American.' Well, what is a Latin American? Somebody who came over four hundred years ago? I mean, what's the difference? What they're saying is we can't respond to somebody from eighteenth-century England, but we can respond to somebody from sixteenth-century Spain."

The real issue was not where he was from, but what he brought. "I'm just a biologist who cares about a living set of resources on the Earth's surface, and what I'm set on doing is trying to get those things embedded in a social consciousness. I could do this in California, I could do it in Brazil, I can do it anywhere. The fact that it's here doesn't make a damn bit of difference."

It just so happened that Dan Janzen had landed in northwestern Costa Rica in the 1960s to do his biology. This was the biology he knew best, and it was the biology he cared most about.

To the list of transcendental items that made the Guanacaste restoration possible must be added a transcendental nation, Costa Rica. Because of its political, economic, and social history, Costa Rica may have been the best possible place to restore a tropical forest in the late twentieth century.

It was an unusual, if not unique, nation: a small, relatively stable democracy without an army and which from time to time could restructure itself to close gaps between the haves and the have-nots, although the gap had been increasing throughout the period of the project. It had a tradition of free speech and protest. Unshackled by the aura of repression and violence common in other Central American countries, nonviolent argument and compromise were the norms for settling conflict.

Conservation was a high priority in part because the national budget wasn't burdened by having to support a standing army. "You can't build anything when you have that kind of lodestone around your neck," León told me. With this "peace dividend," Costa Rica was better able to invest in protecting the environment. The investment the Arias administration made in conservation became self-fulfilling as Sweden and other governments recognized it, to the benefit of the Guanacaste project. That outside money, in fact, was crucial.

Costa Ricans were exposed to nature from childhood. Even if they lived in the city, families frequently took weekend trips to national parks and reserves. "This creates an admiration for nature," León said. "We are exposed to the beauty of the rain forest, to the birds, to the diversity of insects. An irrational yet basic feeling develops when you see these things and when you see the thread that holds them together. Many people feel it's a spiritual experience, or an aesthetic experience."

This political tradition and conservation ethic had given Boza and Ugalde the basis to lobby for national parks in the 1960s and '70s. Against incredible odds, they established a park system that became the underpinning of Costa Rica's international reputation in conservation. "They were the pioneers," Gámez told me. "Those guys actually had to fight, and in a way the Guanacaste initiative was no longer a frontier battle, but rather a development that came after, thanks to the battles that were won by the Alvaro Ugaldes and Mario Bozas." The Boza-Ugalde campaign created a mechanism for channeling international donations to the parks, established a network of contacts within conservation and philanthropic foundations, and helped generate the first waves of publicity about the greening of Costa Rica in *National Geographic*, *Audubon*, and other media. Gámez, León, and other leaders fortified that system with their political, managerial, and financial acumen.

As Costa Ricans and their U.S. colleagues learned more about conservation biology in the 1970s, '80s, and '90s, scientific ideas that shaped the Guanacaste project emerged, such as connecting wildlands over a range of habitats to enable seasonal migrations of insects, birds, and other animals. Ugalde told me, "If I thought I would die happily with 10,000 hectares [39 square miles] in Santa Rosa, it was because I knew *nothing* about conservation biology back then."

The momentum provided by Janzen connected it all. Miller, who became one of ACG's international advisers, may have put it best: "You have moments in history when forces and events come together. If you're lucky, you have somebody with the guts to grab those moments."

A cynic might say the biggest lesson of the Guanacaste project was, "Whatever you do, do it in Costa Rica." Behind that simplistic observation was perhaps the most important fact established by the project: in a world marked by environmental destruction, large-scale restoration *could be achieved*. It was no longer impossible to do. It was no longer outrageous to conceive a plan to do it again, somewhere else. Many observers reminded me, repeating a metaphor originated by Janzen, that the Wright brothers hadn't learned to fly airplanes in a mountain blizzard. They chose a steady breeze on a flat North Carolina beach. Only later did others create craft capable of traversing a stormy sky over Colorado, or flying faster than the speed of sound, or carrying a man to the Moon.

Blanco put it this way: "For us, the lesson is: it's possible. If it's possible for us to make this happen, it's possible for other countries in the world to make this happen."

What were the major lessons learned from Guanacaste? Observers offered a wealth of ideas, from conceptual to practical.

"The lesson in ACG is that running a major conservation project is an exercise in politics," Janzen told me. Nearly everyone else emphasized that point, and they noted that politics meant achieving a goal with people. That, in turn, highlighted the importance of leadership, and the fact that the leaders of the Guanacaste project were strong. They had strong personalities, strong connections in Costa Rica and beyond, and a strong commitment to the notion of taking back the forest.

These leaders handled politics well at many levels. They created a broad coalition, nationally and internationally. The inner circle—people such as Umaña, Gámez, and Janzen—sought assistance from longtime supporters of Costa Rican conservation and science, such as Thomas Lovejoy, Peter Raven, William Reilly, and Murray Gell-Mann, the conservation-minded Nobel physicist on the MacArthur board. At another level, they recruited local landowners, politicians, religious leaders, and others in Guanacaste. "People became committed to the conservation of the area because they love it," León told me. "Now you have a fairly large cadre of people in Guanacaste who feel the park is their patrimony. So I don't see any political power in Costa Rica that could do away with the park."

Several Costa Ricans noted that the leaders succeeded partly thanks to "our idiosyncrasy," what Ticos believe is their special cultural capacity for negotiation and compromise. It was not just leadership, but leadership that was willing to negotiate, and willing to concede.

Working with political leaders—local, national, and international—to educate them about the Guanacaste effort and its dimensions was "absolutely essential," Gámez said. For a foreign scientist to work successfully in Costa Rica, he had to understand the nuances of local and national political culture, and he had to work with politicians.

"Politicians are very legitimate people with very legitimate needs," Janzen said. "If you want to work with the political process, work with politicians, you have to figure out what their agenda is, instead of expecting them to adopt your agenda 100 percent. Rather, find out their agenda, and then see how to marry your agenda to their agenda." It's a simple lesson, yet not obvious to a biologist used to working in academia, which by and large consisted of biologists with like mindsets. Politicians were a different animal. A biologist trying to bring to life a restoration project would learn the meaning of the expression, "Politics makes for strange bedfellows." Janzen admitted that he'd heard that expression all his life but didn't understand what it meant. Fortunately, he encountered Umaña. "He taught me about learning how to work together with somebody totally different from me," Janzen said.

The Guanacaste project leaders also succeeded in what might be called the politics of the individual. That meant at least three things. For one, it meant inspiring individuals to set aside personal gain and comfort, and devote themselves to a vision, "a dream," as many of them put it. "One tends to think what's required is science and money, and one usually forgets above all it's people that you are going to be dealing with," Gámez said. "When you start analyzing the reasons why you fail or succeed, you realize that sharing the dream and working for that dream are absolutely essential. It is not easy. It takes skills." This lesson, of course, already had been well practiced by Boza, Ugalde, and other pioneers.

A second meaning of the politics of the individual had to do with what Janzen called "betting on people who have a vested interest." In running a conservation project, people with a vested interest were people who lived there. They included residents of the area in and around ACG. But they also were people such as Sigifredo Marín, the ACG director, who sold his property in San José, bought land in Liberia, built a house on it, and put his kids in school there. The same was true of several other ACG administrators, fire crew members, education team members, and all the rest. Because they no longer lived in San José and commuted every few weeks to the far northwest, these Park Service employees had, in a very real sense, immigrated to Guanacaste. This psychological transition made them feel more connected to their jobs, responsibilities, and commitments. They had become vested in their new land the same way European immigrants had when they sailed across the Atlantic to build a life in the New World. They cared about it because they were now *of* it. The survival of ACG became important to them not just because of the land; it was important for their survival as well. In short, ACG was not just a place to visit; it was home.

A third meaning of the politics of the individual had to do with the shift of authority from the bureaucrat in San José to the leader in the field. In other words, success derived from giving people on the ground the responsibility to get the job done. That's why, for instance, the fire team became a successful unit. Julio Díaz, originally a park guard and part-time fire fighter, was made head of the Program for Prevention and Control of Forest Fires. That meant the central office was out of the picture, and Díaz had complete control of the fire program. Just like he could determine how to raise his own children, or build his own house, now Díaz could design and run his own team and program for controlling fires. He could decide whom to hire, when to work, when to train, when to do controlled burns, what tactics to use in fire fighting, and when to post lookouts. If he failed, he was gone. If he succeeded, he got the credit. His identity was one and the same with fire control. "That's why Julio Díaz would die to keep the conservation area from burning up," Janzen told me.

Indeed, the success of the fire program had little to do with the technology of fighting fires. "Sure you need a broom," Janzen said. "Sure you need a pump. Sure you need a truck. That's just purchasing. The tricky part is ensuring that there's somebody there [at a fire] at 2 a.m. on Easter Sunday. And that is a people problem."

Because the project succeeded at the politics of the individual, it became staffed with people who combined an idealistic sense of mission with a practical pride in accomplishing it. During my visits to ACG in 1996, I saw that they routinely worked eighteen-hour days. They were enthusiastic and filled with purpose, and they seemed to enjoy the fact that they were always thinking. "We have a common goal," Róger Blanco once said when I asked him why he worked so hard. "We want to make a future. This job is not a job. It is our life."

The success in Guanacaste created a problem. Giving the power to Díaz meant taking it away from the centralized bureaucracy, and that created political enemies. "One person or the other is driving the boat," Janzen explained. "You take it away from one and give it to another, the first one gets upset. That's politics." Gámez put it this way: "You have to be prepared and conscious that any new project, any new initiative, raises internally a lot of resistance by the people or institutions who obviously prefer to maintain their traditions. Consequently, they see in the new initiatives a threat to the status quo. And there is going to be inherent resistance within the system."

Another lesson of the Guanacaste project was the importance of basic research in areas targeted for restoration. Many observers, including some scientists, viewed basic research as fairly esoteric, with little or no immediate usefulness. However, results from basic research in Guanacaste fueled the vision and mechanics of restoring the area. Such knowledge also helped justify government protection and outside funding.

"Think of it another way," said León. "As long as species don't have names, they disappear and nobody even knows about them. But when you put a name on them, then when they disappear, a name has to disappear too. This is much more upsetting." Conducting research in an area is a way of adding value to it, of adding value to the resource and calling it to the attention of the world outside biology. It's a lesson that was not so obvious at first, but it became obvious to all during the Guanacaste project.

"If you want an area protected, make sure you put some scientists in there before you start writing anything about protection," León advised. "Get them to do basic research on *anything they want*. It really doesn't matter what it is. When you find something, now you have a reason to protect it." The Guanacaste project benefited greatly by the proximity of biologists, whose new knowledge was immediately applied. That spontaneity may have seemed chaotic, but it also created a sense of excitement and led to new solutions for pressing problems.

The project showed in a very practical way the need for scientists to become politically active on conservation issues. Janzen's early mistakes suggest that scientists might save themselves a lot of work and trouble if they initially consulted political scientists, who are as expert on the complexities of politics and culture of a region as biologists are on the ecology of a forest. Scientists' views on politics can be superficial and uninformed. However, the Guanacaste story also suggests the success scientists can have in any case, if they just throw themselves into it. Norman Myers, author of *The Primary Source* and other books on tropical nature, once told me that Janzen's leadership in Guanacaste was "a magnificent public gesture of a scientist who could have had an even more outstanding scientific career. He set his academic career aside to go out and do something on the real firing line."

An obvious lesson demonstrated in Guanacaste was that adequate funds are needed to achieve success in a conservation project. A less obvious lesson was the need to build an endowment.

"Serious conservation is expensive," Janzen said. "If conservation is going to last, if it's going to live on past the emotional pulse of a given moment, it's going to cost a lot of money." New land acquisitions must be paid for immediately, and financial resources must be available to care for it. "Mission creep" was not advisable. Rather, conservationists needed to plan big from the beginning and jump in with intensive fund-raising.

Such financing required an international effort, Gámez advised. "If the world wants tropical biodiversity to be preserved, then the world has to pay the bill. There are political, economic, intellectual, and social considerations in all of this, and no country can do it on its own." Government agencies, philanthropies, scientific institutions, and corporations from several countries became involved in the Guanacaste project because of the tireless work of its leaders.

The endowment, an unusual and eminently successful idea borrowed from Janzen's experience as a professor at a private university, was established with the proceeds of the Swedish-Salomon bonanza. "If we had used that money from the debt swap and just spent it, we'd have a much finer park, and it would be dead," Janzen said. "Because when you endow it, you can make mistakes like crazy, and all you lose is that year's interest. Next year, start over. The first of January the check comes in again."

In essence, the endowment guaranteed a certain level of funds to meet operating costs, and a certain level of independence from outside forces. This concept ran against the grain of most conservation funding programs. Funders generally wanted to put money into a park but didn't want to endow it. An endowment took away their control, and they wanted to control how the money was spent. But in the hands of a finance manager for the endowment account, the money became invisible, and the administrator running the project was no longer beholden to the funder. He was beholden to local forces, for example, a board of directors for the conservation area. The advantage was that if he made a major mistake, he was fired, and the project continued because the interest kept rolling in.

The endowment also helped overcome one of the major doubts among ACG critics: that the conservation area would not be sustainable because it was too dependent on outside money. "Many don't see it as the local community taking charge," one scientist told me. "They still see it as dependent on outside expertise."

Corollary to financial success in a conservation project was playing the right combination of politics and diplomacy to win the backing of credit institutions. In the case of ACG, project leaders needed acceptance by not only the central government to initiate the debt-for-nature swaps, but also the board of the Central Bank. Thus the top politicians and economists in the country were made to see the political and economic benefits, short- and long-term, of ACG. The benefits included all the financial ones that came with the debt swap, but also the promise of socioeconomic development of the country from biodiversity prospecting, biocultural education, and related programs.

The Guanacaste project leaders learned and applied several lessons of a social nature. The number one lesson, learned in the Park Service experience in Corcovado and applied assiduously in Guanacaste, was repeated by several observers: "You can't fence parks in."

"If you have an unfriendly population outside, they'll eventually get you," said Umaña. "You cannot protect parks with fences or with guns."

The old Park Service paradigm of barbarians at the gate was replaced with the new model of finding ways to get the local population in. One way was to help them make a sustainable living without destroying the forest. Make them parataxonomists. Make them ecotourism guides. Make them staff and administration. Make the ranchers neighbors and coworkers, not enemies. Another way was to make them leaders on the conservation area board. Despite the broad international financial and intellectual support, it was crucial that the project be viewed internally as a local project, conducted by local people. Now, not all local people are the same, and a question remained whether politicians, large landowners, and other local elites were participating more than poor farmers on the board. Many of their interests, of course, are quite different. Yet the local participation that did exist ensured some measure of credibility necessary for long-term survival.

Still another way of connecting locals was through biocultural education. The local population needed to be convinced, along with national leaders, that preserving biodiversity was a worthwhile goal for socioeconomic development. "It's very difficult to make a project succeed if the society doesn't perceive any direct benefit, particularly in small, poor, developing countries," Gámez said. Consciously or unconsciously, people will raise the question, "How is this going to benefit me, my family, my children?" Unless biodiversity came to be viewed as a beneficial instrument, "it's going to go," Gámez said.

What is the relevance of the Guanacaste project to conservation today and what are the prospects for using its approach in other parts of the world? Was the combination of the Guanacaste leaders and Costa Rica of the 1980s indeed just a fluke, some perfect Wright brothers-Kitty Hawk coincidence so rare that their first flight will never be repeated?

One way to approach these questions is to look at whether tropical forest *restoration*, as opposed to *preservation*, has caught on anywhere else. The answer, generally, is no. Janzen wrote in January 1988 in *Science* that restoration "already occurs throughout the tropics. . . . Succession on a landslide or in a tree-fall gap is natural ecological restoration. The fallow phase of traditional slash-and-burn farming is applied restoration ecology. At least half of Barro Colorado Island, Panama, is restored forest. The farmlands that once supported the Mayan civilizations of the Petén and Yucatan are today unintentionally restored *Manilkara* and *Brosimum* forest, albeit with many fewer species than were originally on the sites."

However, none of these examples derived from a planned mega-conservation project. Tropical restoration appeared to be barely on the radar of major environmental groups. So, too, for biocultural restoration. The popular drumbeat remained to find ways to preserve existing tropical forest.

An indication of how ACG fits into conservation thinking came in February 1992 at the fourth World Congress on National Parks and Protected Areas, in Caracas, Venezuela. A "Global Biodiversity Strategy" released by 500 scientists, conservationists, and government officials from the World Resources Institute, World Conservation Union, and United Nations Environment Programme recommended eighty-five "actions" that local, national, and international leaders could take to protect biodiversity and turn it to human benefit. Action 51 was to restore degraded lands. "As the availability of lands ideally suited for either agriculture or protected area status shrinks under population and production pressures, and as the area of degraded land increases, the widespread adoption of techniques to restore the earth becomes a necessity," the report said. It noted that regenerating natural ecosystems was "gaining popularity in industrialized countries." However, these efforts often involved a trade-off in which pristine land was developed. Many conservationists considered this a Faustian bargain that could only lead to failure. The report listed several programs—in Brazil, Ethiopia, India, and Peru, for example—in which degraded land had been reclaimed for mixed agriculture, timber production, and watershed protection. Its sole example of incorporating degraded land into a protected area and restoring it was ACG.

Interest and expertise in restoration ecology grew in the United States in the late '80s and '90s, and Janzen was widely viewed as a kind of prophet by U.S. scientists and environmental historians. At a 1993 meeting in St. Louis on "Restoring Diversity," sponsored by the Center for Plant Conservation, several hundred biologists from around North America focused on the problems of reintroducing endangered plants to prevent extinction in the face of subdivisions, shopping malls, and other trappings of economic development. But when the organizers discovered that Janzen was in town for another event, they arranged for him to speak early one morning before the preset starting time. The capacity crowd gave him a standing ovation as his introducer said, "Dan is one who needs least to be introduced but deserves most to be."

Interest also grew in dry forest conservation. Scientists who studied tropical dry forests, no doubt at least partly motivated by Janzen, began calling attention to their demise, pointing out that the calamity striking these forests had been overshadowed by the tragedy of the rain forest. In a 1991 paper in *Trends in Ecology and Evolution*, Manuel Lerdau and colleagues from Stanford University, reporting on a dry forest workshop held in Mexico that spring, wrote, "Without immediate action, tropical deciduous forests may soon be of interest only to paleoecologists." The workshop recommended a three-part strategy to conserve dry forests. All involved government action, including efforts to "restore degraded tropical deciduous forests and develop sustainable land-use practices." The authors cited the Guanacaste restoration, saying attempts like it were "the only way to increase the amount of tropical deciduous forests."

As for the Wright brothers-Kitty Hawk combination, if readers of this book come away with the message that every major conservation project needed a Dan Janzen, an Oscar Arias, and a 1980s Costa Rica, that would be unfortunate, and incorrect. There was only one Guanacaste project, but there are many devastated ecosystems around the world in need of restoration. Even so, the project was viewed by some biologists and conservationists as a peculiar story. "It's not clear that there's much hope of reproducing it elsewhere, or that if you had those resources you couldn't use them better," one scientist commented. No one argued that Guanacaste was a formula that could be applied everywhere, but some said there were plenty of places where the politics and people, although not necessarily similar to those of the Guanacaste project, might come together to accomplish something like it.

When pressed, Janzen acknowledged that some conservationists might try to use the Guanacaste formula as a model for establishing tropical national parks in other parts of the world. The irony was that this would be just as erroneous as if ACG were to have been modeled after Yellowstone, Yosemite, or Great Smoky Mountains. "You've got to sit down and look at the circumstance you have and invent a circumstance for it." But it doesn't hurt, he added, to read someone else's recipe first.

If he ever did issue a recipe, it was probably in the January 1988 *Science* paper: "Choose an appropriate site, obtain it, and hire some of the former users as live-in managers. Sort through the habitat remnants to see which can recover. Stop the biotic and physical challenges to those remnants. The challenge is to turn the farmer's skills at biomanipulation to work for the conservation of biodiversity. Explicit and public agreement on management goals is imperative. Is the goal a low-overhead zoo, botanical garden, gene bank, functioning watershed, teaching laboratory, or some combination of these and other goals? Agreement is particularly critical in freewheeling tropical frontier societies, where ownership and administrative responsibility for public goods are poorly defined."

Only time will tell whether the Guanacaste project, or anything like it, is replicable elsewhere. But if nothing else, the project sent a hopeful signal amid the accelerating environmental destruction at the end of the twentieth century. Norman Myers told me that it ranked among several such signs. One was in Thailand, where the birth rate had rapidly dropped. Another was the creation, since the mid–1980s, of more than 200 organizations dedicated to saving the tropical forest. "There's still a very great deal to play for," Myers said.

The successful restoration in Guanacaste symbolized a new attitude. "It's a major turn in the tide, a beginning of a new commitment," Pedro León told me. "In some ways it's an awakening. Not only do we have to protect our environment, but we also have to work to restore the damage that's been done. That is a very powerful notion. How much it will catch on and how much we can really restore is yet to be seen. Maybe in a century we will be able to look back and say, 'Yes, committed people can really restore a lot of their natural environment.'"

Arias said the world needs more Guanacaste projects. He admitted to being "quite disillusioned" about the propensity of developed nations to spend billions of dollars on weapons and only a few million for the protection of the environment, but that was another reason for taking heed of the Guanacaste example. "It is a positive story," he said. "It is a constructive story."

In November 1995, ten members of the Natural Areas Association, a group of U.S. biologists and conservationists, traveled to Costa Rica for a workshop on the history and politics of the conservation areas system. Alvaro Ugalde hosted the group, and among the sites they went to see was ACG. Hank Tyler, president of the association and one of the travelers, found it remarkable that such large-scale restoration was under way in a relatively poor country. "It's a positive demonstration of a can-do attitude," Tyler told me after he returned. "Look around the U.S. and tick off similar activities. There aren't any."

The group was "inspired," he said, to see concepts of conservation biology being put into practice: restoration, biological corridors, links to the marine habitat, and others. It was an inspiration, as well, for Ugalde. Tyler recalled the image of Santa Rosa's first director standing atop the Monument to the Heroes at sunset, looking down at the park for the first time in several years, smiling.

In all this discussion, it is important not to lose sight of three essential facts about the Guanacaste project: (1) It was the first large-scale attempt to restore a tropical forest; (2) the forest was returning to the province—the living dead were being resurrected; and (3) a new generation of scientists and citizens had been inspired by its example and the idea that humans, if they try, can find a way to take nature back. In the crisis discipline that is the science of conservation biology, here was perhaps the best example yet that scientists could overcome "the Nero dilemma." That term, coined by Michael E. Soulé in *Conservation Biology: The Science of Scarcity and Diversity*, referred to a situation biologists often confronted: needing to make immediate tactical decisions on solving a real-world conservation problem before all the data came in on the complex dimensions of the problem and the range of its solutions. Janzen, Hallwachs, and the others achieved this in Guanacaste.

As for inspiration, whether he realized it or not, Janzen's publicity campaign, despite the personal cost, might well have been more successful than any other in the history of Costa Rican parks, perhaps even among the most successful for conservation anywhere. The campaign ignited the project's incredibly successful fund-raising effort, but perhaps just as important, it reached people all over the world with the message that nature could be restored on a grand scale.

In December 1995, in a conversation on the stairway of the Hotel Galilea, in San José, I first began to realize how effective the Guanacaste publicity had been. I was talking with an architect from Long Island, a cordial woman in her forties who had come to Costa Rica to research a paper she was preparing on the country's religious buildings. When I told her about this book, she immediately interjected that she knew about Janzen. She had gone to her public library before the trip and checked out the one and only videotape dealing with Costa Rica, the BBC-PBS documentary, "Costa Rica: Paradise Reclaimed." It had given her a sense that Costa Rica was something special, an uncommonly positive place.

A few months later, on a trip to Barro Colorado Island, I met a young graduate student whose life had literally been changed by the Guanacaste project, although he'd never been to ACG. The student told me that he had graduated a few years earlier with a degree in biology and computer science and had been working as a computer programmer in a Los Angeles suburb. In his spare time, he had been restoring an old Chris Craft, with an eye toward cruising down the West Coast to Central America. As part of preparing for the trip, he sought videotapes on the region at his public library. The only tape available was "Paradise Reclaimed." He watched it—and changed his plans. He returned to school, earning a master's in ecology and then beginning field work for his doctorate.

When I asked Norman Myers about the meaning of the Guanacaste project, he emphasized this point: "I've heard so many students say that it's all doom and gloom and they get very despairing. And yet they are so fired up by the example of Dan Janzen. That's what keeps them at it."

Several of the Latin American students with Loiselle atop Cacao spoke of the inspiration of the Guanacaste project. "We know that Costa Rica is an ideal world, but that doesn't mean you can't do it in your country," said Marco Vinicio Centeno, from Guatemala. "It is a symbol for us of what can be done, and what we have to look for."

Rafael Villegas, from Mexico, added, "It's a good example for all of Latin America."

Perhaps most poignant of all, many in this new generation, like Centeno and Villegas, came from countries on the front lines of the tropical conservation wars. It was entirely possible that they would become the core cadre for the future of tropical conservation biology. The next Ugaldes, Bozas, Gámezes, Leóns, and Umañas. The people in whose hands the future of the forests rested.

It was just after three in the afternoon on August 14, 1996, little more than a decade since I had first climbed the Monument to the Heroes, and since Dan Janzen and Winnie Hallwachs had launched the Guanacaste project. I was again standing atop the monument, looking to the east, toward Orosí, Cacao, and the simmering Rincón de La Vieja. I'd climbed the iron rungs one more time to compare the view with what it had been ten years earlier. As I climbed,

I thought of a story told to me earlier that day by Letty Brown, a biology student from the University of California at San Diego. Brown was at Santa Rosa studying leaf-cutter ants, and she had gone to the monument one night after dark to take in the view. She came down the ladder, jumping past the last few rungs to get to the bottom more quickly. As she landed, the beam of her headlamp caught the eyes of a big brown viper resting on one of the rungs she had skipped. Now ascending, I watched the rungs carefully, then teased myself about the notion that a snake might perch on one in broad daylight. The fear I'd picked up from those ridiculous popular images of the tropical forest before my first trip still held its grip.

The scene at the top of the monument was as panoramic and stunning as ever, only now with the added thrill of witnessing a rebirth. Thomas Belt had written of the drama of such vistas. Commenting about the view from a tropical forest promontory a hundred miles to the northeast, in Nicaragua, he said, "Thoughts arise that can be only felt in their full intensity amid solitude and nature's grandest phases." This view of the Guanacaste Conservation Area was, indeed, an exhibition of one of nature's grandest phases.

Floating down the slope of Cacao toward me was a magnificent thunderhead, dragging a thick tail of rain, like the tentacles of a Portuguese man-of-war. The wind picked up, and I fought to keep my notebook pages from flipping over. The rain tail widened as the storm approached. I switched to pencil when the drizzle began.

I remembered what Ugalde had told me about standing atop the monument with the visitors from the Natural Areas Association and looking out on the reborn forest. "I was so happy to be there," he said, smiling like a man who had just witnessed a dream come true. "I hardly saw any jaragua in the park—in the whole of Santa Rosa! Oh man, that's really fantastic."

Janzen had displayed mixed emotions on the view of the new trees. "You've just got hectare after hectare after hectare of that," he told me once, smiling, even a bit giddy. I'd seen that look before, but where? Maybe it was at the Cafe Relax, after the Distinguished Son of Guanacaste ceremony, the night before the Arias declarations at Santa Rosa that officially launched Guanacaste National Park. However, another time, when I asked him what he felt when he saw the monument view, he snapped.

"I don't think about it. I'm not very much of a backward-looking person. I'm focused on the problems downriver, not the waterfalls we've already been over."

"But many people believe you've accomplished something truly good," I countered.

"Fine. But it doesn't do me any good to think about it."

Maybe you caught him on a bad day, I told myself. Maybe he was tired, or perhaps some political problem was bugging him. Then I realized that no, this was the normal Dan Janzen. This was the man obsessed with saving the forests of Guanacaste, unwilling to allow "idle satisfaction" to seep in, lest his megalo-maniacal drive somehow weaken and the dream slip away.

Thunder clapped above me, and howler monkeys answered from the forest below. Howlers and thunder have been shouting at each other for ages, and it now appeared hopeful that they would continue to do so, there, in northwestern Guanacaste Province, for ages hence. I sensed that this would be my last time atop the monument, my last time at Santa Rosa, my last time in the conserva-tion area. I considered for a moment whether I should say some sort of farewell, but a louder clap of thunder disabused me of that notion. The howlers con-curred, seeming to add a kind of primeval sneer about the futility of human sen-timentality. It was time to climb down.

I paused for a moment near the top rung to gaze at a rainbow that had risen high on the south side of the storm. Stunned by nature's beauty, power, and restorative impulse, I descended in a near daze, not even watching for Letty's viper. It wasn't there anyway, of course.

The rain started to intensify, and I walked quickly down the steps. No snakes, no swarms of bees. Just the bark of the howlers and the shrieks of parrots in distant trees. I reached the shelter of the Casona veranda, and a few moments later the storm opened up. Drops splashed like a million pennies on the orange roof tiles and ran off in streams. The Costa Rican flag flapped on a pole, its red, white, and blue soaked but undaunted. Somehow soothed by the noisy down-pour and the new hope represented by the view from the monument, I won-dered whether there were, at that very moment, more Costa Ricas out there in the world, more madmen and mystics, more forests lying in the ashes, waiting to be reborn.

### Postscripts

On or about the night of August 26, 1999, the roots of the great guanacaste tree in front of the Casona gave out and the tree fell. Because the center and its rings had rotted out, the exact age could not be determined. However, Janzen esti-mated the tree to be 225 years old, give or take 25 years.

On December 2, 1999, the United Nations Educational, Scientific, and Cultural Organization designated the ACG a World Heritage Site, granting it international recognition and protection for its biodiversity and management structure.

In March 2000, ACG leaders began purchasing a 9900–acre patch of rain forest on the volcanic foothills adjoining the conservation area on the northern slope of Rincón de la Vieja. The conservationists had launched a fund-raising drive to acquire the land, called the Rincón Rainforest. Information about its status is available at the following Internet site: http://janzen.sas.upenn.edu.

On February 17, 2000, the arbitration tribunal chosen to resolve the Santa Elena case under the auspices of the International Centre for the Settlement of

Investment Disputes ruled that the government of Costa Rica should pay $16 million for the property. The government promptly did so, and on April 28, Costa Rica took possession of Santa Elena, making it a permanent part of ACG. This fulfilled the final major land-acquisition goal of the Guanacaste project. For all intents and purposes, ACG was now whole.

# Bibliography

Author's note: Much of the information in this book was obtained from observations, speeches, lectures, fact sheets, memos, letters, interviews, and other personal communications. In addition, much came from the following sources.

## Prologue. The *Contrafuego*

Caufield, Catherine. *In the Rainforest: Report from a Strange, Beautiful, Imperiled World*. New York: Alfred A. Knopf, 1984.

Forsyth, Adrian, and Ken Miyata. *Tropical Nature: Life and Death in the Rain Forests of Central and South America*. New York: Charles Scribner's Sons, 1984.

Gómez-Pompa, A., C. Vázquez-Yanes, and S. Guevara. "The Tropical Rain Forest: A Nonrenewable Resource." *Science* 117 (Sept. 1, 1972): 762–765.

Goodland, Robert, and Howard Irwin. *Amazon Jungle: Green Hell to Red Desert?* New York: Elsevier Scientific, 1975.

Gradwohl, Judith, and Russell Greenberg. *Saving the Tropical Forests*. Washington, D.C.: Island Press, 1988.

Janzen, Daniel H. "Tropical Dry Forests: The Most Endangered Major Tropical Ecosystem." In *Biodiversity*, ed. E.O. Wilson, pp. 130–137. Washington, D.C.: National Academy Press, 1988.

Kelly, Brian, and Mark London. *Amazon*. New York: Harcourt Brace Jovanovich, 1983.

Leopold, Aldo. *A Sand County Almanac*. New York: Oxford Univ. Press, 1949.

Leopold, Luna B., ed. *Round River: From the Journals of Aldo Leopold*. New York: Oxford Univ. Press, 1953.

Myers, Norman. *The Primary Source: Tropical Forests and Our Future*. New York: W. W. Norton, 1984.

National Research Council, Committee on Research Priorities in Tropical Biology. *Research Priorities in Tropical Biology*. Washington: National Academy of Sciences, 1980.

Perry, Donald. *Life Above the Jungle Floor*. New York: Simon & Schuster, 1986.

Stone, Roger D. *Dreams of Amazonia*. New York: Viking Penguin, 1985.

### 1. The Nucleus

Brent, Peter. *Charles Darwin: "A Man of Enlarged Curiosity."* New York: Harper & Row, 1981.
Colp, Ralph, Jr. *To Be an Invalid: The Illness of Charles Darwin.* Chicago: Univ. of Chicago Press, 1977.
Janzen, Daniel H., ed. *Costa Rican Natural History.* Chicago: Univ. of Chicago Press, 1983.
Joyce, Frank J. "Nesting success of rufous-naped wrens (*Campylorhynchus rufinucha*) is greater near wasp nests." *Behavioral Ecology and Sociobiology* 32 (1993): 71–77.
McPhaul, John. "Ancient Wisdom, Space-Age Science Join Forces." *Tico Times*, May 16, 1997.
Soriano, George. "Vet Sees Increase in Chagas Disease." *Tico Times*, Nov. 28, 1997.

### 2. Cañón del Tigre

Belt, Thomas. *The Naturalist in Nicaragua.* 1874. Rpt. Chicago: Univ. of Chicago Press, 1985.
*Exploring Costa Rica, 1996.* San José: *Tico Times*, 1996.
Honey, Martha. *Hostile Acts: U.S. Policy in Costa Rica in the 1980s.* Gainesville: Univ. Press of Florida, 1994.
Janzen, Daniel H. "A Host Plant Is More Than Its Chemistry." *Illinois Natural History Survey Bulletin* 33, no. 3 (Sept. 1985): 141–174.
Janzen. *Costa Rican Natural History.*
Leopold. *A Sand County Almanac.*
Lerdau, Manuel, Julie Whitbeck, and N. Michele Holbrook. "Tropical Deciduous Forests: Death of a Biome." *Trends in Ecology and Evolution* 6, no. 7 (July 1991): 201–202.
Murphy, Peter G., and Ariel E. Lugo. "Ecology of Tropical Dry Forest." *Annual Review of Ecology and Systematics* 17 (1986): 67–88.
Terborgh, John. *Diversity and the Tropical Rain Forest.* New York: Scientific American Library, 1992.
Tower, John, Chairman, Edmund Muskie and Brent Scowcroft, members. *The Tower Commission Report: The Full Text of the President's Special Review Board.* New York: Bantam Books and Times Books, 1987.
U.S. Congress. *Report of the Congressional Committees Investigating the Iran-Contra Affair.* U.S. House of Representatives Select Committee to Investigate Covert Arms Transactions with Iran and U.S. Senate Select Committee on Secret Military Assistance to Iran and the Nicaraguan Opposition. Washington, D.C.: U.S. Government Printing Office, 1987.

### 3. The Ant and the Acacia

"Ants and Acacias: An Ecological Study." The Open University and BBC-TV. Film, 1973.
Belt. *The Naturalist in Nicaragua.*
Conniff, Richard. "The King of Sting." *Outside* 21, no. 4 (April 1996): 82–84, 147.
Ehrlich, P. R., and P. H. Raven. "Butterflies and plants: a study in coevolution." *Evolution* 18 (Dec. 1964): 586–608.
Gallagher, Winifred. "Recall of the Wild." *Rolling Stone*, April 21, 1988, 84–96, 116.
Janzen. "A Host Plant Is More Than Its Chemistry."
Janzen, D. H. "Birds and the ant x acacia interaction in Central America, with notes on birds and other myrmecophytes." *The Condor* 71 (1969): 240–256.

Janzen, Daniel H. "Coevolution of Mutualism between Ants and Acacias in Central America." *Evolution* 20 (Sept. 1966): 249–275.

Janzen, Daniel H. "Complexity Is in the Eye of the Beholder." In *Tropical Rainforests: Diversity and Conservation,* ed. Frank Almeda and Catherine M. Pringle, pp. 29–51. San Francisco: California Academy of Sciences and Pacific Division, American Association for the Advancement of Science, 1988.

Janzen. *Costa Rican Natural History.*

Janzen, D. H. "The deflowering of Central America." *Natural History* 83: 48–53.

Janzen, Daniel H. "Degradation of Tropical Forests: A Dialogue." *Bulletin of the Entomological Society of America* 31, no. 1 (Spring 1985): 10–13.

Janzen, Daniel H. "Ecological Distribution of Chlorophyllous Developing Embryos among Perennial Plants in a Tropical Deciduous Forest." *Biotropica* 14, no. 3 (1982): 232–236.

Janzen, D. H. "Euglossine Bees as Long-Distance Pollinators of Tropical Plants." *Science* 171 (Jan. 15, 1971): 203–205.

Janzen, Daniel H. "The Future of Tropical Ecology." *Annual Review of Ecology and Systematics* 17 (1986): 305–24.

Janzen, Daniel H. *Guanacaste National Park: Tropical Ecological and Cultural Restoration.* San José, Costa Rica: Editorial Universidad Estatal a Distancia, 1986.

Janzen, Daniel H. "Herbivores and the Number of Tree Species in Tropical Forests." *American Naturalist* 104, no. 940 (Nov.-Dec. 1970): 501–528.

Janzen, D. H. "How many babies do figs pay for babies?" *Biotropica* 11 (1979): 48–50.

Janzen, Daniel H. "How To Be a Fig." *Annual Review of Ecology and Systematics* 10 (1979): 13–51.

Janzen, Daniel H. "Mice, big mammals, and seeds: it matters who defecates what where." In *Frugivores and Seed Dispersal,* ed. A. Estrada and T. H. Fleming, pp. 251–271. Dordrecht: Dr. W. Junk Publishers, 1986.

Janzen, Daniel H. "The microclimate differences between a deciduous forest and adjacent riparian forest in Guanacaste Province, Costa Rica." *Brenesia* 9 (1976): 29–33.

Janzen, Daniel H. "New horizons in the biology of plant defenses." In *Herbivores: Their Interaction with Secondary Plant Metabolites,* ed. G. A. Rosenthal and D. H. Janzen, pp. 331–350. New York: Academic Press, 1979.

Janzen, Daniel H. "No park is an island: increase in interference from outside as park size decreases." *Oikos* 41 (1983): 402–410.

Janzen, Daniel H. "Notes on the Behavior of Four Subspecies of the Carpenter Bee, *Xylocopa (Notoxylocopa) tabaniformis* in Mexico." *Annals of the Entomological Society of America* 57 (1964): 296–301.

Janzen, D. H. "The Occurrence of the Human Warble Fly (*Dermatobia hominis*) in the Dry Deciduous Forest Lowlands of Costa Rica." *Biotropica* 8, no. 3 (Sept. 1976): 210.

Janzen, Daniel H. "The peak in North American ichneumonid species richness lies between 38 and 42 degrees North latitude." *Ecology* 62 (1981): 532–537.

Janzen, Daniel H. "Seed Predation by Animals." *Annual Review of Ecology and Systematics* 2 (1971): 465–492.

Janzen, Daniel H. "Small terrestrial rodents in 11 habitats in Santa Rosa National Park, Costa Rica." *Brenesia* 17 (1980): 163–174.

Janzen, Daniel H. "Synchronization of Sexual Reproduction of Trees with the Dry Season in Central America." *Evolution* 21 (1967): 620–637.

Janzen, Daniel H. "Two potential coral snake mimics in a tropical deciduous forest." *Biotropica* 12 (1980): 77–78.

Janzen, D. H. "The uncertain future of the tropics." *Natural History* 81 (1972): 80–89.

Janzen, Daniel H. "Weather-Related Color Polymorphism of *Rothschildia lebeau* (Saturniidae)." *Bulletin of the Entomological Society of America* 30, no. 2 (Summer 1984): 16–20.

Janzen, Daniel H. "Why Bamboos Wait So Long to Flower." *Annual Review of Ecology and Systematics* 7 (1976): 347–391.

Janzen, D. H. "Why don't ants visit flowers?" *Biotropica* 9 (1977): 252.

Janzen, D. H., and C. M. Pond. "A Comparison, by Sweep Sampling, of the Arthropod Fauna of Secondary Vegetation in Michigan, England and Costa Rica." *Transactions of the Royal Entomological Society of London* 127 (1975): 33–50.

Janzen, D. H., and C. M. Pond. "Food and feeding behavior of a captive Costa Rican least pigmy owl." *Brenesia* 9 (1976): 71–80.

Janzen, D. H., and T. W. Schoener. "Differences in insect abundance and diversity between wetter and drier sites during a tropical dry season." *Ecology* 49 (1968): 96–110.

Janzen, Daniel H., P. DeVries, D. E. Gladstone, M. L. Higgins, and T. M. Lewinson. "Self- and cross-pollination of *Encyclia cordigera* (Orchidaceae) in Santa Rosa National Park, Costa Rica." *Biotropica* 12 (1980): 72–74.

Janzen, D. H., R. Dirzo, G. C. Green, J. C. Romero, F. G. Stiles, G. Vega, and D. E. Wilson. "Corcovado National Park: a perturbed rainforest ecosystem." Report to the World Wildlife Fund, July 16, 1985.

Janzen, M. J., D. H. Janzen, and C. M. Pond. "Tool-using by the African grey parrot (*Psittacus erithacus*)." *Biotropica* 8 (1976): 70.

Kricher, John C. *A Neotropical Companion: An Introduction to the Animals, Plants, and Ecosystems of the New World Tropics*. Princeton, N.J.: Princeton Univ. Press, 1989.

Langreth, Robert. "The World According to Dan Janzen." *Popular Science*, Dec. 1994, pp. 79–82, 112, 114, 115.

Lessem, Don. "From bugs to boas, Dan Janzen bags the rich coast's life." *Smithsonian* 17, no. 9 (Dec. 1986): 110–119.

Maslow, Jonathan. "Doctor Dry Forest," *BBC Wildlife* 5, no. 12 (Dec. 1987): 630–636.

Stone, Donald E. "The Organization for Tropical Studies (OTS): A Success Story in Graduate Training and Research," in Almeda and Pringle, *Tropical Rainforests: Diversity and Conservation*, pp. 29–51. San Francisco: California Academy of Sciences and Pacific Division, American Association for the Advancement of Science, 1988.

### 4. The Tree with Ears

Biesanz, Richard, Karen Zubris Biesanz, and Mavis Hiltunen Biesanz. *The Costa Ricans*. Englewood Cliffs, N.J.: Prentice-Hall, 1987.

Boza, Mario A. *Costa Rican National Parks*. San José: Incafo, 1996.

Cruz, Vernon. *Administracion del Parque Nacional Santa Rosa*. San José: Universidad de Costa Rica, 1975.

Edelman, Marc. *The Logic of the Latifundio: The Large Estates of Northwestern Costa Rica Since the Late Nineteenth Century*. Stanford, Calif.: Stanford Univ. Press, 1992.

Edelman, Marc, and Joanne Kenen, eds. *The Costa Rica Reader*. New York: Grove Weidenfeld, 1989.

"Expropriate Somoza Land." *Tico Times*, June 3, 1977.

Interview with old Guanacaste man. *Tico Times*, Oct. 2, 1987.

Janzen. *Costa Rican Natural History*.

Janzen. *Guanacaste National Park: Tropical Ecological and Cultural Restoration*.

Janzen. "No park is an island: increase in interference from outside as park size decreases."

Janzen, Daniel H., and Paul S. Martin. "Neotropical Anachronisms: The Fruits the Gomphotheres Ate." *Science* 215 (Jan. 1, 1982): 19–27.

Meléndez, Carlos. "Santa Rosa: Quince minutos bastaron para que se escribiera una página gloriosa de nuestra historia patria." *Museo: Boletin del Museo Nacional* 1, no. 11 (1955): 1–27.

Miller, Kenton R. *Planning National Parks for Ecodevelopment: Methods and Cases from Latin America.* Madrid: Fundación para la Ecologia y para la Protección del Medio Ambiente, 1982.

Rosengarten, Frederic, Jr. *Freebooters Must Die! The Life and Death of William Walker, the Most Notorious Filibuster of the Nineteenth Century.* Wayne, Pa.: Haverford House, 1976.

"Somoza to Sell Guanacaste Ranch." *Tico Times,* July 1, 1977.

Tosi, Joseph A., Jr. *Land Capability as Determined by Climate, Physiography, and Soils Conditions in the Northeastern Part of Guanacaste Province, Costa Rica.* San José: Tropical Science Center, 1967.

Wallace, David Rains. *The Quetzal and the Macaw: The Story of Costa Rica's National Parks.* San Francisco: Sierra Club Books, 1992.

## 5. The Fires of Guanacaste

Janzen. *Costa Rican Natural History.*

Janzen, Daniel H. "The Eternal External Threat." In *Conservation Biology: The Science of Scarcity and Diversity,* ed. Michael E. Soulé, Sunderland, Mass.: Sinauer Associates, 1986.

Janzen. *Guanacaste National Park: Tropical Ecological and Cultural Restoration.*

Janzen, Daniel H. "Management of Habitat Fragments in a Tropical Dry Forest." *Annals of the Missouri Botanical Garden* 75 (1988): 105–116.

Murphy and Lugo. "Ecology of Tropical Dry Forest."

## 6. The Living Dead

Gaston, Kevin J. "The Magnitude of Global Insect Species Richness." *Conservation Biology* 5, no. 3 (Sept. 1991): 283–296.

Hanson, Paul E., and Ian D. Gauld. *The Hymenoptera of Costa Rica.* Oxford: Oxford Univ. Press, 1995.

Janzen. *Costa Rican Natural History.*

Janzen. "The Future of Tropical Ecology."

Janzen. *Guanacaste National Park: Tropical Ecological and Cultural Restoration.*

Janzen. "Management of Habitat Fragments in a Tropical Dry Forest."

Janzen, D. H. "When and when not to leave: Dan Janzen's thoughts from the tropics 8." *Oikos* 49 (1987): 241–243.

LaSalle, J., and I. D. Gauld, eds. *Hymenoptera and Biodiversity.* Wallingford, Oxon, U.K.: C.A.B. International, 1993.

Lerdau et al. "Tropical Deciduous Forests: Death of a Biome."

## 7. The Battle Plan

Goodland and Irwin. *Amazon Jungle: Green Hell to Red Desert?*

Gradwohl and Greenberg. *Saving the Tropical Forests.*

Janzen. *Guanacaste National Park: Tropical Ecological and Cultural Restoration.*

Leopold. *A Sand County Almanac.*

Langreth. "The World According to Dan Janzen."

Lessem. "From Bugs to Boas, Dan Janzen Bags the Rich Coast's Life."

Wille, Chris. "Riches from the Rain Forest: Costa Rica's Parataxonomists Search for 'Green Gold.'" *Nature Conservancy News*, Jan.-Feb. 1993.


### 8. A Tropical Christmas Catalogue

Belt. *The Naturalist in Nicaragua.*

Dyer, Jake. "Clandestine Airfield Occupied." *Tico Times*, Sept. 26, 1986.

Dyer, Jake. "Landings Reported in July." *Tico Times*, Oct. 3, 1986.

Honey. *Hostile Acts: U.S. Policy in Costa Rica in the 1980s.*

Janzen, D. H. "A Tropical Christmas Catalogue." *Bulletin of the Ecological Society of America* 67 (1986): 299.

Janzen. *Guanacaste National Park: Tropical Ecological and Cultural Restoration.*

Janzen, D. H. "Guanacaste National Park—A Tropical Christmas Catalog." *Biotropica* 18, no. 3 (1986): 272.

LeMoyne, James. "Americans Reportedly Supervised Airstrip Project near Nicaragua." *New York Times*, Sept. 29, 1986.

LeMoyne, James. "Costa Rica Closes Airstrip near Nicaraguan Border." *New York Times*, Sept. 25, 1986.

McPhaul, John. "Airstrip Uproar Chills Plan to Expand Park." *Tico Times*, Oct. 24, 1986.

Secord, Richard V. *Honored and Betrayed: Irangate, Covert Affairs, and the Secret War in Laos.* New York: John Wiley & Sons, 1992.

*Tico Times.* Articles: on Green Berets training Ticos at Murciélago, Feb. 21, 1986; on U.S. consulate bombing, April 18, 1986; on fear of increasing border tensions, June 27, 1986; on Sandinistas burning contra buildings in northern Costa Rica, July 18, 1986; and on Guanacaste residents wanting to form a U.S. state, July 25, 1986.

U.S. Congress. *Report of the Congressional Committees Investigating the Iran-Contra Affair.*


### 9. Earthquakes

Conservation International. *The Debt-for-Nature Exchange: A Tool for International Conservation.* Washington, D.C.: Conservation International, Sept. 1989.

Hallwachs, W. "Agoutis (*Dasyprocta punctata*): The Inheritors of Guapinol (*Hymenaea courbaril*: Leguminosae)." In Estrada and Fleming, *Frugivores and Seed Dispersal*, pp. 285–304.

Janzen, Daniel H. "Costa Rican Parks: A Researcher's View." *Nature Conservancy News*, Jan.-Feb. 1984.

"Trogons." In *Birds of the World*, vol. 5, part 6, no. 54. London: IPC Magazines, 1969.

Wallace. *The Quetzal and the Macaw: The Story of Costa Rica's National Parks.*


### 10. Little Dances

Berenbaum, May R. *Ninety-nine Gnats, Nits, and Nibblers.* Urbana, Ill.: Univ. of Illinois Press, 1989.

Edelman. *The Logic of the Latifundio.*

Evans, Howard Ensign. *Life on a Little-known Planet.* 1968. Rpt. Chicago: Univ. of Chicago Press, 1984.

"Green Cosmology." *Outside*, April 1990, pp. 52–53.

Janzen. *Costa Rican Natural History.*

Maslow, Jonathan Evan. "A Dream of Trees." *Philadelphia*, Nov. 1987.

Wheeler, William Morton. *Demons of the Dust.* New York: W. W. Norton, 1930.


### 11. The Clifftop and the Volcano

Hallwachs. "Agoutis (*Dasyprocta punctata*): The Inheritors of Guapinol (*Hymenaea courbaril:* Leguminosae)."

Janzen. *Costa Rican Natural History.*

Janzen. *Guanacaste National Park: Tropical Ecological and Cultural Restoration.*

Janzen. "Management of Habitat Fragments in a Tropical Dry Forest."

Janzen. "Mice, Big Mammals, and Seeds: It Matters Who Defecates What Where."

Kricher. *A Neotropical Companion.*

Leopold. *Round River: From the Journals of Aldo Leopold.*


### 12. Home Runs

Berenbaum, May. *Bugs in the System: Insects and Their Impact on Human Affairs.* Reading, Mass.: Addison-Wesley, 1995.

"Costa Ricans to Plant Trees on Airstrip North Arranged." *New York Times*, July 25, 1987.

Janzen. *Costa Rican Natural History.*

Shannon, Don. "Secret Contra Supply Airstrip in Costa Rica to Become a Park." Los Angeles *Times*, July 23, 1987.

Will, George. "Ergo Schmergo and More Theories on This Season's 'Juiced' Baseballs." *International Herald Tribune*, July 17, 1987.

World Wildlife Fund. "Costa Rica Reduces Debt in Swap for Tropical Forest: Area Added to National Park Includes Secret Iran-Contra Airstrip." News release, July 27, 1987.

"WWF Pushing New Approach." *Tico Times*, Feb. 20, 1987.


### 13. The Green Magician

British Broadcasting Corp. "Costa Rica: Paradise Reclaimed." Documentary film, 1987.

Edelman. *The Logic of the Latifundio.*

Eisner, Thomas, Edward O. Wilson, and Peter Raven. "One Tropical Forest That Can Be Saved." *New York Times*, Oct. 12, 1987.

Maslow. "Doctor Dry Forest."


### 14. Touch of the Money Spider

Anon. "A Debt Swap to Aid Nature." *New York Times*, Jan. 12, 1989.

Asiedu-Akrofi, Derek. "Debt-for-Nature Swaps: Extending the Frontiers of Innovative Financing in Support of the Global Environment." *International Lawyer* 25 (Fall 1991): 557.

Conservation International. *The Debt-for-Nature Exchange: A Tool for International Conservation*.

Curtis, Randall K. "Bilateral Debt Conversion for the Environment." Paper presented at the IUCN World Conservation Congress, Workshop on Debt and Debt Conversion, Montreal, Oct. 17, 1996.

Evangelauf, Jean. "Daniel Janzen: Biology Professor, Ecologist, International Financier." *Chronicle of Higher Education*, May 23, 1990.

Gertsch, Willis J. *American Spiders*. 2nd ed. New York: Van Nostrand Reinhold, 1979.

Harris, Brian. "Wildfire Burns 3,000 Hectares in Park." *Tico Times*, April 2, 1996.

Jacob, Rahul. "Biggie Named in Wall Street Sting." *Fortune*, Jan. 15, 1990.

Janzen, D. H. "Debt Purchase." Undated fact sheet, 1987.

Klee, Kenneth, ed. "A wasp named Gutfreund." *Institutional Investor*, Dec. 1989, p. 12.

Liddell, Jamie. "Arson Suspected in Park Fire." *Tico Times*, April 18, 1997.

Nature Conservancy. "Officially Sanctioned and Funded Debt-for-Nature Swaps to Date with Commercial Bank Debt." Fact sheet, Dec. 1992.

Nature Conservancy. "Swapping Debt for Nature." Undated fact sheet.

Passell, Peter. "Economic Scene: Saving the Forest in Costa Rica." *New York Times*, Feb. 8, 1989.

"Propuesta Plan General de Manejo: Area de Conservacion Guanacaste." Unpublished report by an interdisciplinary group from the University of Costa Rica, Sept. 1994.

Shabecoff, Philip. "Bolivia Protects Land in Debt 'Swap.'" *New York Times*, July 14, 1987.

Sun, Marjorie. "Guanacaste Paves the Way." *Science* 239 (March 18, 1988): 1368.

Tye, Larry. "Winning One for the Forest: Alliance of Bankers, Environmentalists Creates Debt-for-Nature Swaps." *Boston Globe*, April 10, 1989.

U.S. Congress. Tropical Forest Conservation Act of 1998, Public Law No. 105–214, signed July 29, 1998, by President Bill Clinton.

Walsh, John. "Bolivia Swaps Debt for Conservation." *Science*, Aug. 7, 1987, pp. 596–597.

### 15. Area de Conservación Guanacaste

"Biological Stations, Guanacaste Conservation Area." Unpublished report, 1996.

Janzen, Daniel H. "How to Grow a Wildland: The Gardenification of Nature." Paper presented at the National Academy of Sciences forum on "Nature and Human Society," Washington, D.C., Oct. 29, 1997.

Janzen, Daniel. "Tropical Reforestation: The Guanacaste Project." Proceedings of the First Pan-American Furniture Manufacturers' Symposium on Tropical Hardwoods, Grand Rapids, Mich., 1991.

"Propuesta Plan General de Manejo: Area de Conservacion Guanacaste."

### 16. A Time of War

Aldhous, Peter. "Ecologists Draft Plan to Dig in the Dirt." *Science*, Sept. 9, 1994, p. 1521.

Borenstein, Seth. "Scientists to Classify Virtually All Flora, Fauna in Smokies." *Raleigh News and Observer*, Dec. 25, 1998.

Blackmore, Stephen. "Knowing the Earth's Biodiversity: Challenges for the Infrastructure of Systematic Biology." *Science*, Oct. 4, 1996, pp. 63–64.

Gámez, Rodrigo, Sandra Rodríguez, and Ana Elena Valdéz. "Biodiversity and Sustainable Human Development: The Costa Rica Agenda." Paper presented at the National

Academy of Sciences forum on "Nature and Human Society," Washington, D.C., Oct. 30, 1997.

Gámez, Rodrigo, et al. "The Planning and Initiation of the Costa Rican Instituto Nacional de Biodiversidad (INBio)." Unpublished proposal to the U.S. National Science Foundation, April 23, 1989.

Goodland, Robert, ed. *Race to Save the Tropics: Ecology & Economics for a Sustainable Future.* Washington, D.C.: Island Press, 1990.

Janzen, Daniel, and Winnie Hallwachs. "All Taxa Biodiversity Inventory (ATBI) of Terrestrial Systems: generic protocol for preparing wildland biodiversity for non-damaging use." Report of a National Science Foundation Workshop, April 16–18, 1993, Philadelphia, 1994.

Kaiser, Jocelyn. "Great Smokies Species Census Under Way." *Science,* June 11, 1999, pp. 1747–1748.

Kaiser, Jocelyn. "Unique All-Taxa Survey in Cost Rica 'Self-Destructs.'" *Science,* May 9, 1997, p. 893.

Langreth. "The World According to Dan Janzen."

Reid, Walter V., Sarah Laird, Carrie A. Meyer, Rodrigo Gámez, Anna Sittenfeld, Daniel H. Janzen, Michael A. Gollin, and Calestous Juma. *Biodiversity Prospecting: Using Genetic Resources for Sustainable Development.* Washington, D.C.: World Resources Institute, 1993.

Roberts, Leslie. "Chemical Prospecting: Hope for Vanishing Ecosystems?" *Science,* May 22, 1992, pp. 1142–1143.

Solloway, Nicky. "INBio Signs New Pacts for Drug Research." *Tico Times,* Feb. 28, 1997.

"Tropical Biologist Wins Kyoto Prize." *Science,* July 4, 1997, p. 41.

Wille. "Riches from the Rain Forest: Costa Rica's Parataxonomists Search for 'Green Gold.'"

Yoon, Carol Kaesuk. "Counting Creatures Great and Small." *Science,* April 30, 1993, pp. 620–622.

## 17. The Vigilant

Allen, William H. "The varied bats of Barro Colorado Island: What roles do tropical bats play in forest regeneration?" *BioScience,* Oct. 1996, pp. 639–642.

"Arson Suspected in Park Fire." *Tico Times,* April 18, 1997.

"Biological Stations, Guanacaste Conservation Area."

Brennan, Peter. "Helms Battles Costa Rica Over Ranches' Expropriation: Drug Trafficker Owned One, Illegal Airstrip Was on Another." *San Francisco Examiner,* Dec. 15, 1994.

Donovan, Brian. "Big-Stick Diplomacy? War's Legacy in Costa Rica." *Newsday,* Feb. 1, 1993.

Farah, Douglas. "Helms Makes Managua Skittish: Incoming Panel Chairman's Old Antagonists Hustle Off the Record." *Washington Post,* Dec. 21, 1994.

"Guanacaste's Wildfires: Man-Made Threat." *Tico Times,* April 5, 1991.

Janzen. *Costa Rican Natural History.*

Kornbluh, Peter, and Martha Honey. "The Case of Ollie's Airstrip." *The Nation,* Feb. 22, 1993.

McPhaul, John. "Mystery Firm Cited in Aid Delay." *Tico Times,* March 15, 1991.

Mears, Teresa. "Animal-Alien Legend Enters Miami Lore." *Boston Globe,* Sept. 2, 1996.

Ministry of Environment and Energy, Guanacaste Conservation Area. "Administrative Structure." Internal ACG report, 1996.

"New Action on Environment." *Tico Times,* June 7, 1991.

"Park Director Defends Areas' Privileged Status." *Tico Times,* March 15, 1991.

Pratt, Christine. "Tribunal Named to Rule on Expropriation Dispute." *Tico Times,* June 6, 1997.

Tenenbaum, David. "The Greening of Costa Rica." *Technology Review,* Oct. 1995.

"Wildfire Burns 3000 Hectares in Park." *Tico Times*, April 2, 1996.

Wilkinson, Tracy. "Central America: Costa Ricans Baffled, Angry at Strained Relations with U.S." *Los Angeles Times*, Feb. 26, 1994.

U.S. Senate, Committee on Foreign Relations. "Confiscated Property of American Citizens Overseas: Cases in Honduras, Costa Rica, and Nicaragua." Republican Staff Report. Washington, D.C.: U.S. Government Printing Office, March 1994.

### 18. Life Jackets

Belt. *The Naturalist in Nicaragua.*

"Biological Stations, Guanacaste Conservation Area."

Camargo, Camilo. "Control biologíco, enemigos naturales y plagas en agropaisajes localizados junto a un área silvestre." Report to the ACG Biological Education Program, 1996.

Castro Salazar, Rene, Alejandro Esquivel Gerli, and Norman Warren. "Convenio de cooperacion de beneficios ambientales entre el Ministerio del Ambiente y Energia, Fundación de Parques Nacionales y Del Oro, S.A., un subsidiario de CDC/Inversiones Guanaranja, S.A., en el noroeste de Costa Rica." Unpublished contract proposal, Aug. 18, 1997.

Escofet, Guillermo. "Minister Defends Forest Policy." *Tico Times*, Aug. 29, 1997.

Escofet, Guillermo. "Orange-Peel Experiment Under Fire." *Tico Times*, Oct. 23, 1998.

Escofet, Guillermo. "Paying for the 'Bichos.'" *Tico Times*, Sept. 4, 1998.

Escofet, Guillermo. "Study Backs Claims of British Citrus Firm." *Tico Times*, Feb. 19, 1999.

Janzen. "How to Grow a Wildland: The Gardenification of Nature."

Jukofsky, Diane, and Chris Wille, eds. "Linking Mesoamerica's Greenways." *Eco-Exchange*, Tropical Conservation Newsbureau, Rainforest Alliance, May-June 1996.

Marín, Sigifredo, and D. H. Janzen. "The Original 1992 Proposal for the Development of the Bridge." Unpublished paper, 1992.

Morales, David. "Descripcion de la metodologia de cultivo de bosques, a implementar en el proyecto de restauracion de bosque húmedo en pastizales dentro del Corredor Biologico Rincón-Cacao." Internal ACG report, 1996.

Morales, David. "Evaluacion de los resultados del primer año del proyecto de restauracion de bosques en pastizales dentro del Corredor Biologico Rincón-Cacao." Internal ACG report, 1996.

Reid et al. *Biodiversity Prospecting: Using Genetic Resources for Sustainable Development.*

### Epilogue: Lessons from a Tropical Kitty Hawk

Belt. *The Naturalist in Nicaragua.*

Berger, Jonathan. *Restoring the Earth: How Americans Are Working to Renew Our Damaged Environment.* New York: Anchor Press, Doubleday, 1987.

Janzen. "The Future of Tropical Ecology."

Janzen, Daniel H. "Tropical Ecological and Biocultural Restoration." *Science* 239 (Jan.15, 1988): 243–244.

Lerdau et al. "Tropical Deciduous Forests: Death of a Biome."

Soulé, Michael E. "Conservation Biology in the 'Real World.'" In Soulé, *Conservation Biology: The Science of Scarcity and Diversity.*

Stiles, F. Gary, and Alexander Skutch. *A Guide to the Birds of Costa Rica*. Ithaca, N.Y.: Cornell Univ. Press, 1989.

World Resources Institute, World Conservation Union, and United Nations Environment Programme. *Global Biodiversity Strategy: Guidelines for Action to Save, Study, and Use Earth's Biotic Wealth Sustainably and Equitably*. Washington: World Resources Institute, World Conservation Union, United Nations Environment Programme, 1992.

# Index